DECISION SCIENCE AND OPERATIONS MANAGEMENT OF SOLAR ENERGY SYSTEMS

DECISION SCIENCE AND OPERATIONS MANAGEMENT OF SOLAR ENERGY SYSTEMS

VIKAS KHARE

School of Technology, Management and Engineering, NMIMS, Indore, Madhya Pradesh, India

CHESHTA J. KHARE

Department of Electrical Engineering, SGSITS, Indore, Madhya Pradesh, India

SAVITA NEMA

Department of Electrical Engineering, Maulana Azad National Institute of Technology (MANIT), Bhopal, India

PRASHANT BAREDAR

Energy Department, Maulana Azad National Institute of Technology (MANIT), Bhopal, India

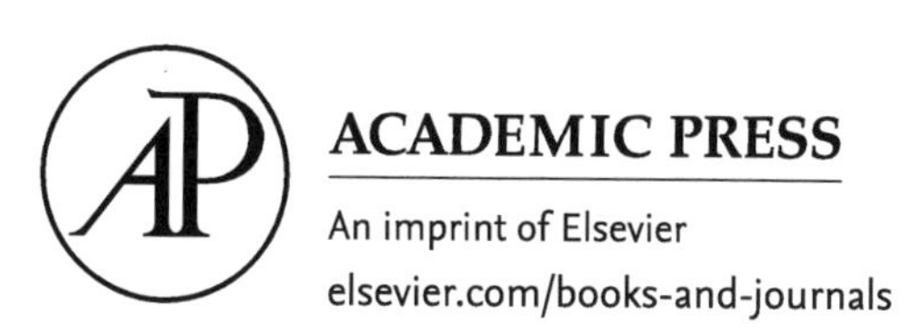

Academic Press is an imprint of Elsevier
125 London Wall, London EC2Y 5AS, United Kingdom
525 B Street, Suite 1650, San Diego, CA 92101, United States
50 Hampshire Street, 5th Floor, Cambridge, MA 02139, United States
The Boulevard, Langford Lane, Kidlington, Oxford OX5 1GB, United Kingdom

ISBN: 978-0-323-85761-1

For Information on all Academic Press publications visit our website at https://www.elsevier.com/books-and-journals

Publisher: Charlotte Cockle
Acquisitions Editor: Lisa Reading
Editorial Project Manager: Moises Carlo P. Catain
Production Project Manager: Prasanna Kalyanaraman
Cover Designer: Greg Harris

Typeset by Aptara, New Delhi, India

Contents

CHAPTER ONE

Fundamental and basic principles

Learning objective

- Understand the basics concepts of solar energy system.
- Know how the photo-voltaic system is used for electricity generation.
- Learn worldwide and Indian scenario of solar energy system.
- Understand the basics of decision science.
- Discuss about the fundamental concept of operational management process.

1.1 Introduction

Growing demand of electricity cannot be met through the conventional energy sources, renewable energy sources are emerging options to fulfill the additional energy demand but are unpredictable in nature due to the stochastic character of their occurrence, Renewable energy system play very vital role to mitigate the environment problem and lower the emission of greenhouse gases. There are several types of renewable energy system, such as solar energy system, wind energy system, biomass energy system, wave energy system, tidal energy system and geo thermal energy system, which are a good option to replace conventional energy system. Solar energy system plays very crucial role in the smart city concept, by making effective utilization of solar radiation for the generation of electrical energy. The electrical energy generation through the solar energy system mainly depends on the amount of solar radiation at a particular location which in turn depends on the quantity of temperature, longitude and latitude of the particular location.

Lot of researchers is already working in the field of solar energy system. Earlier Rahman et al. explore the feasibility study of PV-Fuel cell hybrid energy system and conquer the intermittent difficulty with PV and provide new application of the fuel cell technology and it offers the advantage of reducing the fuel cell consumption. Khan et al. [1] gave a meticulous description about pre-feasibility study of stand-alone solar-wind hybrid energy system for application in Newfoundland with a drop of fuel cell cost to 65%, a wind-diesel fuel cell battery system would be feasible. The paper demonstrates hybrid energy system with hydrogen as an energy transporter

Decision Science and Operations Management of Solar Energy Systems.
DOI: https://doi.org/10.1016/B978-0-323-85761-1.00005-6

for application in Newfoundland Canada. Sizing, recital and various cost indexes were also analyzed in this article. Graditi et al. [2] developed scientific, ecological, and practical aspects of hybrid systems and nonconventional energy sources. Nayar et al. [3] describe a case study of a PV/Wind/Diesel hybrid energy system and show how a resourceful hybrid system is being implemented in three isolated islands in the Maldives republic.

This chapter includes the principles of solar energy system and solar photo-voltaic system. Optimum design of solar energy system is also included in this chapter. Further Indian and worldwide scenario of solar energy system is also part of this chapter. Chapter also includes the fundamental of decision science and operational management, where decision science is the part of data analysis and data analysis is a process of inspecting, cleansing, transforming, and modeling data with the goal of discovering useful information, informing conclusions, and supporting decision-making. Operations management is mainly concerned with planning, organizing and supervising in the contexts of production, manufacturing or the provision of services. As such, it is delivery-focused, ensuring that an organization successfully turns inputs to outputs in an efficient manner.

1.2 Principles of solar energy system

Solar energy is the most environmentally friendly and widely available renewable energy source. This energy may be used by modern technology for a multitude of purposes, including generating electricity, giving light, and heating water for household, commercial, and industrial use. We can also use solar energy to meet our electricity needs. Solar radiation is directly transformed to DC power using solar photovoltaic (SPV) cells. This electricity can be used immediately or saved in the battery. A photovoltaic (PV) system is made up of one or more solar panels, an inverter, and other electrical and mechanical components that use the sun's energy to generate power. PV systems come in a wide range of sizes, from small rooftop or portable systems to large utility-scale power plants.

1.2.1 How do these system works?

The photovoltaic effect occurs when light from the Sun, which is made up of packets of energy called photons, falls upon a solar panel and generates an electric current. Each panel produces a little quantity of energy on its own, but when combined with other panels in a series parallel arrangement, a big

Table 1.1 Parameters for decision science and data analysis of solar panel.

Array nominal power at standard test condition
Operating mode through direct coupling
Operating mode through MPPT converter
Operating mode through dc-dc converter
Value of plane irradiance
Number of modules and string
Tilt/Azimuth angle
Module area
Solar energy fraction
Module and string mismatch losses

amount of energy is produced. The electricity produced from a solar panel (or array) is in the form of direct current (DC). Therefore DC power so produced must be converted to AC using an inverter. This AC electricity from the inverter can then be used to feed power locally, or be fed to the electrical grid.

1.2.2 System component

There are additional crucial components of a photovoltaic system that are generally referred to as the "balance of system" or BOS, in addition to the solar panels. Power conditioning units, inverters, racks, wiring, combiners, disconnects, circuit breakers, and electric meters, as well as batteries, are among these components. The primary system components of a solar energy system are as follows:

Solar panel: A solar panel is made up of several solar cells that have semiconductor qualities and are encased in a material that protects them from the elements. These characteristics allow the cell to capture light, or more precisely, photons from the sun, and transform their energy into useable power via a process known as the photovoltaic effect. A layer of conducting material is present on both sides of the semiconductor, which "collects" the electricity generated. An anti-reflection coating is also applied on the lit side of the panel to reduce reflection losses. The vast majority of solar panels are built of crystalline silicon, which has a theoretical efficiency limit of 33% for converting sunlight into electricity. Greater-efficiency semiconductor materials and solar cell technologies have been developed, however they come at a higher manufacturing cost. Table 1.1 shows the number of parameters, which is utilized for data analysis and decision science of solar panel.

Table 1.2 Parameters for decision science and data analysis of battery.

Basic parameters	Number of cells in series and parallel
	Nominal voltage
	Capacity at C10
	Internal resistance at reference temperature
	Reference temperature
	Columbic efficiency
Behaviour at limits	Charge cut-off voltage
	Discharge cut-off voltage
	Maximum charging current
	Maximum discharging current
	Minimum charging temperature
	Minimum discharging temperature
Full battery indicators	Stored capacity at depth of discharge
	Total stored energy
	Specific energy
	Specific weight
Graphical parameters	Charge/discharge v/s state of charge
	Charge voltage v/s state of charge
	Discharge voltage v/s depth of discharge
	Charge voltage v/s time
	Discharge voltage v/s time

Inverters: An inverter is an electrical device that takes direct current (DC) electrical electricity and converts it to alternating current (AC). This means that the DC current from the solar array is sent through an inverter, which transforms it to AC for solar energy systems. To operate most electrical devices or connect to the electrical grid, this conversion is required. Inverters are essential for practically all solar energy systems and, after the solar panels, are usually the most expensive component. The majority of inverters have conversion efficiency of 90% or greater, as well as crucial safety features such as ground fault circuit interruption and anti-islanding.

Battery: Solar batteries are often referred to as solar panel batteries, solar power batteries, and solar battery storage. It is a term used to describe equipment that store solar energy for later use. For power backup, a solar battery is designed to link with a solar charger controller or a solar inverter. With off grid solar systems and hybrid solar systems, all types of solar systems power the linked load during the daytime during sunlight and export the excess electricity to solar batteries.

Table 1.2 shows the parameters, to be utilized for data analysis and decision science of Battery.

Racking: The mounting gear that secures the solar array to the ground or rooftop is referred to as racking. These devices, which are usually made of steel or aluminum, mechanically fix the solar panels with a high level of precision. Extreme weather events, such as hurricane or tornado-force winds and/or large snow accumulations, should be designed into racking systems. To avoid electrocution, racking systems must also electrically link and ground the solar array. Flat roof racking systems and pitched roof racking systems are the two most common types of rooftop racking systems.

1.2.3 Other components

Combiners, disconnects, breakers, meters, and wire are the remaining components of a conventional solar PV system. As the name implies, a solar combiner combines two or more electrical wires into one bigger one. Combiners are used on all medium to large and utility-scale solar arrays, and they often feature fuses for protection. Disconnects are electrical gates or switches that allow an electrical wire to be manually disconnected. These devices, known as the "DC disconnect" and "AC disconnect," are typically used on either side of an inverter to provide electrical isolation when the inverter needs to be installed or removed. Overcurrent or surge protection is provided via circuit breakers. Breakers are designed to automatically activate when the current reaches a certain level, but they can also be operated manually to serve as an extra disconnect. Electric meters, on the other hand, monitor the quantity of energy that travels through them and are often employed by utility companies to measure and charge customers. A specific bi-directional electric meter is used for solar PV systems to monitor both incoming energy from the utility and outgoing energy from the solar PV system. Finally, the wiring or electrical cables must be correctly sized to carry the current and convey the electrical energy from and between each component.

1.3 Optimum design of solar energy system

The foremost concern in the unit size of a hybrid renewable energy system is the accurate selection of a system component that can economically satisfy the load demand. First Chedid et al. [4] explored unit sizing and control of integrated wind solar power system to determine the precise design for either autonomous or grid linked application with the help of a computer added design tool. This tool uses linear programming scheme to diminish the cost of electricity. Kellog et al. [5] developed unit sizing solar wind hybrid renewable energy system using the concept of the simple arithmetical algorithm. Generation and storage system are appropriately sized in order to meet the annual load and reduce the total annual cost to the customer. Nelson et al. [6] analyses most favourable sizing and economic analysis of a hybrid renewable energy system with the addition of Fuel Cell (FC) stack, an electrolyser and hydrogen storage tank which has worked as an energy storage arrangement. The study was performed using a graphical user interface programmed in MATLAB and break-even distance analysis is also calculated for each configuration. Lopez et al. [7] presents the influence of power converter losses assessment in the sizing of a Hybrid Renewable Energy System (HRES). An improvement to the precision of loss calculation in static converter used in HRES is presented in this article with the objective to assess energy losses in the system during the unit sizing process. Khatibi et al. [8] presents an analysis for escalating the saturation of renewable energies by the optimal sizing of pumped storage power plant. Belfkira et al. [9] developed, intended and optimize a grid independent wind/PV/diesel system. In this paper a tactic of sizing optimization of the standalone hybrid energy system is presented. Paudel et al. [10] presented a feasibility assessment and unit size of solar, wind energy system and battery bank for an isolated hybrid renewable energy system at the remote location of Nepal. Nagabhusana et al. [11] calculated the optimal sizes of the module of a solar-wind HRES by a linear programming idea that minimizes the present worth of capital and operating cost, whereas Kumar et al. [12] used probabilistic reliability analysis to find out the most favourable size of solar wind HRES. The stochastic belongings in all the parameter in the optimization models are taken into description for estimating power reliability. Laterra et al. [13] explains automatic procedure to develop the unit sizing of a grid connected solar wind HRES by fuzzy logic based multi-objective optimization idea. Both scientific and cost-effective objective functions are used to find out the best configuration. Further Erdinc et al. [14] explore a novel perspective component performance squalor issue in optimum sizing configuration. Observe and focus algorithm based optimization is used to compute the degradation and meticulous configuration of each integrated system component performance is studied. Angel et al. [15] analyse the influence of some unit sizing structure of grid connected solar-wind integrated non-conventional energy system with energy storage and load consumption on their crossing point with the electrical network. In this paper sizing constraint are used as sizing factor which is the ratio of yearly energy produced by the non-conventional energy system and yearly energy demanded by an end user. Erdinc et al. [16] explain unit sizing criteria of solar wind integrated renewable energy system by real time routine analysis and scrutinize the effectiveness of the optimum sized approach to entire generated system.

To identify, the unit sizing of solar energy system, it is necessary to calculate the load demand of the consumer. The load and its operating time vary for different appliances; therefore special care must be taken during energy demand calculations. The energy consumption of the load can be determined by multiplying the power rating (W) of the load by its number of hours of operation. Thus, the unit can be written as watt × hour or simply Wh.

Energy demand Watt-hour = Power rating in Watt × Duration of operation in hours

Thus, the daily total energy demand in Wh is calculated by adding the individual load demand of each appliance per day.

Total energy demand Watt-hour = Σ (Power rating in Watt × Duration of operation in hours)

A system should be designed for the worst-case scenario, that is, for the day when the energy demand is highest. A system designed for the highest demand will enhance the reliability of the system. But designing the system for the highest demand will increase the overall cost of the system.

Solar PV system sizing

1. **Determine the amount of energy required**

   The first stage in constructing a solar PV system is to determine the total power and energy consumption of all loads that the solar PV system must supply, as follows:

   - **Calculate the total number of Watt-hours utilized per day for each appliance:** To calculate the total Watt-hours per day that must be delivered to the appliances, add the Watt-hours required for all appliances together.
   - **Calculate the total daily Watt-hours required from the PV modules:** Add the entire number of appliances together. The total watt-hours per day that must be given by the panels are calculated by multiplying the watt-hours per day by 1.3 (the energy lost in the system).

2. **Size the PV modules:** PV modules of various sizes create varied amounts of power. The total peak watt produced is required to determine the PV module sizing. The peak watt (Wp) generated is determined by the size of the PV module and the climate of the site. We must take into account the panel generation factor, which varies by site. The panel generation factor, for example, is 3.43. PV module sizing is calculated as follows:
   - **Calculate the total watt-peak rating needed for PV modules:** To get the total watt-peak rating required from the PV panels to operate the appliances, multiply the total Watt-hours per day to be taken from the PV modules by 3.43.
   - **Calculate the number of PV panels for the system:** Divide the total watt-peak rating required for PV modules by the rated output watt-peak of the PV modules to get the result. Any fractional element of the result is multiplied by the next largest full number, which is the number of PV modules required.

     The computation yields the smallest number of PV panels. The system will work better and the battery life will be extended if additional PV modules are fitted. The system may not work at all during overcast periods if fewer PV modules are employed, and battery life will be lowered.
3. **Inverter sizing:** When AC power output is required, an inverter is utilised in the system. The inverter's input rating should never be less than the total wattage of the appliances. The nominal voltage of the inverter and the battery must be the same. Inverters for stand-alone systems must be large enough to handle the complete amount of watts that will be needed at any given moment. The inverter should be 25-30% larger than the overall wattage of the appliances. If the appliance includes a motor or compressor, the inverter size should be at least three times the capacity of the appliance, with additional capacity added to manage surge current during startup.
4. **Battery sizing**: Deep cycle batteries are designed to be promptly recharged after being discharged to a low energy level, or to be cycle charged and discharged for years on end. The battery should be large enough to store enough energy for the appliances to function at night and on cloudy days. The following formula is used to calculate the battery size:
   - Calculate total Watt-hours per day used by appliances.
   - Divide the total Watt-hours per day used by 0.85 for battery loss.
   - Divide the above answer by 0.6 for depth of discharge.

- Divide the above answer by the nominal battery voltage.
- Multiply the answer obtained in above case with days of autonomy (the number of days that you need the system to operate when there is no power produced by PV panels) to get the required ampere-hour capacity of deep-cycle battery.

  Battery capacity (Ah) = [Total Watt-hours per day used by appliances x Days of autonomy]/(0.85 × 0.6 x nominal battery voltage)

5. **Solar charge controller sizing**

   The capacity of a solar charge controller is usually measured in amperes and voltages. Choose a solar charge controller that matches the voltage of the PV array and the batteries, and then choose which sort of solar charge controller is best for a given application. Ascertain that the solar charge controller has sufficient capacity to handle the current generated by the PV array. The controller size for a series charge controller is determined by the total PV input current given to the controller, as well as the PV panel layout (series or parallel configuration).

The short circuit current (Isc) of the PV array is multiplied by 1.3 when sizing a solar charge controller, according to conventional practice. Solar charge controller rating = Total short circuit current of PV array x 1.3

Example: A house has the following electrical appliance usage:
4 hours every day, one 30 watt fluorescent bulb with electronic ballast

Each day, one 60-watt fan is utilized for four hours.

One 80-watt refrigerator that works 24 hours a day, with the compressor running 12 hours and then turning off 12 hours.

The system will be powered by 12 Vdc, 110 Wp PV module.

1. **Determine power consumption demands**

   Total appliance use = (30 W x 4 hours) + (60 W x 4 hours) + (80W x 24 × 0.5 hours) = 1320 Wh/day

   Total PV panels energy needed = 1320 × 1.3 = 1716 Wh/day.

2. **Size the PV panel**

   2.1 Total Wp of PV panel capacity needed = 1716 / 3.4 = 504.7 Wp

   2.2 Number of PV panels needed = 504.7 / 110 = 4.58~ =5 modules

   Actual requirement = 5 modules

   so this system should be powered by at least 5 modules of 110 Wp PV module.

3. **Inverter sizing**

   Total Watt of all appliances = 30 + 60 + 80 = 170 W

   For safety, the inverter should be considered 25-30% bigger size.

   The inverter size should be about 220 W or greater.

4. **Battery sizing**
   Total appliances use = (30 W x 4 hours) + (60 W x 4 hours) + (80 W x 12 hours)
   Nominal battery voltage = 12 V
   Days of autonomy = 3 days
   Battery capacity = [[(30 W x 4 hours) + (60 W x 4 hours) + (80 W x 12 hours)] x 3]/(0.85 × 0.6 × 12)
   Total Ampere-hours required 647.05Ah
   So the battery should be rated 12 V 650 Ah for 3 day autonomy.
5. **Solar charge controller sizing**
   PV module specification
   Pm = 110 Wp
   Vm = 16.7 Vdc
   Im = 6.6 A
   Voc = 20.7 A
   Isc = 7.5 A
   Solar charge controller rating = (4 strings x 7.5 A) x 1.3 = 39 A
   So the solar charge controller should be rated 40 A at 12 V or greater.

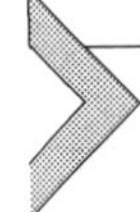

1.4 Worldwide and Indian scenario of solar energy system

1.4.1 Worldwide scenario of solar energy system

According to the International Renewable Energy Agency's latest data, the globe installed 176 GW of new renewable energy capacity in 2019. (IRENA). At the end of December, total clean energy capacity was 2,536.8 GW, with hydropower and wind continuing to be the leading sources, with 1,310.2 GW and 622.7 GW, respectively. Solar installations, which include photovoltaic (PV) and concentrated solar power (CSP), continue to trail wind, with a total installed capacity of 586.4 GW. Grid-connected PV accounted for 580.1 GW, while CSP accounted for 6.27 GW. With 330.1 GW of cumulative installed PV capacity, Asia is the region of the globe with the most PV capacity. China has the most cumulative installations in the area, with 205.7 GW, followed by Japan (61.8 GW), India (34.8 GW), and South Korea (10.5 GW). By the end of 2019, Europe had 138.2 GW of solar power installed, with 129.8 GW in the European Union. With 49.9 GW, Germany remains the continent's largest market, followed by Italy (20.9 GW), the

United Kingdom (13.3 GW), France (10.5 GW), and Spain (8.6 GW). At the end of December, total grid-connected PV capacity in North America had reached 68.2 GW. The United States installed 60.5 GW, followed by Mexico with 4.8 GW and Canada with 3.3 GW. The total grid-connected PV capacity in Central America and the Caribbean surpassed 2.1 GW. Honduras (511 MW), Dominican Republic (293 MW), Panama (242 MW), and El Salvador are the region's main markets (237 MW). Chile and Brazil are the largest markets in South America, with cumulative installations of 2.6 GW and 2.4 GW, respectively. By the end of December, the continent's total installed PV capacity had reached 6.46 GW. Total PV capacity in the Middle East reached 5.14 GW. The United Arab Emirates and Israel are the region's solar leaders, with 1.7 GW and 1.1 GW of total installations, respectively. According to IRENA, Africa's cumulative total reached 6.36 GW by the end of 2019. Armenia, Azerbaijan, Russia, Georgia, and Turkey contributed a total of 7.14 GW to the Eurasian region. Oceania's total hit 16.23 GW, with Australia leading the way with 15.9 GW.

1.4.2 Indian scenario of solar energy system

India is the fourth largest energy consumer in the world after the United States, China, and Russia. India's energy policy focuses on securing energy sources to meet the need of its growing economy. The Ministry of New and Renewable Energy has started a project to track the changing solar power regulatory environment and create a centralized library of data. This exercise should assist participants better grasp the changing nature of solar energy regulations and related concerns, as well as provide a platform for sharing information on relevant topics.

India has an abundance of solar energy. With the available commercially proven technology, solar radiation of roughly 5,000 trillion kWh per year is incident over its land mass, with an average daily solar power potential of 0.25 kWh per m2 of usable land area. The installed capacity was 21.65 GW as of March 31, 2018, accounting for 2% of total utility energy generation.

Solar power plants require nearly 2.4 hectares (0.024 km2) of land per MW capacity, which is comparable to coal-fired power plants when life cycle coal mining, consumptive water storage, and ash disposal areas are considered, as well as hydro power plants when the submergence area of a water reservoir is considered. On India's 1% land, solar plants with a capacity of 1.33 million MW can be constructed (32,000 square km). In all areas of India, there are enormous swaths of land appropriate for solar power that are unproductive, barren, and devoid of vegetation, accounting for more than 8% of the country's total area. When solar power plants are installed on waste land (32,000 square kilometers), they can generate 2000 billion kWh of electricity (twice the total generation in 2013-14), with land annual productivity/yield of 1.0 million (US$14,000) per acre (at 4 Rs/kWh price), which is comparable to many industrial areas and many times more than the best productive irrigated agriculture lands. Furthermore, these solar power plants are self-produced and do not rely on the supply of any raw materials. If all marginally productive lands are taken by solar power plants in the future, solar electricity has an endless potential to replace all fossil fuel energy requirements (natural gas, coal, lignite, nuclear fuels, and crude oil). India's solar power potential can fulfil per capita energy consumption levels comparable to those of the United States and Japan for the peak population in its demographic transition. In May 2017, India's solar PV electricity tariff fell to $2.44 (US$3.5) per kWh, the lowest of any source of electricity generation in the country.

When compared to fossil fuel power plants, solar thermal power plants with thermal storage are cheaper (US $5/kWh) and cleaner. They can precisely meet load/demand around the clock and also serve as base load power plants when solar energy is harvested in excess throughout the day. Without the use of expensive battery storage or non-solar power plants with dispatch ability and reliability, a correct blend of solar thermal and solar PV can fully match load changes. Land acquisition in Kadi Gujarat is a hurdle for solar farm projects in India. Some state governments are experimenting with innovative ways to manage land availability, such as deploying solar capacity above their massive irrigation canal projects, capturing solar energy while lowering irrigation water loss due to sun evaporation. Gujarat was the first to undertake the Canal Solar Power Project, which uses the state's 19,000 km (12,000 mph) long network of Narmada canals to install solar panels and create electricity. It was India's first initiative of its kind.

1.4.1.1 Solar potential site assessment in India

A pre-feasibility study of a solar energy system is required prior to installation and operation. An initial research is conducted in solar energy projects to establish whether it is worthwhile to proceed to the feasibility study stage. A detailed feasibility assessment should provide a timeline of the projects' development. To determine the optimal place to create a solar energy system, factors such as the climate of the application site, the availability of solar energy sources, the potential of solar energy sources, and the load demand of application sites are considered. In most cases, technical development and project implementation come before feasibility. It must therefore be conducted with a balanced approach to provide information upon which decisions can be based. In India there are a number of locations, which is fulfilled from the feasible solar radiation and that amount of solar radiation is capable to produce electrical energy from solar energy system. 115 automatic solar and meteorological monitoring stations, known as SRRA stations, have been set up across the country by the Ministry of New and Renewable Energy. Because solar radiation fluctuates throughout the day, precise measurements are critical in the design, development, and performance analysis of solar power facilities. Satellite-based irradiation estimates are commonly utilised for a continuous spatial coverage of vast regional areas, although the best quality data is provided by ground-based observations, which are also utilised for validating, benchmarking, and improving satellite-derived data. The average intensity of solar radiation received in India is 200 MW/km square (megawatt per kilometre square). With a geographical area of 3.287 million km square, this amounts to 657.4 million MW. However, solar energy is a dilute source. The India Meteorological Department (I.M.D.) maintains a network of radiation monitoring stations throughout India to fulfil this need. The network, which began with four stations in 1957 and now has around 40, has grown throughout time. Table 1.3 shows the coordinates of a solar station in India. The most basic parameter, the incident solar radiant energy directly and scattered is measured at all the stations.

At 24 locations across India, the scattered component diffuse solar irradiance is measured. Solar mapping of the entire country based on satellite images and carefully corroborated by ground truth data will provide information on both Direct Normal Irradiance (DNI) and Global Horizontal Irradiance (GHI) on a continuum basis with 15% accuracy. Fig. 1.1 shows the global only, global & diffuse, global-diffuse & direct, global-diffuse-direct & net terrestrial, net radiation only, global & direct and direct radiation location in all over the India. Fig. 1.2 shows the month-wise global solar radiation in all over the India. Figure demonstrate in the month of April, May, June solar radiation is highest and in that month maximum chance to generate electricity from the solar energy system, but in July and August due to minimum solar radiation, we always prefer a hybrid renewable energy system and also follow generator as a backup supply. It is feasible to find places with more sun radiation and set up ground stations to measure solar radiation and

Table 1.3 Coordinates of solar station in India.

City name	Latitude (N)	Longitude (E)	Height (m) amsl
Minicoy	08 18	73 09	1
Thiruvananthapuram	08 29	76 57	60
Port blair	11 40	92 43	13
Bangalore	12 58	77 35	921
Chennai	13 00	80 11	10
Goa	15 29	73 49	58
Hyderabad	17 27	78 28	530
Visakhapatnam	17 41	83 18	7
Pune	18 32	73 51	555
Mumbai	19 07	72 51	8
Nagpur	21 06	79 03	308
Bhavnagar	21 45	72 11	5
Kolkata	22 39	88 27	5
Ahmedabad	23 04	72 38	55
Bhopal	23 17	77 21	523
Ranchi	23 19	85 19	652
Varanshi	25 18	83 01	90
Shillong	25 34	91 53	1598
Patna	25 36	85 10	51
Jodhpur	26 18	73 01	217
Jaipur	26 49	75 48	390
New Delhi	28 29	77 08	273
Srinagar	34 05	74 50	1585

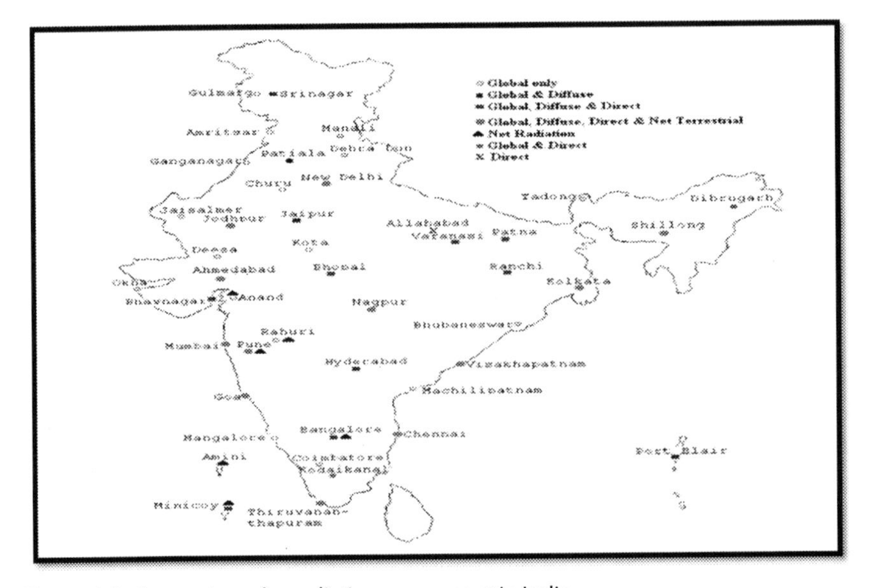

Figure 1.1 State-wise solar radiation assessment in India.

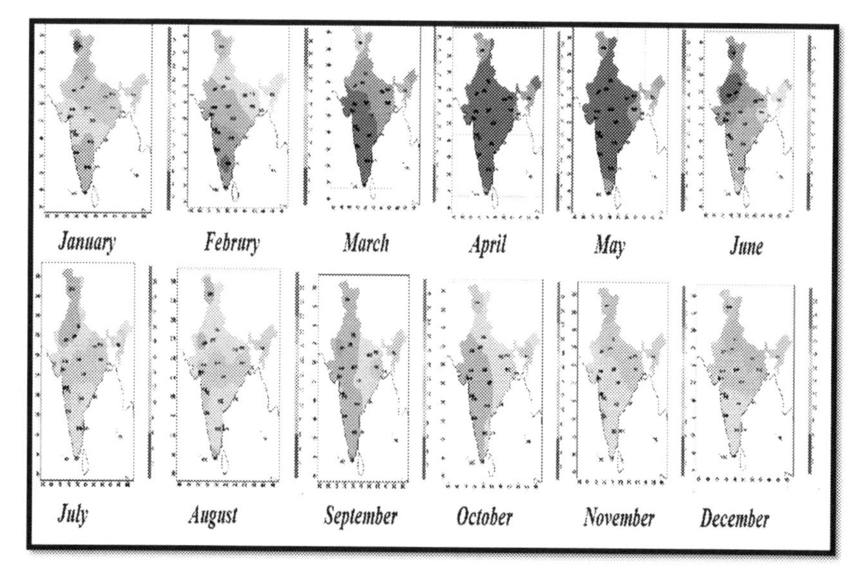

Figure 1.2 Month-wise solar radiation in all states of India.

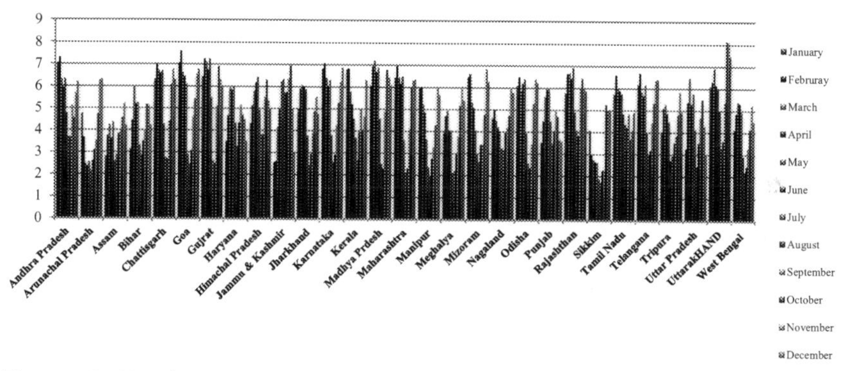

Figure 1.3 Month-wise Solar Radiation of all the States of India.

other climatic data more accurately. State of Gujarat, Rajasthan, Karnataka and Madhya Pradesh possesses highest solar radiation in almost eight months in the year and all of these states have large potential of electricity generation through solar energy system. The Fig. 1.3 shows month wise average solar radiation of all states of India. It will be able to avoid the costly task of establishing a big number of ground stations across the country. Investors, on the other hand, are required to set up their own ground measurement equipment at the project site in order to get a more exact estimate of data for their investment selections.

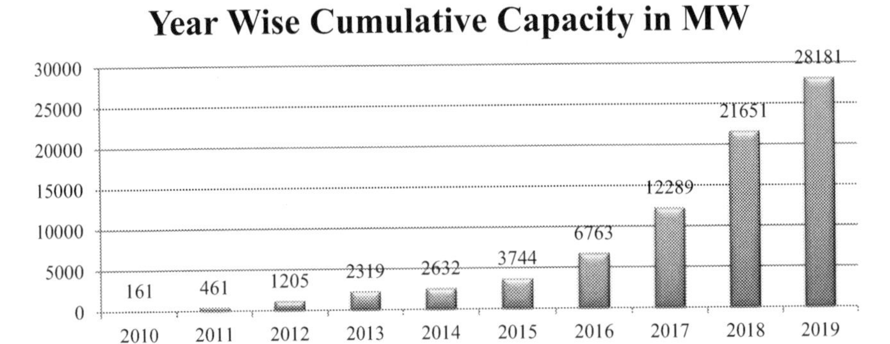

Figure 1.4 Year wise cumulative capacity in MW.

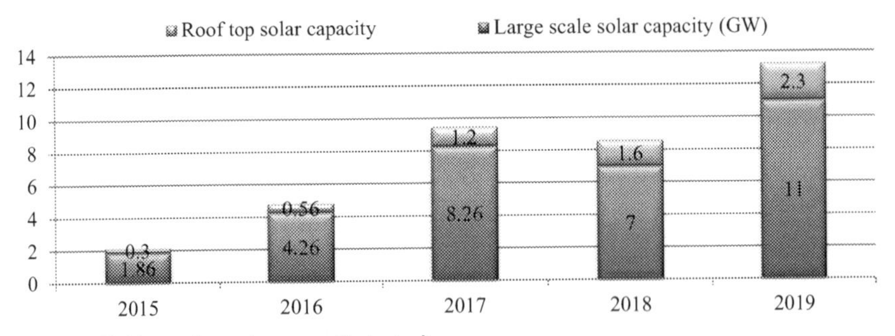

Figure 1.5 Year wise solar capacity in India.

1.4.1.2 Solar energy capacity in India

Solar energy is a rapidly growing business in India. As of March 31, 2019, the country's solar installed capacity was 28.18 GW. The Indian government set a goal of 20 GW of capacity by 2022, which was met four years ahead of schedule. The goal was expanded in 2015 to 100 GW of solar capacity by 2022, with a target investment of $100 billion. India's solar-generation capacity has increased eightfold from May 26, 2014 from 2650 MW to over 20 GW as of January 31, 2018. In 2015–16, the country built 3 GW of solar capacity, 5 GW in 2016-17, and over 10 GW in 2017-18, with solar electricity's average current price falling to 18% below that of coal-fired energy. Fig. 1.4 shows year wise cumulative capacity of solar energy system in India. This Figure shows solar cumulative capacity in India is increased from 161MW to 28181MW. In the last 5 year solar energy capacity is increasing at a very high level and from 2010 to 2019 it increases directly 175 times. This type of data shows the actual present status of India in the field of solar energy system. Fig. 1.5 shows year wise roof top and large scale solar

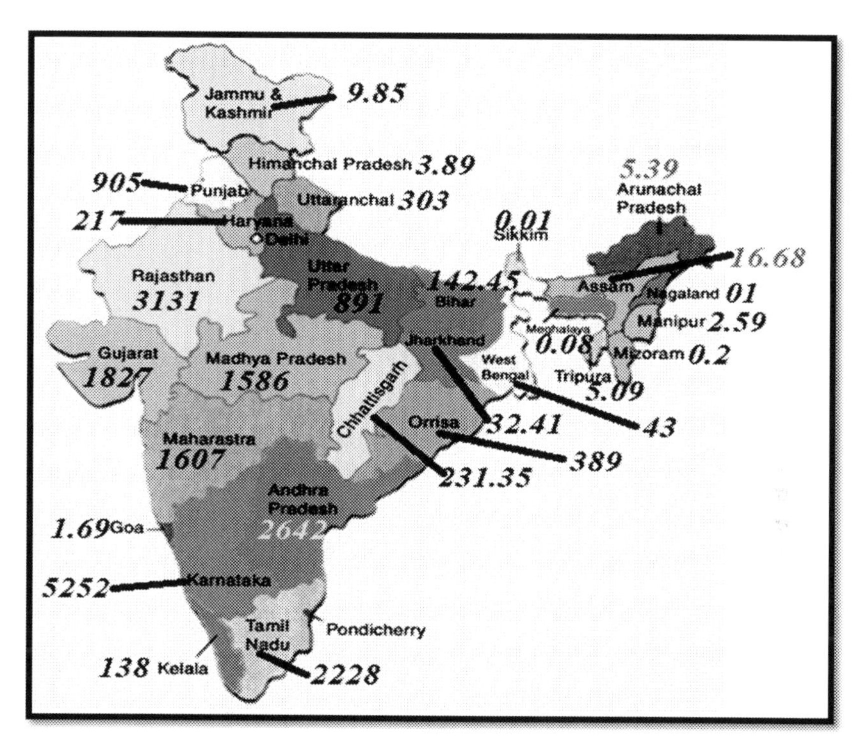

Figure 1.6 State-wise solar capacity in India.

capacity in India. Rooftop solar power accounts for 3.4 GW, of which 70% is industrial or commercial.

The Fig. 1.6 shows state-wise solar capacity in India. According to the figure state of Rajasthan, Gujarat, Maharashtra, and Madhya Pradesh have highest solar capacity 3131MW, 1827MW, 1607MW, and 1586MW, respectively.

As of November 30, 2018, Andhra Pradesh had more than 2,590 MW of installed solar capacity. The 250-MW NP Kunta Ultra Mega Solar Power Project near Kadiri in Anantapur district was agreed to by NTPC and APTransCo in 2015. Kurnool Ultra Mega Solar Park, the world's largest solar power facility at the time, was commissioned in October 2017 with a capacity of 1000 MW. Greater Visakhapatnam commissioned a 2 MW grid-connected floating solar plant in August 2018, making it India's largest operating floating solar PV facility. On its water supply reservoir, NTPC Simhadri plans to build a 25 MW floating solar PV facility. The Ananthapuram - II solar park near Tadipatri was commissioned in February 2019 with a capacity of 200 MW. Gujarat is one of India's most solar-developed states, having 1,637 MW of total photovoltaic power by the end of January 2019. Due to its high solar-power potential, availability of vacant land, connectivity, transmission and distribution facilities, and utilities, Gujarat has been a leader in solar-power generation in India. These characteristics are reinforced by political will and investment, according to a report by the Low Emission Development Strategies, Global Partnership (LEDS GP). Gujarat's solar power policy framework, financing mechanisms, and incentives from 2009 have contributed to the state's greener investment climate and grid-connected solar power targets. Near the village of Charanka in Patan district, the state has opened Asia's largest solar park. The park has generated 345 MW of its entire intended capacity of 500 MW as of March 2016, and the Confederation of Indian Industry has praised it as a creative and environmentally sustainable project. In December 2018, the Raghanesda solar park contracted a 700 MW solar PV facility at 2.89 Rs/unit regulated pricing.

The state government has launched a rooftop solar-power generation programme to help Gandhinagar become a solar-powered metropolis. Gujarat expects to create 5 MW of solar power under the project by installing solar panels on roughly 50 state-owned buildings and 500 private structures. Haryana has set a target of 4.2 GW solar power (including 1.6 GW solar root top) by 2022, citing the state's great potential of at least 330 sunny days. With a total installed and commissioned capacity of 73.27MW, Haryana is one of the fastest developing states in terms of solar energy. In FY 2016/17, 57.88MW of this capacity was put into service. Haryana's solar power strategy, which was launched in 2016, provides farmers with a 90% subsidy for solar powered water pumps, as well as subsidies for solar street lighting, home lighting solutions, solar water heating schemes, and solar cooking programmes. Solar capacity of 3% to 5% is required for new residential buildings greater than 500 square yards, with no construction plan approval necessary and a loan of up to Rs. 10 lakhs made accessible to residential property owners. For rooftop solar installations in Haryana, all electricity, taxes, electricity duty, wheeling charges, cross subsidy charges, transmission and distribution charges, and other fees are waived completely. Karnataka is India's leading solar state, with 5,000 MW of installed capacity at the end of the fiscal year 2017-18. The Pavagada solar park now has a capacity of 1400 MW, with a total capacity of 2,050 MW expected by the end of 2020.

Madhya Pradesh is one of India's most solar-developed states, having 1,117 MW of total photovoltaic power by the end of July 2017. The Welspun solar MP project, the state's largest solar-power plant, was developed on

305 ha (3.05 km2) of land for a cost of 1,100 crore (US$160 million) and will supply power at a rate of 8.05 (12 US) per kWh. Prime Minister Narendra Modi unveiled a 130 MW solar power plant project in Bhagwanpura, a village in Neemuch district. Welspun Energy is India's largest solar generator and one of the top three renewable-energy companies in the country. In the Rewa district, a 750 MW solar power facility is also being developed and is projected to be finished in 2021. Rajasthan is another solar-friendly state, with a total photovoltaic capacity of 2289 MW as of June 30, 2018. The Dhirubhai Ambani solar complex in Rajasthan also houses the world's largest Fresnel type 125 MW CSP plant. With about 1,500 MW of installed capacity, Jodhpur district tops the state, followed by Jaisalmer and Bikaner. The Bhadala solar park is being built in four phases, with a total capacity of 2,255 MW. NTPC Limited has already commissioned 260 MW capacity. At the end of June 2018, the total installed capacity was 745 MW, with the remaining capacity projected to be operational by March 2019. India is developing off-grid solar power for local energy requirements in addition to its large-scale grid-connected solar photovoltaic (PV) effort. Solar products are increasingly being used to address rural requirements; by the end of 2015, the country had sold slightly over one million solar lamps, eliminating the need for kerosene. In India, 118,700 solar household lighting systems and 46,655 solar street lighting systems were erected as part of a national programme, while little over 1.4 million solar cookers were supplied. Table 1.4 shows state-wise cumulative capacity of installation of solar photovoltaic system in India. Table 1.5 shows statistical analysis of Lanterns & Lamp numbers and Home Lights numbers. Table 1.6 shows statistical analysis of Street Light numbers and Pump numbers. Table 1.7 shows the statistical analysis of standalone power plants numbers.

1.5 Fundamental of decision science

Solar energy companies around the world make decisions that decide whether they will be profitable and expand or whether they will stagnate and die in the field of renewable energy systems. In a solar energy system, business statistics is the tool that collects, analyses, summarizes, and presents solar energy data to aid decision-making, and business statistics plays a vital role in the on-going saga of decision-making within the dynamic world of solar business. Fig. 1.7 shows the different parameters of decision science.

A descriptive statistic is a summary statistic that quantitatively describes or summarizes features from a collection of information, and also is the

Table 1.4 State wise cumulative installation of SPV system 2019.

State	Lanterns & lamps nos.	Home light nos.	Street light nos.	Pumps nos.	Stand alone power plants
Andhra Pradesh	51360	22972	7812	19526,8907	3786
Arunachal Pradesh	14433	18945	1671	22	650
Assam	13379,12258	6926	318	45	1605
Bihar	210391,160274	12303	955	1882	4168,200
Chhattisgarh	3311	7754	2042	26673,15203	28660,216
Delhi	4807	0	22018	90	1269
Goa	1093	393	58718	15	32
Gujarat	31603	9253	5806	8010	13576
Haryana	98853	56727	22018	1243,700	2321
Himachal Pradesh	33909	29342	58718,6860	6	1905,52
Jammu & Kashmir	51224	65319	5806	39	7720
Jharkhand	138723,115349	9450	787	3598,452	3640
Karnataka	7334	52638	2694	4118,641	7754
Kerala	54367	41912	1735	818,8	15825,1931
Madhya Pradesh	529101,519657	4016	9378	5584,1771	3654
Maharashtra	239297,170614	3497	10420	3315,1287	3857

(continued on next page)

Manipur	4787	3900	1888	40	1241
Meghalaya	24875	7844	4900,3627	19	1084,200
Mizoram	9589	6801	5056	37	2019,300
Nagaland	6766	1045	6235	3	1506
Odisha	99843,89961	5274	5834	8570,1491	568
Punjab	17495	8626	42758	1857	2066
Rajasthan	225851,221135	166978,11709	6852	41377,187	10850
Sikkim	23300	15059	504	0	850
Tamil Nadu	16818	273015,42695	39235,2433	4459	12752
Telangana	0	0	351,107	424	6643,1269
Tripura	64282	32723	1199	151	662
Uttar Pradesh	104791,42776	235909	185091	10877	10041
Uttarakhand	93927	91595	21905	26	2365,826
West Bengal	17662	145332	8726	653	1730

Table 1.5 Statistical analysis of lanterns & lamp nos. and home lights nos.

Parameters	Lanterns lamps number	Home light number
Count	30	30
Sum	1.365166E + 12	4.400025E + 10
Mean	4.550554E + 10	1.466675E + 09
Standard deviation	1.159396E + 11	5.752532E + 09
Standard error	2.116757E + 10	1.050264E + 09
Lower 95% CL mean	2.212995E + 09	-6.813557E + 08
Upper 95% CL mean	8.879809E + 10	3.614706E + 09
Median	32756	10876.5
Minimum	0	0
Maximum	5.291015E + 11	2.730154E + 10
Range	5.291015E + 11	2.730154E + 10
Variance	1.344198E + 22	3.309162E + 19
Coefficient of dispersion	1389227	134847.6
Skewness	2.909501	3.811641
Kurtosis	11.46576	16.26893

Table 1.6 Statistical analysis of street light number and pump number.

Parameters	Street light number	Pumps number
Count	30	30
Sum	1.029332E + 09	3.087653E + 09
Mean	3.431108E + 07	1.029218E + 08
Standard deviation	1.26738E + 08	4.85948E + 08
Standard error	2.313908E + 07	8.872157E + 07
Lower 95% CL mean	-1.301366E + 07	-7.853419E + 07
Upper 95% CL mean	8.163581E + 07	2.843778E + 08
Median	6543.5	1255
Minimum	318	0
Maximum	5.871869E + 08	2.667315E + 09
Range	5.871866E + 08	2.667315E + 09
Variance	1.606251E + 16	2.361455E + 17
Coefficient of dispersion	5243.102	82009.3
Skewness	3.683633	5.146163
Kurtosis	15.19568	27.66511

process of using and analyzing those statistics. Descriptive statistics are used to characterize the fundamental characteristics of a study's data. They give quick summaries of the sample and the metrics. They are the foundation of practically every quantitative data analysis, along with simple graphical analysis. Inferential statistics are sometimes distinguished from descriptive statistics. Descriptive statistics are used to describe what the data is or

Table 1.7 Statistical analysis of standalone power plants numbers.

Parameters	Stand alone power plants number
Count	30
Sum	2.632636e + 08
Mean	8775454
Standard deviation	3.107066e + 07
Standard error	5672701
Lower 95% cl mean	-2826523
Upper 95% cl mean	2.037743e + 07
Median	3821.5
Minimum	32
Maximum	1.582519e + 08
Range	1.582519e + 08
Variance	9.653862e + 14
Coefficient of dispersion	2295.891
Skewness	4.118966
Kurtosis	19.5424

indicates. With inferential statistics, you're attempting to draw conclusions that go beyond the facts at hand. Predictive analytics encompasses a variety of statistical techniques from data mining, predictive modeling, and machine learning, which analyse current and historical facts to make predictions about future or otherwise unknown events. Predictive analytics is the use of data, statistical algorithms and machine learning techniques to identify the likelihood of future outcomes based on historical data.

Prescriptive analytics is the third and final phase of business analytics, which also includes descriptive and predictive analytics. Prescriptive analytics is a type of data analytics which is used to help businesses make better decisions through the analysis of raw data. Specifically, a prescriptive analytics factor informs about possible situations or scenarios, available resources, past performance, and current performance, and suggests a course of action or strategy. It can be used to make decisions on any time horizon, from immediate too long term. Regression analysis, on the other hand, is a collection of statistical methods for estimating relationships between a dependent variable and one or more independent variables. It can be used to determine the strength of a relationship between variables and to predict how they will interact in the future. There are various types of regression analysis, including linear, multiple linear, and nonlinear. Nonlinear regression analysis is commonly used for more complicated data sets in which the dependent and independent variables show a nonlinear relationship.

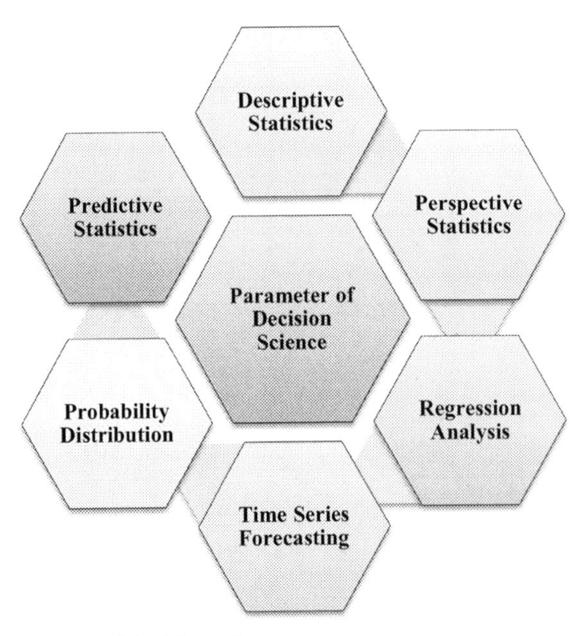

Figure 1.7 Parameters of decision science.

1.6 Fundamental of operational management process

Operations management is a well arranged approach and process to addressing issue in the transformation process that convert inputs into revenue generating outputs. In other words operations management is also a process of manage the operations of any project in a convenient manner. Following are the four aspects of operations management:

- Operations management is a continuous and systematic process.
- Operations management solves various issues of an organization.
- Conversion process is the intermediate process of operations management.
- The main aim of operations management is to increase the system performance and decrease the overall costs.

Operations management is a continuous and systematic process to understanding the nature of different affecting factors and problems to be analysed; measurement of performance assessment; collection of huge number of operations; different measurement tools, technology and different assessment techniques for analysis and evaluation. Operations management solves various issues of an organization and different issues according to the time interval, nature of the problems and commitment of the different types of the resources and at that point resources may be tangible, intangible and human resources. The conversion process ensures that different input factors are converted into the well-defined output parameters through the different types of intermediate operations management processes. The main aim of operations management is to reduce the costs and increase the performance of an organization at the external and internal levels.

Finance, operations, marketing and human resource management are the four key functions of an organization. The decision made in each of these functional areas could form an important input in another functional area. If top management wants to increase the performance level of an organization, then it is necessary to adopt systematic operations management process at individual level in financial, operational, marketing and human resource management department. The marketing department is responsible for understanding the requirement of customers, creating a demand for the products and services produced, and satisfying customer requirements by delivering the right products and services to customer at the right time. Operations management is the system perspective, where input is converted into the revenue generated output. Different types of input factors in the operation management are labor, material and capital or investment cost. The availability of labour and their cost is the important factor for the successful adaptation of operational management. The second input factor is the material and all the organizations process raw materials and convert them into useful products. For the example, in solar energy project raw material is the material for solar panel and battery such as crystalline, monocrystalline, lithium-ion, etc. Processing includes the different activities that an operating system undertakes to convert the raw material into useful products for the customers. In the operation management process, another important factor is the capital, which is categorized into two forms fixed and variable cost. The output factors of an operation management are goods and services, where goods are related to tangible items and services are related to the intangible items. In between input and output factors processing includes the various activities that an operating system undertakes to convert the raw material into useful products for customers. Fig. 1.8 shows the process of operations management.

1.6.1 Operations management functions

From the assessment of the various factors of operations management in the system perspective, it is also assess the important functions of operations management. The important functions can be assessed by two factors,

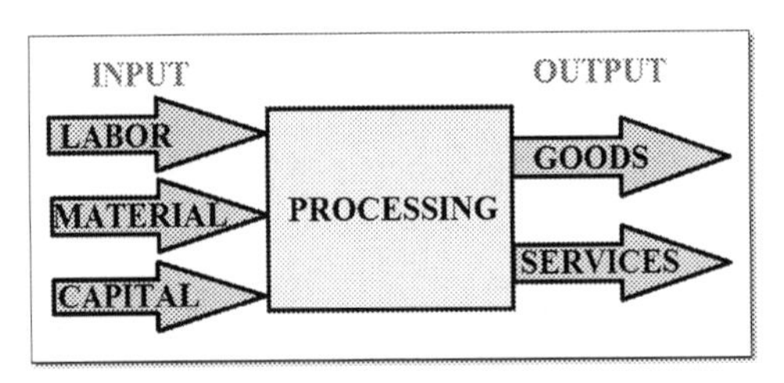

Figure 1.8 Operation management as a systematic approach.

Table 1.8 Functions of operations management.

Design issues	Operational issues
Design of goods and services	Predictive analysis
Systematic process design	Planning of operations
Total quality management	Supply chain management
Location assessment	Operation and maintenance system

"design v/s operational control issues" and long-term v/s short-term issues. Design concerns the setup of the operations system and provides an overall framework within which the operations system will operate. Designing products, services, and processes, establishing a quality assurance system, deciding on the layout and location of facilities, and capacity planning are all operations management responsibilities related to design. Another helpful way to comprehend the various operations management functions is to look at them from the perspective of the planning horizon. Concept of operations management functions in the field of solar energy system is often used. Optimum design of overall standalone or grid connected solar energy system is related to the design issue, where design of photo-voltaic system, design of battery, design of converter are the key point to be given due consideration. The continuous electricity generation through the designed solar energy system is the part of operations issues. Day-wise, month-wise and year-wise electricity generation through the solar energy system is related to the short term and long term issues of the functions of operation management. Table 1.8 shows the function of operation management.

In the reality, operations management is done through the step by step process, which includes facilities location, plant layout, productivity and production, manufacturing economics, inventory management and models, total quality management and theory of constraints. Fig. 1.9 shows the operations management trends and issues.

Facility Location pertains to the choice of appropriate geographical sites for locating various manufacturing and/or service facilities of an

Figure 1.9 Operations management trends and issues.

organization. Location assessment is the prefeasibility assessment of the process of operations management. In the process of location assessment for an organization number of factors to be kept in mind are labor cost, transportation facilities; connectivity with other areas, marketing condition. In the current scenario information technology infrastructure is an important factor for facility location assessment. A location planning exercise requires three steps; analyze different factors that affect the location decision, connectivity between the different parameters, and methodology to analyze the impacts of different input factors, which directly or indirectly effect the decision of location assessment. Fig. 1.10 shows the factor affecting the location decision.

Cost related issues capture the desirability or otherwise of competing locations on the basis of the cost of operating the system in alternative locations. Typically fixed and variable costs are often considered for the analysis, because these costs are direct, tangible and easy to evaluate. However based on the availability of huge number of information, such as wages, tax and other tariff benefits could also be included.

Plant layout is the optimum design of a plant considering all the direct and indirect parameters. In an organization the framework and design of

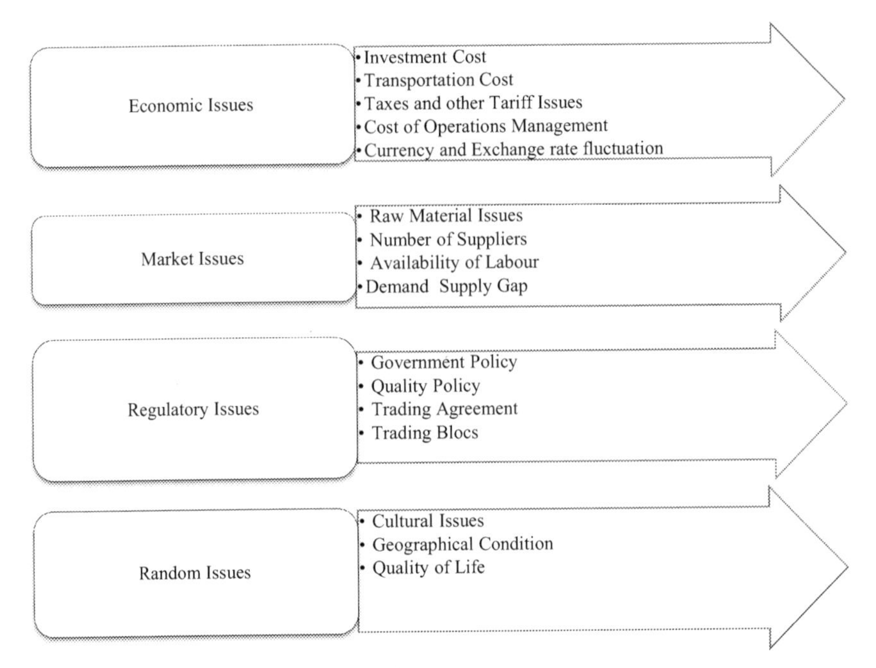

Figure 1.10 Factor affecting location decisions.

different internal department is also a part of plant layout. Layout planning in manufacturing and service organization involves the physical arrangement of various tangible and intangible resources available in the system to improve the performance of the operating system, thereby providing better customer service. Layout planning provides a set of technology that helps an operating manager decide where to assign the resources to get 100% output in the future. Fig. 1.11 shows the types of layout. There are different types of layout, such as process layout, product layout, group technology and fixed position layout. Design of plant layout is complete with the help of following three steps:

I. Identify the number of department required.
II. Estimate the magnitude of flow.
III. Full layout of the department.

- A process layout is a systematic framework of different tangible and intangible resources on the basis of operational characteristics. Example includes a maintenance department, quality control department, procurement, stores and production planning.

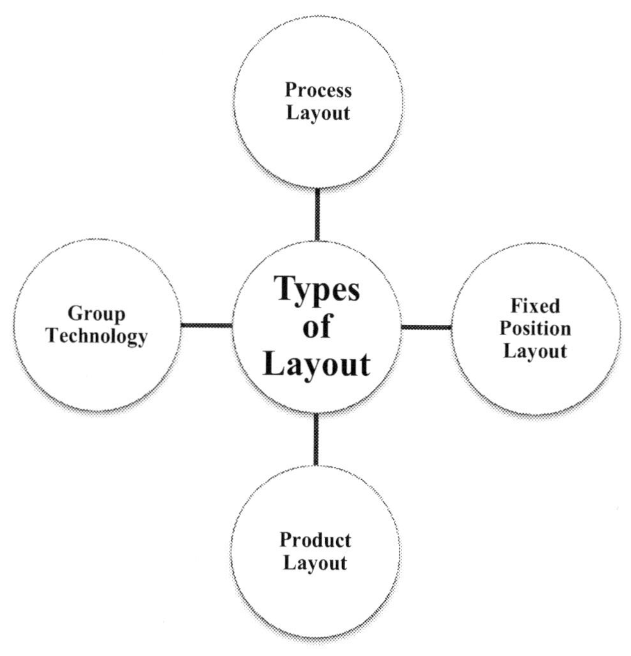

Figure 1.11 Types of layout.

- A product layout is an alternative design for the arrangement of resources. In this case, the order in which the resources are placed exactly follows the process sequence dictated by a product.
- Group technology layout is a philosophy that seeks to exploit the commonality in manufacturing and uses this as the basis for grouping components and resources. The implementation of group technology is often known as cellular manufacturing.
- Fixed position layouts are typically employed in large project type organization.

Productivity and Production in the operating system are characteristics by complex relationship between people, resources, market and material and often involve lead time affecting several of the processes. Production is the process of creating, growing, manufacturing, or improving goods and services. It also refers to the quantity produced. In economics productivity is used to measure the efficiency or rate of production. It is the amount of output per unit of the input.

Manufacturing economics is related to the material requirement planning, capacity requirement planning and distribution requirement planning.

Material requirement planning is the information technology based system for the planning of required quantities of material for manufacturing processes. Following factors affects the manufacturing economics directly and indirectly:

I. Multiple levels of dependencies
II. Product structure and bill of material
III. Determine the quantity of material
IV. Project time duration
V. Tangible, intangible and human resources

Capacity requirement planning is a technique that applies the material requirement planning logic to address the capacity issues in an organization on the other hand a distribution requirement planning exercise will help organizations and their supply chain partners to jointly plan and reduce investment in inventory in the supply chain.

1.7 Conclusion

Solar energy systems are the best alternative of conventional power plants. Recent technology such as artificial intelligence, machine learning, cloud computing and block-chain must be used to find out the different parameters of solar energy system. This chapter provide clear view, how operations management and decision science can be applied in the field of solar energy system. Different parameters of solar panel, battery and inverter shows that data analysis is must to identify the proper functioning of solar energy system. From the future perspective of solar energy system, it is necessary to implement different methodologies of operations management and decision science to enhance the efficiency and use of solar energy system.

1.8 Exercise/question 1

1. What is the basic principle of solar energy system?
2. How the solar energy system does work?
3. What are the different system components of solar energy system?
4. Explain the function of following elements w.r.t. the solar energy system:
   a. Solar Panel
   b. Inverter
   c. Battery

5. What is the meaning of optimum design of solar energy system?
6. How you can consider inverter and battery sizing for solar energy system.
7. Write short note on the following:
   a. Worldwide scenario of solar energy system
   b. Indian scenario of solar energy system
8. Explain the basic principle of decision science.
9. Explain the different parameters of decision science.
10. Write the short note on the following function w.r.t. the decision science:
    a. Descriptive statistics
    b. Predictive statistics
    c. Perspective statistics
    d. Regression analysis
    e. Probability distribution
    f. Time series forecasting
11. Why operation management as a systematic approach? Justify with example.
12. Explain the four aspects of operations management.
13. What are the key functions of an organization, explain in details.
14. What are the different trends in the operational management?
15. What are the different factor which effect directly and indirectly locations decision of an organization?
16. What are the different types of plant layout?

References

[1] M.J. Khan, M.T. Iqbal, Pre-feasibility study of standalone hybrid energy system for application in Newfoundland, Renew. Energy 30, 2005, pp. 835–854.
[2] G. Graditi, S. Favuzzas, E.R. Sanseverino, Technical, environmental and economical aspects of hybrid system including renewable and fuel cells, SPEED-IEEE (2006) 531–536.
[3] C. Nayar, M. Tang, W. Suponthana. A case study of a PV/Wind/diesel hybrid energy system for remote island in the republic of Maldives, AUPEC-IEEE:1-7.
[4] R. Chedid, S. Rahman, Unit sizing and control of hybrid wind solar power system, IEEE Trans. Energy Convers. 12 (1997) 79–85.
[5] W.D. Kellog, M.H. Nehrir, G. Venkataramanan, V. Gerez, Generation unit sizing and cost analysis for standalone wind, PV and hybrid wind PV system, IEEE Trans. Energy Convers. 13 (1998) 70–75.
[6] D.B. Nelson, M.H. Nehrir, C. Wang, Unit sizing and cost analysis of standalone hybrid Wind/PV/Fuel power generation system, Renew. Energy 31 (2006) 1641–1656.
[7] M. Lopez, D. Morales, J.C. Vannier, C.D. Sadarnac, Influence of power converter losses evaluation in the sizing of a HRES, ICCEP-IEEE (2007) 249–254.

[8] M. Khatibi, M. Jazaeri, An analysis for increasing the penetration of renewable energies by optimal sizing of pumped storage power plants, EPEC-IEEE (2008) 1–5.
[9] R. Belfkira, C. Nichita, P. Reghan, G. Barakat, Modeling and optimal sizing of hybrid renewable energy system, IEEE (2008) 1834–1839.
[10] S. Paudel, J.N. Shrestha, F.J. Neto, J.A.F. Ferreira, Optimization of hybrid PV/Wind power system for remote telecom station, ICPS-IEEE (2011) 1–6.
[11] A.C. Nagabhusana, J. Rohini, A.B. Raju, Economic analysis and comparison of proposed HRES for standalone applications at various places in Karnataka state, IEEE PES (2011) 380–385.
[12] A. Kumar, M. Zaman, N. Goel, Probabilistic reliability assessment in the optimization of hybrid power generating units, EPEC-IEEE (2012) 75–79.
[13] G. Laterra, G. Salvina, G.M. Tina, Optimal sizing procedure for hybrid solar wind power system by fuzzy logic, MELECON-IEEE (2006) 865–868.
[14] O. Erdinc, M. Uzunoglu, A new perspective in optimum sizing of hybrid renewable energy system: consideration of component performance degradation issue, Int. J. Hydrog. Energy 37 (2012) 10479–10488.
[15] ÁA. Bayod-Rújula, ME. Haro-Larrodé, A. Martínez-Gracia, Sizing criteria of hybrid photovoltaic–wind systems with battery storage and selfconsumption considering interaction with the grid, Sol. Energy 98 (2013) 582–591.
[16] O. Erdinc, O. Elma, M. Uzunoglu, U.S. Selamogullari, Real-time performance analysis of an optimally sized conversion unit, Energy Build. 75 (2014) 419–429.

CHAPTER TWO

Data visualization and descriptive statistics of solar energy system

Learning objective

- Understand the basics of data visualization and descriptive statistics.
- Know how frequency distribution play effective role to understand the prefeasibility analysis of solar energy system.
- Learn qualitative and quantitative analysis of solar energy system.
- Understand the concept of cross tabulation and scatter plot of solar radiation data.
- Discuss about the measurement of central tendency and variability of solar energy data.

2.1 Introduction

In the recent scenario it is necessary to develop data assessment of every phenomenon and every product. In the field of electrical engineering, statistical analysis plays very vital role to identify the future demand of electrical consumer. It is also necessary to data analysis to find out the perfect location for installation of conventional and non-conventional energy source. For the transition to an energy trading system that prioritizes local resources, improves dependability, produces green jobs, and promotes energy independence, high-quality solar energy prefeasibility analysis data, as well as other technical and management information system data, is critical. Data visualization and descriptive statistics is essential tool for making accurate judgments ranging from government policy and investment decisions to dependable solar power sector planning. Data-driven decisions reflect proper ambition, maximize cost effectiveness, and enable successful solar energy project execution. Using high-quality solar energy resource and other geospatial data as inputs into decision-making processes can help decision-makers improve the effect and outcome of their judgments. Policymakers, investors, and system operators, as well as the universities, non-governmental organizations, and other institutions that assist them, all rely on solar energy statistics to make sound judgments. Solar energy targets can provide market signals to investors and entrepreneurs, as well as inform policymakers about

Decision Science and Operations Management of Solar Energy Systems.
DOI: https://doi.org/10.1016/B978-0-323-85761-1.00002-0

the scope of the issue. Government agencies, utilities, regulators, system operators, technical institutions that support government decisions, and other public and private sector stakeholders are all involved in creating solar energy targets. Geographic factors such as atmospheric phenomena (solar radiation), transportation infrastructure, and political boundaries are represented by geospatial data sets. These data sets can be combined in an unlimited number of ways to generate new information on solar energy potential and to help people make better solar energy decisions. Solar energy resource data that has been created using modeling approaches is known as modelled data. To estimate solar irradiance, for example, atmospheric scientists utilize a range of methodologies, including empirical, semi-empirical, and, more recently, physics-based methods. Atmospheric scientists use downscaled reanalysis data sets and quantitative weather models (e.g., Weather Research and Forecasting Model) to obtain solar resource data. Point-specific solar energy data obtained by skilled persons or meteorological measuring equipment is referred to as measured data. These statistics can provide a reasonable insight of a region's solar resource if numerous points are accessible across the region. These data can also be utilised to evaluate projects on a site-by-site basis. Data visualization is usually done at the time when solar energy resource data is created. Comparing modeled data to "ground truth" data to quantify how closely the model fits reality is referred to as visualization. Visualization can take numerous forms, such as comparing diurnal and annual cycles of solar radiation from a meteorological position, or comparing power output ramps and frequency distributions from a power standpoint. If the validation measurements reveal systemic biases, visualization can play a crucial role in analytic application, sometimes necessitating the usage or acquisition of another data set. Meteorological data, often known as weather data, is frequently modelled or measured in conjunction with wind or solar data, and it is included in or later combined with those data sets. These data can be used to help in solar energy analysis. Air temperature, for example, has an effect on solar photovoltaic (PV) efficiency, increasing or lowering energy output and so altering downstream analyses.

Lots of researchers already apply different concept of statistics in the field of solar energy system. Kumar et al. [1] presented the solar energy potential evaluation for India's southern states based on satellite data. Due to the overall lack of in-situ meteorological data sets, this study relied on satellite-derived simulated solar insolation datasets (especially solar radiation data). The ability to visualize solar insolation components such as direct normal irradiance (DNI) and global horizontal irradiance (GHI) allowed researchers to better understand energy combinations and their potential. The average DNI variance was 5.21 kWh/m^2 and the GHI variation was 5.72 kWh/m^2. A detailed map of southern Indian states with a grid resolution of 100 km x 100 km was presented in the supplemental study. Yeom et al. [2] explored COMS MI geostationary satellite data combined with numerical weather prediction reanalysis variables were used to investigate solar and wind energy resources in North Korea. Under instantaneous (87.90 W m^2 and 16.84 W m^2, respectively) and daily 'all sky circumstances' (624.98 Wh m^2 d1 and 13.89 Wh m^2 d1, respectively), the root mean square error (RMSE) and mean bias error (MBE) were used to compare pyrometer- and satellite-sourced solar radiation. These low results show that satellite-based solar irradiance is accurate enough to be utilised to forecast future solar energy on North Korean land surfaces. Noussan et. al. [3] described data analysis of solar collector energy performance The results demonstrate the great reliability of such systems, which can convert 40 to 60 percent of available radiation and produce annual output yields of more than 400 kWh/m^{2}y. The conversion efficiency varies by season, with the winter months being the least favorable, owing to lesser direct radiation. Other characteristics such as slope, azimuth, and operating temperatures could be the sources of these variances in the district heating systems studied in this study. Pashiardis et al. [4] presented statistical technique for characterization of solar energy utilisation and solar radiation inter-comparison at two sites in Cyprus. For the period January 2013–December 2015, a statistical analysis and inter-comparison of solar radiation at two places in Cyprus representing two different climate regimes of the island is presented. The global, horizontal beam, and diffuse radiation mean yearly and mean monthly daily totals, as well as their frequency distribution for both sites, are estimated and discussed. On a monthly basis, the skewness and kurtosis coefficient values are utilized to determine the frequency distribution type of the aforesaid radiation parameters.

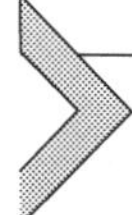

2.2 Basics of data visualization and descriptive statistics

Descriptive statistics are a set of short descriptive coefficients that summarize a data collection, which might be a representation of the complete population or a sample of it. Measures of central tendency and measures of variability are two types of descriptive statistics. The mean, median, and mode are examples of central tendency measures, while standard deviation, variance, minimum and maximum variables, as well as kurtosis and skewness, are examples of variability measurements. The prefeasibility study of a solar energy system is the starting point for data visualization and descriptive statistics. A prefeasibility analysis of a solar energy system must be completed prior to installation and operation. An initial research is conducted in any solar energy project to assess whether it is worthwhile to proceed to the viability study stage. A detailed feasibility assessment should provide a timeline of the projects' development. To determine the optimal place to create a solar energy system, factors such as the climate of the application site, the availability of solar energy sources, the potential of solar energy sources, and the load demand of application sites are considered. In most cases, technical development and project implementation come before a feasibility assessment. As a result, it must be undertaken objectively in order to produce information on which judgments can be made. In this case, descriptive statistics can be used to examine a solar energy system at the primary level. Following are the different primary level factor, which is necessary to assess through descriptive statistics for proper functioning of solar energy system.

- Availability of solar radiation
- Load capacity
- Size of the battery
- Inverter capacity
- Cost analysis of overall project

If any solar energy company want to establish 50kW solar power plant at a particular location, top level management visit number of sites and decide which location will be more compatible for solar energy system, so in that case one question keep in mind, amount of solar radiation at that place. So now it is necessary to data analysis in terms of descriptive statistics of solar radiation of a number of locations. Here first you can assess measures of central tendency of solar radiation in terms of ungrouped data. Measures of central tendency yield information about the centre, or middle part, of a group of number.

2.2.1 3M concept of measurement of central tendency of solar energy system data

When you will apply the concept of central tendency to assess the solar radiation, then assessment is done through the "3M", which means MODE, MEDIAN and Mean. The mode is the most frequently occurring value of solar radiation in a given set of data. Median is the middle value of solar

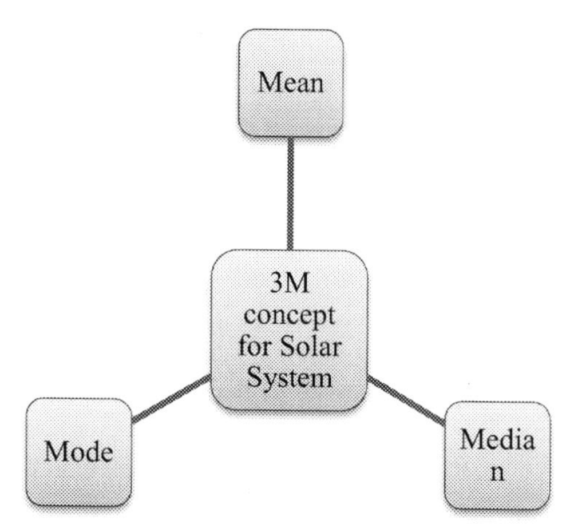

Figure 2.1 3M concept of solar energy system.

radiation in an ordered array data. For an odd number of values of solar radiation, find the middle value of the ordered array as a median and for an even number of values of solar radiation, find the average of the middle two terms. The arithmetic mean is the average of a group of numbers and is computed by summing all numbers and dividing by the number of numbers. Here assessment of solar energy system is also done through the R language, where R language a programming language and free software environment for statistical computing and graphics supported by the R Foundation for Statistical Computing. The R language is widely used among statisticians and data miners for developing statistical software and data analysis. Fig. 2.1 shows the 3M concept of solar energy system.

So again for establishment of 50kW solar energy system, company identify three possible locations and collect month-wise average solar radiation (Kwh/m^2/day) of an individual location in Table. Table shows data of solar radiation in ordered array. Table 2.1 shows the month-wise average solar radiation in kWh/m^2/day and Table 2.2 shows month-wise average solar radiation in ascending order.

Then according to the given data set of solar radiation, find out the value of mean, mode and median. Figs. 2.2–2.4 shows the mean of solar radiation for location 1, location 2 and location 3 through R language respectively. Figs. 2.5–2.7 shows the mode of solar radiation for location 1, location 2 and location 3 through R language respectively. Figs. 2.8–2.10 shows the median of solar radiation for location 1, location 2 and location 3 through R language respectively.

Table 2.1 Month-wise average solar radiation (Kwh/m^2/day).

Month	Location 1	Location 2	Location 3
January	4.5	6.3	3.1
February	5.1	7.1	4.2
March	6.3	8.4	4.7
April	6.5	8.6	4.9
May	6.8	8.8	5.3
June	6.8	9.0	6.1
July	6.3	7.6	5.8
August	6.1	7.1	4.9
September	5.2	6.5	3.9
October	5.4	6.6	4.7
November	5.7	6.3	4.3
December	5.0	6.3	6.7

Table 2.2 Month-wise average solar radiation (ascending order) (Kwh/m^2/day).

Location 1	Location 2	Location 3
4.5	6.3	3.1
5	6.3	3.9
5.1	6.3	4.2
5.2	6.5	4.3
5.4	6.6	4.7
5.7	7.1	4.7
6.1	7.1	4.9
6.3	7.6	4.9
6.3	8.4	5.3
6.5	8.6	5.8
6.8	8.8	6.1
6.8	9	6.7

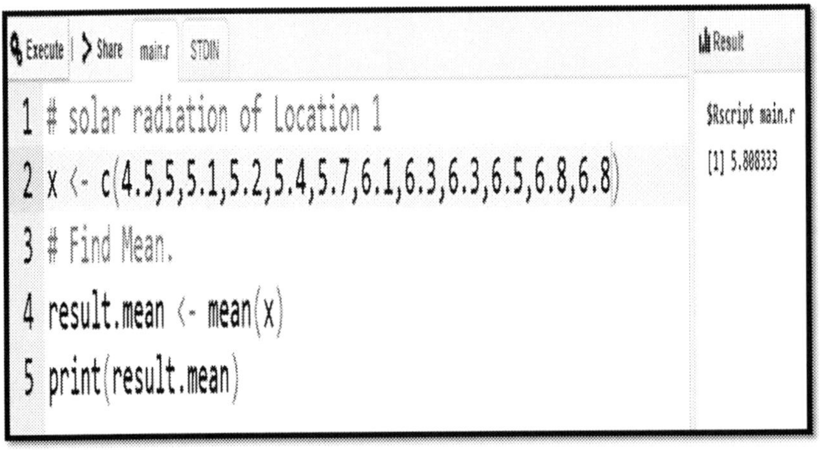

```
Execute | Share   main.r   STDIN                          Result

1 # solar radiation of Location 1                         $Rscript main.r
2 x <- c(4.5,5,5.1,5.2,5.4,5.7,6.1,6.3,6.3,6.5,6.8,6.8)   [1] 5.808333
3 # Find Mean.
4 result.mean <- mean(x)
5 print(result.mean)
```

Figure 2.2 Mean of solar radiation for location 1 through R language.

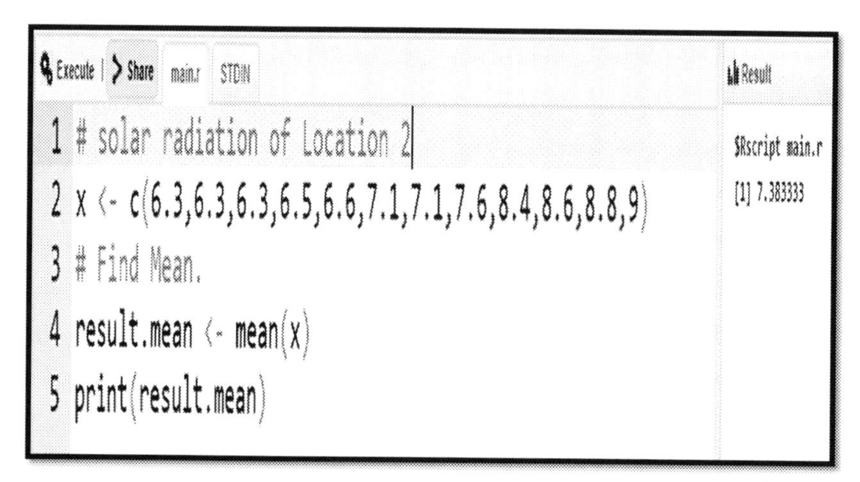

Figure 2.3 Mean of solar radiation for location 2 through R language.

```
1 # solar radiation of Location 3
2 x <- c(3.1,3.9,4.2,4.3,4.7,4.7,4.9,4.9,5.3,5.8,6.1,6.7)
3 # Find Mean.
4 result.mean <- mean(x)
5 print(result.mean)
```

```
$Rscript main.r
[1] 4.883333
```

Figure 2.4 Mean of solar radiation for location 3 through R language.

```
1 # Create the function.
2 getmode <- function(v) {
3    uniqv <- unique(v)
4    uniqv[which.max(tabulate(match(v, uniqv)))]
5 }
6 # Solar radiation for location 1.
7 v <- c(6.3,6.3,6.3,6.5,6.6,7.1,7.6,8.4,8.6,8.8,9)
8 # Calculate the mode using the user function.
9 result <- getmode(v)
10 print(result)
```

```
$Rscript main.r
[1] 6.3
```

Figure 2.5 Mode of solar radiation for location 1 through R language.

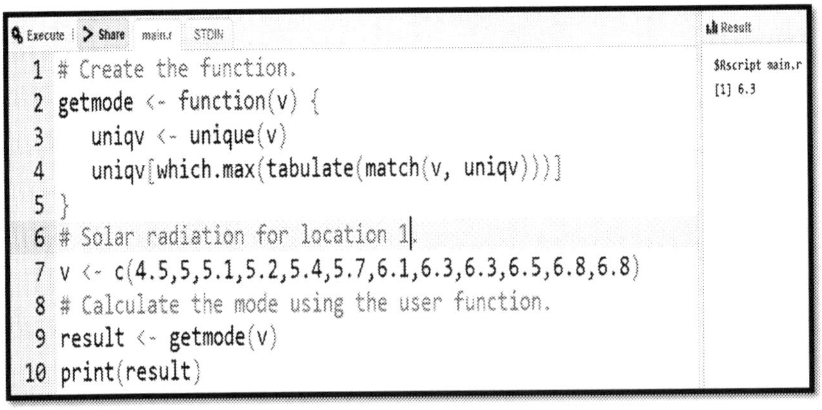

```
Execute | > Share   main.r   STDIN                                   Result

 1 # Create the function.                                             $Rscript main.r
 2 getmode <- function(v) {                                           [1] 6.3
 3    uniqv <- unique(v)
 4    uniqv[which.max(tabulate(match(v, uniqv)))]
 5 }
 6 # Solar radiation for location 1.
 7 v <- c(4.5,5,5.1,5.2,5.4,5.7,6.1,6.3,6.3,6.5,6.8,6.8)
 8 # Calculate the mode using the user function.
 9 result <- getmode(v)
10 print(result)
```

Figure 2.6 Mode of solar radiation for location 2 through R language.

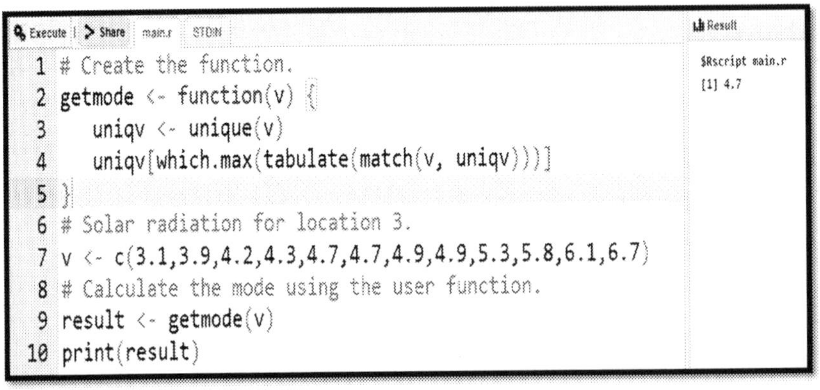

```
Execute | > Share   main.r   STDIN                                   Result

 1 # Create the function.                                             $Rscript main.r
 2 getmode <- function(v) {                                           [1] 4.7
 3    uniqv <- unique(v)
 4    uniqv[which.max(tabulate(match(v, uniqv)))]
 5 }
 6 # Solar radiation for location 3.
 7 v <- c(3.1,3.9,4.2,4.3,4.7,4.7,4.9,4.9,5.3,5.8,6.1,6.7)
 8 # Calculate the mode using the user function.
 9 result <- getmode(v)
10 print(result)
```

Figure 2.7 Mode of solar radiation for location 3 through R language.

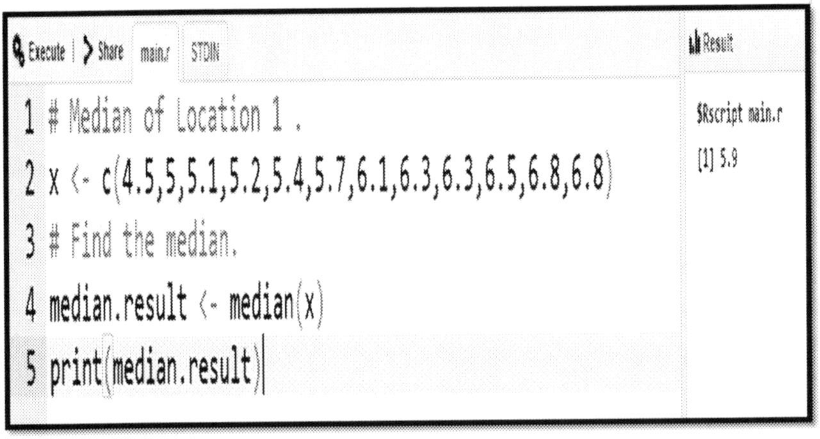

```
Execute | > Share   main.r   STDIN                                   Result

1 # Median of Location 1 .                                            $Rscript main.r
2 x <- c(4.5,5,5.1,5.2,5.4,5.7,6.1,6.3,6.3,6.5,6.8,6.8)               [1] 5.9
3 # Find the median.
4 median.result <- median(x)
5 print(median.result)
```

Figure 2.8 Median of solar radiation for location 1 through R language.

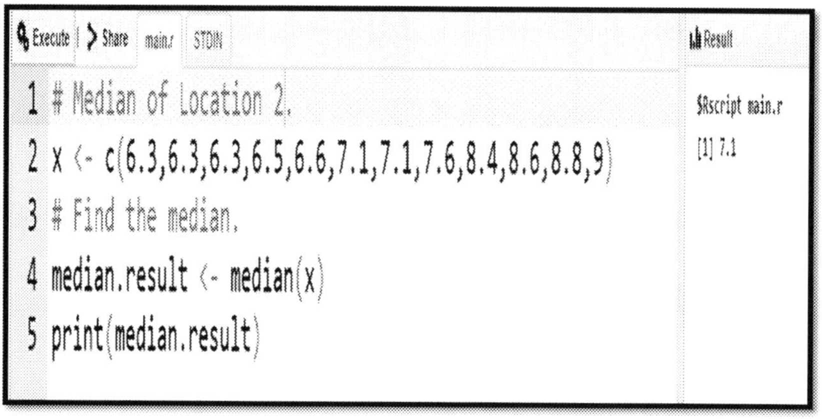

```
Execute | Share  main.r  STDIN                                   Result
1 # Median of Location 2.                                          $Rscript main.r
2 x <- c(6.3,6.3,6.3,6.5,6.6,7.1,7.1,7.6,8.4,8.6,8.8,9)            [1] 7.1
3 # Find the median.
4 median.result <- median(x)
5 print(median.result)
```

Figure 2.9 Median of solar radiation for location 2 through R language.

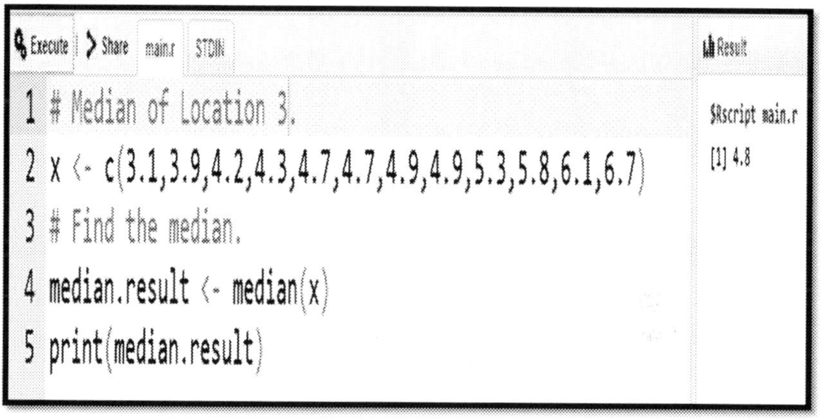

```
Execute | Share  main.r  STDIN                                   Result
1 # Median of Location 3.                                          $Rscript main.r
2 x <- c(3.1,3.9,4.2,4.3,4.7,4.7,4.9,4.9,5.3,5.8,6.1,6.7)          [1] 4.8
3 # Find the median.
4 median.result <- median(x)
5 print(median.result)
```

Figure 2.10 Median of solar radiation for location 3 through R language.

Mean of location 1 = (4.5 + 5+ 5.1 + 5.2 + 5.4 + 5.7 + 6.1 + 6.3 + 6.3 + 6.5 + 6.8 + 6.8)/12 = 69.7/12 = 5.80

Mean of location 2 = (6.3 + 6.3 + 6.3 + 6.5 + 6.6 + 7.1 + 7.1 + 7.6 + 8.4 + 8.6 + 8.8 + 9)/12 = 88.6/12 = 7.38

Mean of location 3 = (3.1 + 3.9 + 4.2 + 4.3 + 4.7 + 4.7 + 4.9 + 4.9 + 5.3 + 5.8 + 6.1 + 6.7)/12 = 58.6/12 = 4.88

Mode of location 1 = Most frequent value of solar radiation = 6.3 (Number of Occurrence 2)

Mode of location 2 = Most frequent value of solar radiation = 6.3 (Number of Occurrence 3)

Table 2.3 Descriptive statistics.

	Location 1	Location 2	Location 3
Mean	5.80	7.38	4.87
Mode (No. of occurrence)	6.3 (2)	6.3 (3)	4.7 (2)
Median	5.9	7.1	4.8

Mode of location 3 = Most frequent value of solar radiation = 4.7 (Number of Occurrence 2)

Median of location 1 = Odd number of values, so median is the average of 6th and 7th number of the value of solar radiation = (5.7 + 6.1)/2 = 5.9

Median of location 2 = (7.1 + 7.1)/2 = 7.1

Table 2.3 shows that descriptive statistics of all the location and result according to the data of solar radiation, location 2 is suitable rather than location 1 and location 3. Because the mean of location 2 is 7.38 kwh/m^2/day, which is greater than the mean of other location, so easily understand 7.38 is suitable value of solar radiation for electricity generation. The mode of location 2 is 6.3 which is greater than mode of location 3 and equal to mode of location 1, but in location 1 number of occurrence is 2 times and for location 2 is number of occurrence is 3. The number of occurrence 3 is shows that at location 3 solar radiation is more stable compare to the other location in addition almost 3 month in a year, value of solar radiation is 6.3 kwh/m^2/day, that's shows stability of location 2 with solar radiation amount is much better than other location.

So that central tendency is useful to find out the descriptive statistics of solar radiation and also useful to find out suitable location for establishment of solar power plant. After the descriptive assessment now you can develop data visualization of solar radiation data through bar chart. Fig. 2.11 comparative graphical assessment of month-wise solar radiation for all the locations.

Now it is necessary to identify the descriptive statistics of load demand of location 2, which is finalize to establish the solar energy system. Table 2.4 shows the month-wise average load capacity of the location 2. Figs. 2.12–2.14 shows mean, mode, median of load demand for location 2 through R Language, respectively.

Mean of location 2 = (26 + 28 + 29 + 29 + 29 + 32 + 32 + 34 + 35 + 36 + 37 + 41)/12 = 388/12 = 32.33

Mode of location 2 = Most frequent value of demand capacity = 29 (Number of Occurrence 3)

Median of location 2 = (32 + 32)/2 = 32

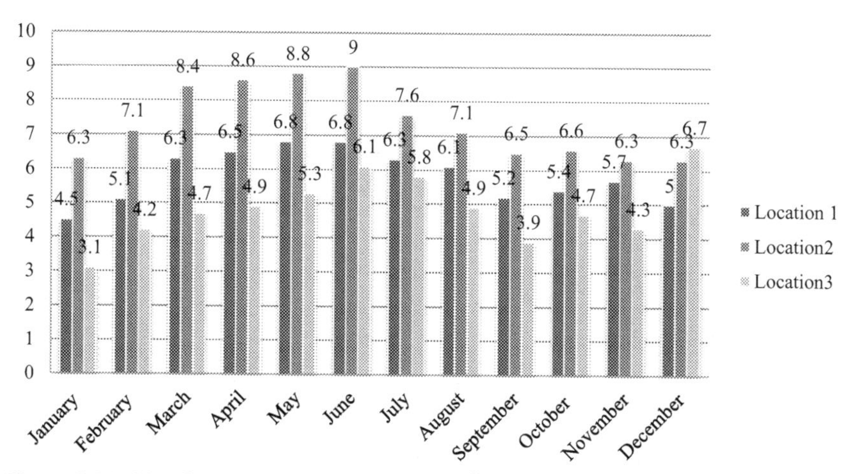

Figure 2.11 Month-wise solar radiation (Kwh/M²/Day) for all the locations.

Table 2.4 Month-wise average load capacity kW.

Month	Load capacity (kw)
January	32
February	29
March	35
April	37
May	36
June	41
July	32
August	34
September	28
October	26
November	29
December	29

Above descriptive statistics shows that mean of load demand is 32.33kW and median is 32KW. Value of mode for the load demand of location 2 for solar power plant is 29 and number of occurrence is 3, which shows that load demand is saturated or steady state. Now you can develop data visualization of load demand and also comparative data visualization between solar radiation of location 2 and load demand of location 2. Figs. 2.15 and 2.16 show graphical representation month wise load capacity and month wise load capacity and solar radiation of location 2 respectively.

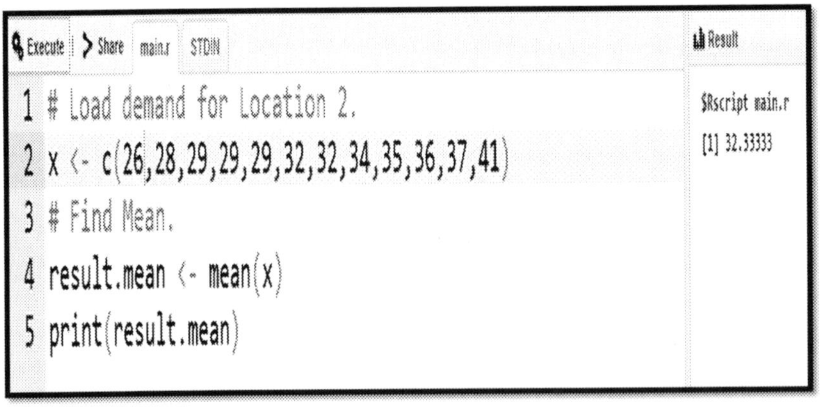

```
Execute  > Share  main.r  STDIN
1 # Load demand for Location 2.
2 x <- c(26,28,29,29,29,32,32,34,35,36,37,41)
3 # Find Mean.
4 result.mean <- mean(x)
5 print(result.mean)
```

```
Result
$Rscript main.r
[1] 32.33333
```

Figure 2.12 Mean of load demand for location 2 through R language.

```
Execute  > Share  main.r  STDIN
1 # Mode of Load Demand
2 getmode <- function(v) {
3    uniqv <- unique(v)
4    uniqv[which.max(tabulate(match(v, uniqv)))]
5 }
6
7 # Create the vector with numbers.
8 v <- c(26,28,29,29,32,32,34,35,36,37,41)
9 # Calculate the mode using the user function.
10 result <- getmode(v)
11 print(result)
```

```
Result
$Rscript main.r
[1] 29
```

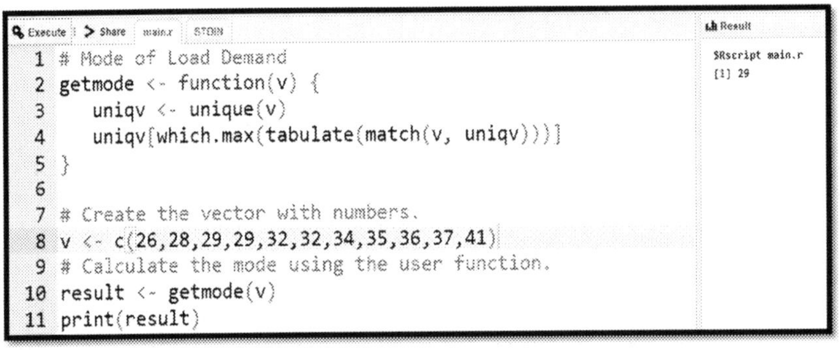

Figure 2.13 Mode of load demand for location 2 through R language.

```
Execute  > Share  main.r  STDIN
1 # Create the vector.
2 x <- c(26,28,29,29,29,32,32,34,35,36,37,41)
3
4 # Find the median.
5 median.result <- median(x)
6 print(median.result)
```

```
Result
$Rscript main.r
[1] 32
```

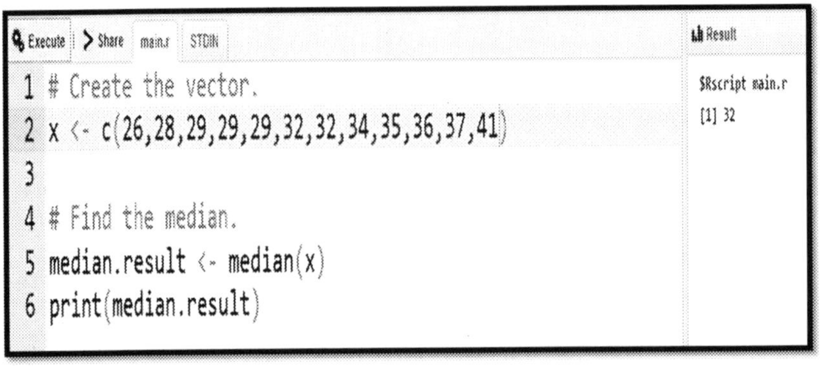

Figure 2.14 Median of load demand for location 2 through R language.

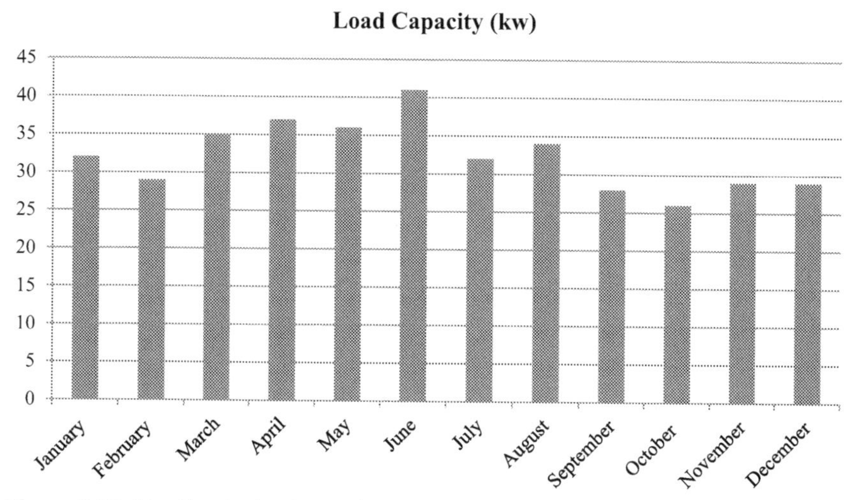

Figure 2.15 Month-wise load capacity.

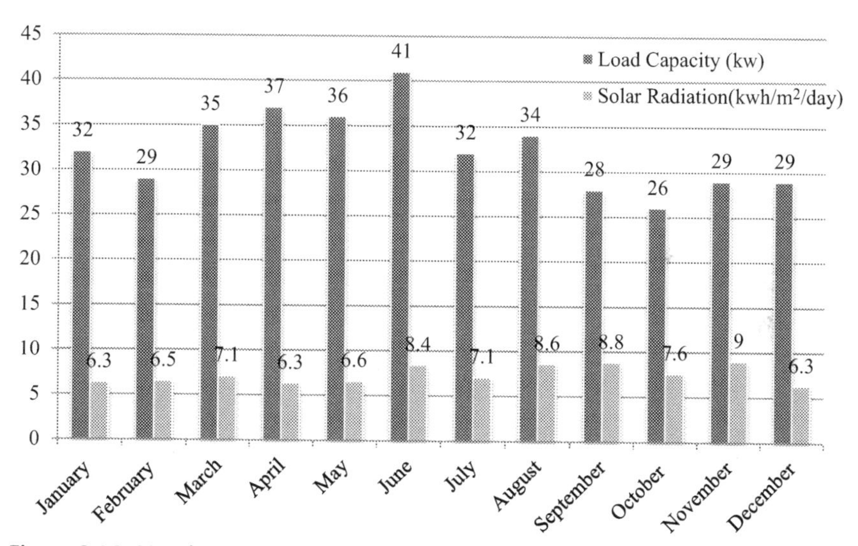

Figure 2.16 Month-wise solar radiation and load capacity.

2.2.2 PQ Assessment of solar energy system data

In the previous analysis, assessment of solar energy system data is evaluated through the "3M" concept, which means Mean, Mode and Median, now assessment is done through the "PQ" concept. PQ is the percentile and quartiles measurement of solar energy system data. Percentiles are measures of central tendency that divide a group of data into 100 parts. There are 97 percentiles, because it takes 97 dividers to separate a group of data into

Table 2.5 Month-wise solar radiation for location 2.

Month	Location 2
January	6.3
February	7.1
March	8.4
April	8.6
May	8.8
June	9.0
July	7.6
August	7.1
September	6.5
October	6.6
November	6.3
December	6.3

100 parts. In the percentiles calculation first arrange the solar energy system data into an ascending order, and then identify the percentile location (k) by

$$k = \left(\frac{P}{100}\right)n$$

Where P is the percentile of interest, k is the percentile location and n is the number in the data set. If k is a whole number, the Pth percentile is the average of the value at the kth and the value at the $(k + 1)^{st}$ location and if k is not a whole number, the Pth percentile value is located at the whole number part of $j + 1$.

Quartiles are measure of central tendency that divide a group of data into four subgroups or parts. The three quartiles are denoted as Q1, Q2, and Q3. The first quartile Q1 separates the first or lowest, one fourth of the data from the upper three fourths and is equal to the 25th percentile. The second quartile Q2 is located at the 50th percentile and equals the median of the data. The third quartile is equal to the value of the 75th percentile.

In the "PQ" assessment if you can determine 40th percentile of the following 12 months (Table 2.5) average solar radiation (kwh/m^2/day) data of location 2:

First organize the data into ascending order:

6.3, 6.3, 6.3, 6.5, 6.6, 7.1, 7.1, 7.6, 8.4, 8.6, 8.8, 9.0

$$k = \left(\frac{40}{100}\right)12 = 4.8$$

Because k is not a whole number, the value of $k+1$ is $4.8 + 1 = 5.8$. The whole number of 5.8 is 6. The 40th percentile is located at the 5th value. The 5th value is 6.6, so 6.6 is the 40th percentile of solar radiation data set.

In the quartile measurement, you want to determine the values of Q1, Q2, Q3 for the above number set of 12th month average solar radiation of location 2.

The value of Q1 is found at the 25th percentile, P_{25} by:

$$\text{For } n = 12k = \left(\frac{25}{100}\right)12 = 3$$

Because k is a whole number, P_{25} is found as the average of third and fourth value of solar radiation $P_{25} = (6.3 + 6.5)/2 = 6.4$

The value of Q1 is $P_{25} = 6.4$.

The value of Q2 is equal to the median. $n = 12$, so median average of two middle values. $Q2 = (7.1 + 7.1)/2 = 7.1$

The value of Q3 is found at the 75th percentile, P_{75} by:

$$\text{For } n = 12k = \left(\frac{75}{100}\right)12 = 9$$

Because k is a whole number, P_{25} is found as the average of ninth and tenth value of solar radiation $P_{75} = (8.4 + 8.6)/2 = 8.5$

2.2.3 MVS assessment of solar energy data

MVS is another measure of variability in terms of "M = mean absolute deviation," "V = variance," and "S = standard deviation."

The mean absolute deviation is the average of the absolute values of the deviations around the mean for a set of numbers. The variance is the average of the squared deviations about the arithmetic mean for a set of numbers. The population variance is denoted by σ^2. The standard deviation is the square root of the variance. The population standard deviation is denoted byσ. Table 2.6 shows deviation from the mean for solar radiation data. Table 2.7 shows variance and standard deviation of solar radiation data.

Now the value of variance and standard deviation is:

$$\text{Variance} = \sigma^2 = \frac{\sum (x-\mu)^2}{n} = \frac{12.25}{12} = 1.02$$

$$\text{Standard Deviation} = \sigma = \sqrt{\frac{\sum (x-\mu)^2}{n}} = \sqrt{1.02} = 1.01$$

Table 2.6 Deviation from the mean for solar radiation data.

Month	Solar radiation (x)	Deviation from mean (x-μ)
January	6.3	6.3-7.38 = −1.08
February	7.1	7.1-7.38 = −0.28
March	8.4	8.4-7.38 = 1.02
April	8.6	8.6-7.38 = 1.22
May	8.8	8.8-7.38 = 1.42
June	9.0	9.0-7.38 = 1.62
July	7.6	7.6-7.38 = 0.22
August	7.1	7.1-7.38 = −0.28
September	6.5	6.5-7.38 = −0.88
October	6.6	6.6-7.38 = −0.78
November	6.3	6.3-7.38 = −1.08
December	6.3	6.3-7.38 = −1.08
	$\Sigma x = 88.6$ Mean $\mu = 88.6/12 = 7.38$	$\Sigma(x - \mu) = 0$

Table 2.7 Variance and standard deviation of solar radiation data.

Month	Solar radiation (x)	Deviation from mean (x-μ)	$(x - \mu)^2$
January	6.3	6.3-7.38 = −1.08	1.1664
February	7.1	7.1-7.38 = −0.28	0.0784
March	8.4	8.4-7.38 = 1.02	1.0404
April	8.6	8.6-7.38 = 1.22	1.4884
May	8.8	8.8-7.38 = 1.42	2.0164
June	9.0	9.0-7.38 = 1.62	2.6244
July	7.6	7.6-7.38 = 0.22	0.0484
August	7.1	7.1-7.38 = −0.28	0.0784
September	6.5	6.5-7.38 = −0.88	0.7744
October	6.6	6.6-7.38 = −0.78	0.6084
November	6.3	6.3-7.38 = −1.08	1.1664
December	6.3	6.3-7.38 = −1.08	1.1664
	$\Sigma x = 88.6$ Mean $\mu = 88.6/12 = 7.38$	$\Sigma(x - \mu) = 0$	$\Sigma(x - \mu)^2 = 12.25$

The standard deviation is the separate entity and as a part of other analysis of solar energy system

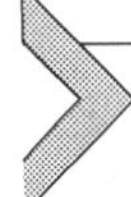

2.3 Frequency distribution of prefeasibility data of solar energy system

One of the first steps in studying and analysing data is to reduce significant and sometimes costly data to a graphic image that is clear, concise, and consistent with the original data's meaning. Frequency distribution is a summary of data presented in the form of class interval and frequencies. Frequency distribution is the assessment of grouped and ungrouped data. Raw data, or data that have not been summarized in any way, are sometimes referred to as "ungrouped data". Data that have been organized into a frequency distribution are called "grouped data". The distinction between ungrouped and grouped data is significant because statistics calculation changes between the two. Table shows that solar radiation data of 365 days for a particular location. Further this data is utilized for quantitative analysis of solar radiation data. Table 2.8 shows the solar radiation data of 365 days for a particular location. Table 2.9 shows frequency distribution of solar radiation data of 365 days. Table 2.10 shows class midpoints, relative frequencies, and cumulative frequency.

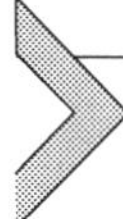

2.4 Quantitative and qualitative analysis of solar radiation data

Quantitative data is defined as the value of data expressed in counts or numbers, with each data set having a distinct numerical value. This data is any quantifiable information that may be utilised for mathematical computations and statistical analysis, with the goal of making real-world decisions based on the results. To answer questions like "how many?" "How often?" and "How much?" quantitative data is used. This information can be validated, and mathematical tools can be used to easily evaluate it.

The gathering of data is an important aspect of the research process. However, in order to make sense of this data, it must be examined. Quantitative data collected in surveys can be analyzed in a variety of ways. They are as follows:

2.4.1 Crosstabulation

The most extensively utilised quantitative data analysis approach is cross-tabulation. It is a recommended strategy since it draws inferences between multiple data sets in the research study using a simple tabular form. It contains information that is either mutually exclusive or has some relationship to one another.

In descriptive statistics, it's fairly typical to want to look at two variables at the same time to see if there's a possible relationship between them. Cross tabulation and scatter plot are two of the more basic tools for observing the

Table 2.8 Solar radiation (kwh/m^2/day) data of 365 days for a particular location.

3.8	3.1	4.9	6.7	8.5	3.3	5.1	6.9	7.8	4.1	5.9	7.7	7.6	4.3	6.1	3.7	5.5	7.3	9.1	3.9
4.5	3.2	5.0	6.8	8.6	3.4	5.2	7.0	7.3	4.2	6.0	7.8	4.3	4.4	6.2	3.8	5.6	7.4	9.2	3.9
6.7	3.3	5.1	6.9	8.7	3.5	5.3	7.1	7.6	4.3	6.1	7.9	5.6	4.5	6.3	3.9	5.7	7.5	9.3	4.1
8.9	3.4	5.2	7.0	8.8	3.6	5.4	7.2	4.3	4.4	6.2	8.0	7.3	4.6	6.4	3.9	5.8	7.6	9.4	4.2
9.1	3.5	5.3	7.1	8.9	3.7	5.5	7.3	5.6	4.5	6.3	8.1	7.8	4.7	6.5	4.1	5.9	7.7	7.6	4.3
3.2	3.6	5.4	7.2	9.0	3.8	5.6	7.4	7.3	4.6	6.4	8.2	9.1	4.8	6.6	4.2	6.0	7.8	4.3	4.4
6.9	3.7	5.5	7.3	9.1	3.9	5.7	7.5	7.8	4.7	6.5	8.3	3.1	4.9	6.7	4.3	6.1	7.9	5.6	4.5
6.9	3.8	5.6	7.4	9.2	3.9	5.8	7.6	9.1	4.8	6.6	8.4	3.2	5.0	6.8	4.4	6.2	8.0	7.3	4.6
8.7	3.9	5.7	7.5	9.3	4.1	5.9	3.8	3.1	4.9	6.7	8.5	3.3	5.1	6.9	4.5	6.3	8.1	7.8	4.7
8.3	3.9	5.8	7.6	9.4	4.2	6.0	4.5	3.2	5.0	6.8	8.6	3.4	5.2	7.0	4.6	6.4	8.2	9.1	4.8
7.8	4.1	5.9	7.7	7.6	4.3	6.1	6.7	3.3	5.1	6.9	8.7	3.5	5.3	7.1	4.7	6.5	8.3	3.1	4.9
7.3	4.2	6.0	7.8	4.3	4.4	6.2	8.9	3.4	5.2	7.0	8.8	3.6	5.4	7.2	4.8	6.6	8.4	3.2	5.0
7.6	4.3	6.1	7.9	5.6	4.5	6.3	9.1	3.5	5.3	7.1	8.9	3.7	5.5	7.3	4.9	6.7	8.5	3.3	5.1
4.3	4.4	6.2	8.0	7.3	4.6	6.4	3.2	3.6	5.4	7.2	9.0	3.8	5.6	7.4	5.0	6.8	8.6	3.4	5.2
5.6	4.5	6.3	8.1	7.8	4.7	6.5	6.9	3.7	5.5	7.3	9.1	3.9	5.7	7.5	5.1	6.9	8.7	3.5	5.3
7.3	4.6	6.4	8.2	9.1	4.8	6.6	6.9	3.8	5.6	7.4	9.2	3.9	5.8	3.4	5.2	7.0	8.8	3.6	5.4
7.8	4.7	6.5	8.3	3.1	4.9	6.7	8.7	3.9	5.7	7.5	9.3	4.1	5.9	3.5	5.3	7.1	8.9	3.7	5.5
9.1	4.8	6.6	8.4	3.2	5.0	6.8	8.3	3.9	5.8	7.6	9.4	4.2	6.0	3.6	5.4	7.2	9.0	3.8	5.6
5.7	5.8	5.9	6.0	6.1															

Table 2.9 Frequency distribution of solar radiation (kwh/m^2/day) data of 365 days for a particular location.

Class interval	Frequency
3 under 4	61
4 under 5	61
5 under 6	65
6 under 7	58
7 under 8	62
8 under 9	58

Table 2.10 Class midpoints, relative frequencies, and cumulative frequency.

Class interval	Frequency	Class midpoints	Relative frequencies	Cumulative frequencies
3 under 4	61	3.5	0.167	61
4 under 5	61	4.5	0.167	122
5 under 6	65	5.5	0.179	187
6 under 7	58	6.5	0.159	245
7 under 8	62	7.5	0.169	307
8 under 9	36	8.5	0.098	343
9 under 10	22	9.5	0.060	365
Total	365			

relationship between two variables. Cross tabulation is a method of creating a two-dimensional matrix that shows frequency counts for two variables at the same time. As an example of prefeasibility analysis of solar energy system, gather a sample of 365 days data of solar radiation (kwh/m^2/day) and clearness index. The minimum and maximum value of clearness index is 0.401 and 0.678 respectively and same factors for solar radiation are 4.3 and 6.9 respectively. Table 2.11 shows solar energy system data by solar radiation and clearness index. Table 2.12 shows cross-tabulation table of solar radiation data.

By tallying the frequencies of responses for each combination of categorize between solar radiation and clearness index and data are cross tabulated according to the two variables. In this example, there is a tally of how many solar energy system data, each possible combination of solar radiation and clearness index data, such as according to the Table, 12 frequency responses, when clearness index is 0.401 and solar radiation is 4.5.

2.4.2 Scatter plot

A scatter plot is a two-dimensional graph plot of two numerical variables' pairs of points. A scatter plot is a graphical tool that is frequently used to

Table 2.11 Solar energy system data by solar radiation and clearness index.

Days	Clearness index	Solar radiation (kwh/m^2/day)
1	0.401	4.5
2	0.407	5.7
3	0.582	6.3
4	0.555	6.9
5	0.674	7.1
6	0.573	6.3
7	0.407	5.2
8	0.582	6.3
9	0.555	6.9
10	0.674	7.1
11	0.407	6.3
12	0.582	6.9
.	.	.
.	.	.
.	.	.
365	0.401	6.3

Table 2.12 Cross-tabulation table of solar radiation data.

		Solar radiation (kWh/m^2/day)						Total
Clearness		4.5	5.2	5.7	6.3	6.9	7.1	
index	0.401	12	8	20	7	3	10	60
	0.407	10	5	5	15	10	10	55
	0.555	14	10	21	9	10	6	70
	0.573	20	15	5	17	13	15	85
	0.582	7	6	3	11	8	5	40
	0.674	5	10	10	10	13	7	55
Total		68	54	64	69	57	53	365

investigate potential relationships between two variables. Table 2.13 shows the assessment of solar energy system.

Data that approximates and characterizes is defined as qualitative data. It is possible to observe and record qualitative data. The nature of this data type is nonnumerical. Observations, one-on-one interviews, focus groups, and other methods are used to acquire this type of information. Data that may be organized categorically based on the features and features of an object or a phenomenon is known as qualitative data in statistics. Fig. 2.17 shows the plot between solar radiation and clearness index.

Table 2.13 Assessment of solar energy system data.

Solar radiation (kWh/m²/day)	Clearness index	Solar radiation (kWh/m²/day)	Clearness index
4.3	0.659	4.5	0.711
4.7	0.678	5.3	0.489
3.2	0.688	6.2	0.322
5.1	0.654	6.8	0.745
5.8	0.632	7.0	0.812
4.9	0.532	6.9	0.367
3.9	0.438	5.3	0.589
5.5	0.476	5.8	0.70

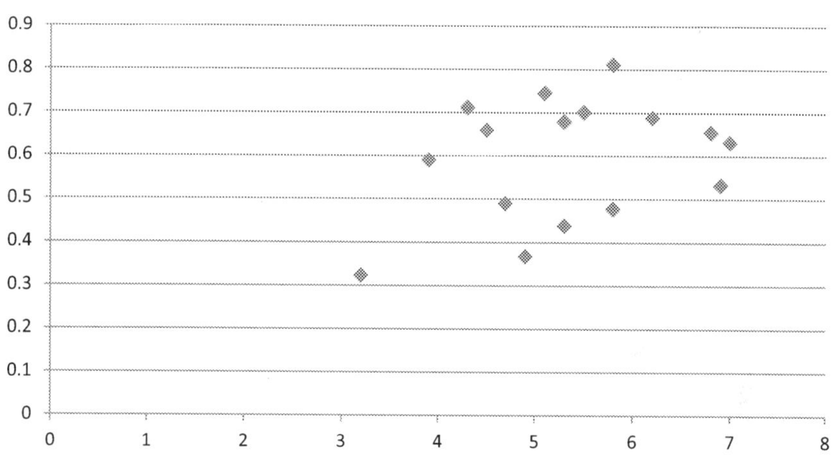

Figure 2.17 Solar radiation v/s clearness index.

2.4.3 Two main approaches to qualitative data analysis

2.4.3.1 Deductive approach

The deductive technique entails examining qualitative data using a predetermined structure devised by the researcher. The questions can be used as a guide for interpreting the data by the researcher. This method is quick and simple, and it can be utilised when a researcher has a good sense of what kind of replies he or she will get from the sample population.

2.4.3.2 Inductive approach

On the other hand, the inductive approach is not reliant on a preset structure or framework. It takes more time and effort to conduct a complete qualitative

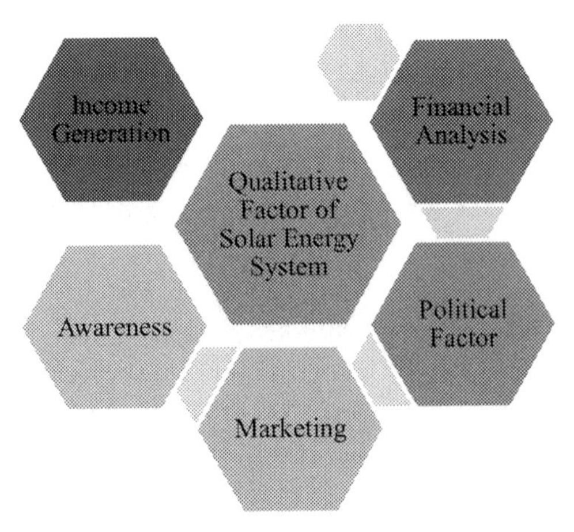

Figure 2.18 Qualitative factor of solar energy system.

data analysis. When a researcher has little or no knowledge of the investigated phenomenon, an inductive methodology is frequently applied. Fig. 2.18 shows the qualitative factor of solar energy system [9].

The term "income production" refers to a profitable investment or commercial activity. Solar energy projects can be a lucrative source of revenue. Land is an important component of any solar energy project and can help landowners generate additional cash. During both the development and operational stages of a solar project, land leases, direct and indirect hiring of locals account for roughly 10-15 percent of capital expenditure and roughly 70%–80% of operational expenditures, according to solar developer Azure Power Global Limited.

Solar energy projects can generate jobs both during and after the development stage, as well as in the operation and maintenance of the projects after they are completed. For the course of the development phase, local laborers, electricians, and others are engaged. Azure also stated that during the construction phase of the project, it employs people from nearby local villages and later provides direct and indirect employment through jobs such as module cleaning, grass cutting, security guards, and technicians, among others, for the project's operations and maintenance. Long-term jobs, on the other hand, are more likely to be produced indirectly. "One such example is of Jagir Singh, a lessor for Azure 2 MW project in Punjab, where he was employed as a security guard to look after the plant, and now, a decade later, he heads security for the plant," according to the source of that company.

Understanding your solar production resource, PV system cost, value of power, and available subsidies is required for a robust financial analysis of a solar energy project. Investors must understand the essential components of a PV proposal and how to decide if the system is a good investment in order to make an informed decision. Another important aspect is to ensure that the proposal provides for the tax benefits as well as any tax rises resulting from the lower utility prices. Many proposals list the system cost after all tax benefits are taken into account, while the electric savings are listed on a pre-tax basis.

When we identify the political factor in the field of solar energy system, so in that case it is necessary to identify three primary objectives, which is to theorize and explore the relationships between solar energy sector and political power and to critically assess tensions associated with an agenda to democratize solar energy systems. It is also necessary, draw out the implications for democratizing solar energy development in practice. If we go through the market analysis of solar energy system, in that case result shows that the global solar energy market was valued at $52.5 billion in 2018 and is projected to reach $223.3 billion by 2026, growing at a CAGR of 20.5% from 2019 to 2026. This report breaks down the global solar energy market by technology, application, component, marketing channel, and region. The market is divided into two categories based on technology: photovoltaic systems and concentrated solar power systems. Monocrystalline, polycrystalline, cadmium telluride, amorphous silicon cells, and other solar modules are available. It is divided into three categories according on its use: residential, commercial, and industrial. The market is divided into four categories based on end use: electricity generation, lighting, warmth, and charging. It is examined in North America, Europe, and Asia (Germany, France, Italy, Spain, UK and rest of Europe).

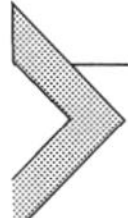

2.5 Measurement of central tendency and variability of solar energy data

Individual values are not revealed when data is grouped. As a result, for grouped data, measures of central tendency and variability must be computed differently than for ungrouped or raw data.

2.5.1 Mean

The mean for grouped data is then computed by summing the products of the class midpoint and the class frequency for each class and dividing that sum by the total number of frequencies. The formula for the mean of grouped data follows.

$$\text{Mean of grouped data } \mu_{\text{Grouped}} = \frac{\sum F_i M_i}{n} = \frac{\sum F_i M_i}{F_i}$$

Where F_i is the class frequency, M_i is the class midpoint and n is total frequency.

2.5.2 Median

The median for ungrouped or raw data is the middle value of an ordered array of numbers. For grouped data, solving for the median is considerably more complicated. The calculation of the median for grouped data is done by using the following formula.

$$\text{Median of the grouped data} = \text{Median} = \left(l + \frac{\frac{n}{2} - CF_P}{F_{MED}} \right) D$$

Where l is the lower limit of the median class interval, CF_P are a cumulative total of the frequencies up to but not including frequency of the median class, F_{MED} is the frequency of the median class, D is the width of the median class interval.

2.5.3 Mode

The mode for grouped data is the class midpoint of the model class. The modal class is the class interval with the greatest frequency.

2.5.4 Measures of variability

The measures of variability for grouped data are presented here: the variance and standard deviation. Again, the standard deviation is the square root of the variance.

"Original formulas" for population variance and standard deviation of grouped data:

$$\sigma^2 = \frac{\sum F_i (M_i - \mu)^2}{n}$$

$$\sigma = \sqrt{\sigma^2}$$

Formula for "computational version" is

$$\sigma^2 = \frac{\sum F_i M_i^2 - \frac{(\sum F_i M_i)^2}{n}}{n}$$

Table 2.14 Calculation of grouped mean.

Class interval	Frequency (F_i)	Class midpoints (M_i)	Cumulative frequencies	F_i M_i
3 under 4	61	3.5	61	213.5
4 under 5	61	4.5	122	274.5
5 under 6	65	5.5	187	357.5
6 under 7	58	6.5	245	377
7 under 8	62	7.5	307	465
8 under 9	36	8.5	343	306
9 under 10	21	9.5	364	199.5
Total	364			$\Sigma F_i M_i = 2193$

Where F_i is the frequency, M_i is the class midpoint, n is the total frequencies of the population and μ is grouped mean for the population.

Refer to the Tables 2.8 and 2.9: Solar radiation (kwh/m^2/day) data of 365 days for a particular location and frequency distribution of solar radiation (kwh/m^2/day) data of 365 days for a particular location.

$$\mu_{\text{Grouped}} = \frac{\sum F_i M_i}{n} = \frac{2193}{364} = 6.02$$

Mean = 6.02

The first step in calculating a grouped median is to determine the value of n/2, which is the location of the median term. Since there are 364 values (n), the value of n/2 is 364/2 =182. The median is 182th term. The question to ask is: where does the 182th term fall? This can be answered by determining the cumulative frequencies for the data, as shown in Table 2.14. An examination of these cumulative frequencies reveals that the 182th term falls in the third class interval because there are only 122 values in first two class intervals. Thus the median value is in the third class interval between 5 and 6. The class interval containing the median value is referred to as the median class interval. Since the 182th value is between 5 and 6, the value of median must be at least 5. How much more than 5 is the median? The difference between the location of the median value, n/2 = 182, and the cumulative frequencies up to but not including the median class interval, 122, the frequency of median class is 65. Width of the class interval is 1. Now we put the value in the formula of median

$$\text{Median}=\left(5+\frac{\frac{364}{2}-122}{65}\right)2 = 5+\left(\frac{182-122}{65}\right)2 = 5+\left(\frac{60}{65}\right)1 = 5.923$$

Median = 5.923

The mode for grouped data is the class midpoint of the modal class. The modal class is the class interval with the greatest frequency. Modal class is 5 under 6, and then class midpoint is 5.5.

Mode = 5.5

2.5.5 Measures of variability

Table 2.15 shows the calculating grouped variance and standard deviation with the original formula. Table 2.16 shows calculating grouped variance and standard deviation with the computational formula.

"Original formulas" for population variance and standard deviation of grouped data:

$$\sigma^2 = \text{Variance} = \frac{\sum F_i(M_i - \mu)^2}{n} = \frac{1170.78}{364} = 3.216$$

$$\text{Standard deviation} = \sigma = \sqrt{\sigma^2} = \sqrt{3.216} = 1.79$$

Formula for "computational version" is

$$\sigma^2 = \frac{\sum F_i M_i^2 - \frac{(\sum F_i M_i)^2}{n}}{n} = \frac{14383 - \frac{(2193)^2}{364}}{364} = \frac{14383 - 13212}{364} = 3.21$$

$$\text{Standard deviation} = \sigma = \sqrt{\sigma^2} = \sqrt{3.21} = 1.79$$

2.5.6 Z score

A Z score represents the number of standard deviations a value (x) is above or below the mean of a set of numbers when the data are normally distributed. Using Z scores allows translation of a value's raw distance from the mean into units of standard deviation.

$$Z\,\text{Scores} = Z = \frac{X - \text{Mean}}{\text{Standard Deviation}}$$

If a Z score is negative, the raw value (x) is below the mean. If the z score is positive, the raw value (x) is above the mean. For the example, the data set of solar radiation (kwh/m^2/day), that is normally distributed with a mean of solar radiation of 11 kwh/m^2/day and standard deviation of 6.32 kwh/m^2/day, suppose energy manager wants to determine the Z score of the value of solar radiation at 13 kwh/m^2/day.

$$Z\,\text{Scores} = Z = \frac{13 - 11}{6.32} = 0.316$$

Table 2.15 Calculating grouped variance and standard deviation with the original formula.

Class interval	Frequency (F_i)	Class midpoints (M_i)	$F_i M_i$	$(M_i - \mu)$ $\mu = 6.02$	$(M_i - \mu)^2$	$F_i (M_i - \mu)^2$
3 under 4	61	3.5	213.5	−2.52	6.3504	387.3744
4 under 5	61	4.5	274.5	−1.52	2.3104	140.9344
5 under 6	65	5.5	357.5	−0.52	0.2704	17.576
6 under 7	58	6.5	377	0.48	0.2304	13.3632
7 under 8	62	7.5	465	1.48	2.1904	135.8048
8 under 9	36	8.5	306	2.48	6.1504	221.4144
9 under 10	21	9.5	199.5	3.48	12.1104	254.3184
Total	364		$\Sigma F_i M_i = 2193$		$\Sigma(M_i - \mu)^2 = 29.61$	$\Sigma \mathrm{Fi}(M_i - \mu)^2 = 1170.78$

Table 2.16 Calculating grouped variance and standard deviation with the computational formula.

Class interval	Frequency (F_i)	Class midpoints (M_i)	$F_i M_i$	$F_i M_i^2$
3 under 4	61	3.5	213.5	747.25
4 under 5	61	4.5	274.5	1235.25
5 under 6	65	5.5	357.5	1966.25
6 under 7	58	6.5	377	2450.5
7 under 8	62	7.5	465	3487.5
8 under 9	36	8.5	306	2601
9 under 10	21	9.5	209	1895.25
Total	364		$\Sigma F_i M_i = 2193$	$\sum F_i M_i^2 = 14383$

2.5.7 The empirical rules in terms of Z score

Between Z = −1 and Z = +1 are approximately 68% of the values.

Between Z = −2 and Z = +2 are approximately 95% of the values.

Between Z = −3 and Z = +3 are approximately 99.7% of the values.

2.5.8 Coefficient of variation

The coefficient of variation is a statistic that is the ratio of the standard deviation to the mean expressed in percentage and is denoted by CV

$$\text{Coefficient of Variations} = \frac{\text{Standard Deviation}}{\text{Mean}} \times 100$$

In the Z score data set mean of solar radiation of 11 kwh/m^2/day and standard deviation of 6.32 kwh/m^2/day. The coefficient of variation essentially is a relative comparison of a standard deviation to its mean.

$$\text{Coefficientof Variations} = \frac{6.32}{11} \times 100 = 57.45\%$$

2.6 Measures of shapes of solar energy data

Measures of shape are tools that can be used to describe the shape of a distribution of data. In this section, we examine two measures of shape, skewness and kurtosis.

2.6.1 Skewness

A distribution of data in which the right half is mirror image of the left half is said to be symmetrical. The skewed portion is the long, thin part of the curve. Many researchers use a skewed distribution to denote that the

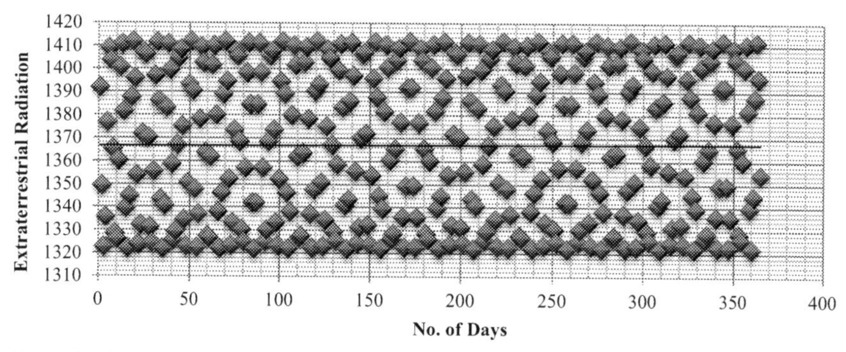

Figure 2.19 Extra-terrestrial radiation of 400 days.

data are sparse at one end of the distribution and piled up at the other end. The concept of skewness helps to understand the relationship of the mean, median and mode. In a unimodal distribution that is skewed, the mode is the apex of the curve and the median is the middle value. The mean tends to be located towards the tail of the distribution, because mean is particularly affected by the extreme values. A bell shaped or normal distribution with the mean, median and mode all at the center of the distribution has no skewness.

2.6.2 Extraterrestrial and terrestrial radiations

The intensity of solar radiation keeps on attenuating as it propagates away from the surface of the sun, through the wavelength remain unchanged. Solar radiation incident on the outer atmosphere of the earth is known as Extra-terrestrial radiation I_{ETR}. The solar constant I_{scons} is defined as the energy received from the sun per unit time, on a unit area of surface perpendicular to the direction of propagation of the radiation at the top of the atmosphere and at the surface mean distance from the sun [10–12].

The Extra-terrestrial I_{ETR} = Isc [1.0 + 0.033cos(360n/365)] W/m^2

Where n is the day of year, starting from 1st January.

Now the data of extra terrestrial radiation of all the 365 days shown in Table 2.17. Table 2.18 shows extra-terrestrial radiation from 211 to 365 days. Fig. 2.19 shows extra-terrestrial radiation of 400 days. Table 2.19 shows frequency distribution of extra-terrestrial radiation (w/m^2) data of 365 days for a particular location. Table 2.20 class midpoints, relative frequencies and cumulative frequency. Table 2.21 shows calculation of grouped mean.

$$\mu_{\text{Grouped}} = \frac{\sum F_i M_i}{n} = \frac{499125}{365} = 1367.46$$

Table 2.17 Extra-terrestrial radiation from 1 to 210 days.

1391.891	1323.805	1366.976	1410.209	1342.15	1338.097	1408.489	1372.017	1322.622	1387.531
1349.358	1332.317	1404.609	1380.032	1321.889	1379.938	1404.663	1332.38	1349.268	1411.828
1322.64	1371.92	1408.528	1338.173	1342.068	1410.181	1367.073	1323.777	1391.81	1395.94
1335.688	1407.113	1375.22	1322.155	1384.597	1401.714	1329.418	1353.921	1412.111	1354.109
1376.806	1406.346	1334.543	1346.339	1411.351	1362.128	1325.453	1395.79	1391.973	1323.834
1409.133	1370.308	1322.962	1389.044	1398.347	1326.91	1358.732	1411.85	1349.448	1332.255
1403.691	1331.305	1350.859	1411.988	1357.242	1327.63	1399.423	1387.704	1322.658	1371.823
1365.357	1324.3	1393.225	1394.603	1324.884	1363.643	1411.048	1344.999	1335.618	1407.068
1328.496	1355.573	1412.082	1352.473	1330.281	1402.666	1383.187	1322.016	1376.71	1406.394
1326.152	1397.09	1390.525	1323.366	1368.594	1409.716	1340.815	1339.359	1409.098	1370.406
1360.426	1411.633	1347.88	1333.375	1405.478	1378.474	1321.916	1381.48	1403.747	1331.364
1400.593	1386.165	1322.374	1373.526	1407.869	1336.946	1343.433	1410.621	1365.455	1324.268
1410.646	1343.516	1336.873	1407.827	1373.623	1322.36	1386.076	1400.658	1328.547	1355.479
1381.573	1321.92	1378.379	1405.529	1333.44	1347.791	1411.619	1360.522	1326.111	1397.017
1339.436	1340.735	1409.684	1368.692	1323.342	1390.442	1397.163	1326.194	1360.329	1411.647
1322.009	1383.095	1402.725	1330.338	1352.381	1412.078	1355.668	1328.445	1400.528	1386.253
1344.913	1411.027	1363.741	1324.849	1394.525	1393.305	1324.331	1365.259	1410.671	1343.6
1387.618	1399.491	1327.678	1357.147	1411.995	1350.95	1331.245	1403.634	1381.665	1321.924
1411.839	1358.829	1326.865	1398.277	1389.129	1322.984	1370.211	1409.168	1339.514	1340.656
1395.865	1325.491	1362.031	1411.369	1346.426	1334.475	1406.298	1376.901	1322.002	1383.004
1354.015	1329.364	1401.652	1384.687	1322.166	1375.123	1407.157	1335.758	1344.828	1411.006

Table 2.18 Extra-terrestrial radiation from 211 to 365 days.

1399.559	1378.285	1323.391	1398.417	1379.844	1323.005	1397.236	1381.388	1322.676	1396.015
1358.925	1409.653	1333.309	1357.338	1410.152	1334.407	1355.763	1410.596	1335.547	1354.202
1325.53	1402.785	1373.429	1324.919	1401.776	1375.027	1324.363	1400.723	1376.615	
1329.311	1363.838	1407.786	1330.224	1362.226	1408.451	1331.185	1360.619	1409.063	
1366.878	1327.726	1405.58	1368.496	1326.955	1404.716	1370.113	1326.235	1403.804	
1404.555	1326.82	1368.79	1405.427	1327.582	1367.171	1406.25	1328.395	1365.552	
1408.566	1361.934	1330.395	1407.91	1363.546	1329.473	1407.202	1365.162	1328.598	
1375.316	1401.589	1324.815	1373.719	1402.606	1325.415	1372.115	1403.577	1326.069	
1334.611	1410.237	1357.051	1333.505	1409.747	1358.636	1332.443	1409.203	1360.232	
1322.941	1380.125	1398.206	1323.317	1378.568	1399.355	1323.749	1376.997	1400.462	
1350.768	1338.248	1411.387	1352.288	1337.019	1411.069	1353.828	1335.829	1410.695	
1393.145	1322.145	1384.777	1394.448	1322.346	1383.278	1395.714	1322.605	1381.758	
1412.085	1346.252	1342.231	1412.002	1347.703	1340.894	1411.86	1349.178	1339.591	
1390.609	1388.959	1321.889	1389.214	1390.358	1321.913	1387.791	1391.728	1321.995	
1347.968	1411.98	1341.987	1346.513	1412.075	1343.35	1345.084	1412.111	1344.743	
1322.389	1394.68	1384.507	1322.177	1393.384	1385.987	1322.023	1392.054	1387.443	
1336.801	1352.566	1411.333	1338.022	1351.042	1411.604	1339.282	1349.538	1411.817	

Table 2.19 Frequency distribution of extra-terrestrial radiation (w/m²) data of 365 days for a particular location.

Class interval	Frequency
1320 under 1330	70
1330 under 1340	38
1340 under 1350	30
1350 under 1360	25
1360 under 1370	26
1370 under 1380	27
1380 under 1390	28
1390 under 1400	34
1400 under 1410	51
1410 under 1420	36

Table 2.20 Class midpoints, relative frequencies, and cumulative frequency.

Class interval	Frequency	Class midpoints	Relative frequencies	Cumulative frequencies
1320 under 1330	70	1325	0.191781	70
1330 under 1340	38	1335	0.10411	108
1340 under 1350	30	1345	0.082192	138
1350 under 1360	25	1355	0.068493	163
1360 under 1370	26	1365	0.071233	189
1370 under 1380	27	1375	0.073973	216
1380 under 1390	28	1385	0.076712	244
1390 under 1400	34	1395	0.093151	278
1400 under 1410	51	1405	0.139726	329
1410 under 1420	36	1415	0.09863	365

Mean = 1367.46

The first step in calculating a grouped median is to determine the value of n/2, which is the location of the median term. Since there are 365 values (n), the value of n/2 is 365/2 =182.5. The median is the average 182th and 183th term.

An examination of these cumulative frequencies reveals that the 182th and 183th term falls in the fifth class interval because there are only 163 values in first four class intervals. Thus the median value is in the fifth class interval between 1360 and 1370. The class interval containing the median value is referred to as the median class interval. Since the 182th and 183th value is between 1360 and 1370, the value of median must be at least 1360. How much more than 1360 is the median? The difference between the location of the median value, n/2 = 182 and 183, and the cumulative frequencies up

Table 2.21 Calculation of grouped mean.

Class interval	Frequency (F_i)	Class midpoints (M_i)	Cumulative frequencies	F_i M_i
1320 under 1330	70	1325	70	92750
1330 under 1340	38	1335	108	50730
1340 under 1350	30	1345	138	40350
1350 under 1360	25	1355	163	33875
1360 under 1370	26	1365	189	35490
1370 under 1380	27	1375	216	37125
1380 under 1390	28	1385	244	38780
1390 under 1400	34	1395	278	47430
1400 under 1410	51	1405	329	71655
1410 under 1420	36	1415	365	50940
Total	365			$\sum F_i M_i = 499125$

to but not including the median class interval, 163, the frequency of median class is 26. Width of the class interval is 10. Now we put the value in the formula of median

$$\text{Median} = \left(1360 + \frac{\frac{365}{2} - 163}{26}\right)2 = 1360 + \left(\frac{182.5 - 163}{26}\right)10$$

$$= 1360 + \left(\frac{19.5}{26}\right)10 = 5.923$$

Median $= 1367.5$

The mode for grouped data is the class midpoint of the modal class. The modal class is the class interval with the greatest frequency. Modal class is 1360 under 1370, and then class midpoint is 1365.

Mode $= 1365$

2.6.2.1 *Measures of variability*

Table 2.22 shows calculating grouped variance and standard deviation with the original formula.

Original formulas" for population variance and standard deviation of grouped data:

$$\sigma^2 = \text{Variance} = \frac{\sum F_i(M_i - \mu)^2}{n} = \frac{374580}{365} = 1026.24$$

$$\text{Standard Deviation} = \sigma = \sqrt{\sigma^2} = \sqrt{1026.24} = 32.07$$

Table 2.23 shows the calculating grouped variance and standard deviation with the computational formula.

Table 2.22 Calculating grouped variance and standard deviation with the original formula.

Class interval	Frequency (F_i)	Class midpoints (M_i)	F_i M_i	($M_i - \mu$) $\mu = 1367.46$	$(M_i - \mu)^2$	F_i $(M_i - \mu)^2$
1320 under 1330	70	1325	92750	−42.46	1802.8516	126199.612
1330 under 1340	38	1335	50730	−32.46	1053.6516	40038.7608
1340 under 1350	30	1345	40350	−22.46	504.4516	15133.548
1350 under 1360	25	1355	33875	−12.46	155.2516	3881.29
1360 under 1370	26	1365	35490	-2.46	6.0516	157.3416
1370 under 1380	27	1375	37125	7.54	56.8516	1534.9932
1380 under 1390	28	1385	38780	17.54	307.6516	8614.2448
1390 under 1400	34	1395	47430	27.54	758.4516	25787.3544
1400 under 1410	51	1405	71655	37.54	1409.2516	71871.8316
1410 under 1420	36	1415	50940	47.54	2260.0516	81361.8576
Total	364		$\sum F_i M_i =$ 499125		$\Sigma(M_i - \mu)^2 =$ 8314.51	$\Sigma \mathrm{Fi}(M_i - \mu)^2$ $= 374580$

Table 2.23 Calculating grouped variance and standard deviation with the computational formula.

Class interval	Frequency (F_i)	Class midpoints (M_i)	F_i M_i	F_i M_i^2
1320 under 1330	70	1325	92750	122893750
1330 under 1340	38	1335	50730	67724550
1340 under 1350	30	1345	40350	54270750
1350 under 1360	25	1355	33875	45900625
1360 under 1370	26	1365	35490	48443850
1370 under 1380	27	1375	37125	51046875
1380 under 1390	28	1385	38780	53710300
1390 under 1400	34	1395	47430	66164850
1400 under 1410	51	1405	71655	100675275
1410 under 1420	36	1415	50940	72080100
Total	365		$\Sigma F_i M_i =$ 499125	$\sum F_i M_i^2 =$ 682910925

Formula for "computational version" is

$$\sigma^2 = \frac{\sum F_i M_i^2 - \frac{\left(\sum F_i M_i\right)^2}{n}}{n} = \frac{682910925 - \frac{(499125)^2}{364}}{365}$$

$$= \frac{682910925 - 682536344}{365} = 1029$$

$$\text{Standard Deviation} = \sigma = \sqrt{\sigma^2} = \sqrt{1029} = 32.07$$

Solar time: The solar time measured with reference to solar noon, which is the time when the sun is crossing the observer's meridian.

Solar time = standard time ∓ 4 $(l_{st} - l_{loc}) + e$

Where l_{st} and l_{loc} are the standard longitude used for measuring standard time of the country and the longitude of the observer's location respectively [5–8].

Where e = 9.87sin2b – 7.53cosb – 1.5 sinb min.

Where b = (360/364) × (n-81)

n = day of the year, starting from 1st January. Fig. 2.20 shows the assessment of extra-terrestrial radiation. Table 2.24 shows the frequency distribution of equation of time of 364 days for a particular location. Table 2.25 shows the class midpoints, relative frequencies and cumulative frequency. Table 2.26 shows calculation of grouped mean.

$$\mu_{\text{Grouped}} = \frac{\sum F_i M_i}{n} = \frac{525}{364} = 1.44$$

Mean = 1.44

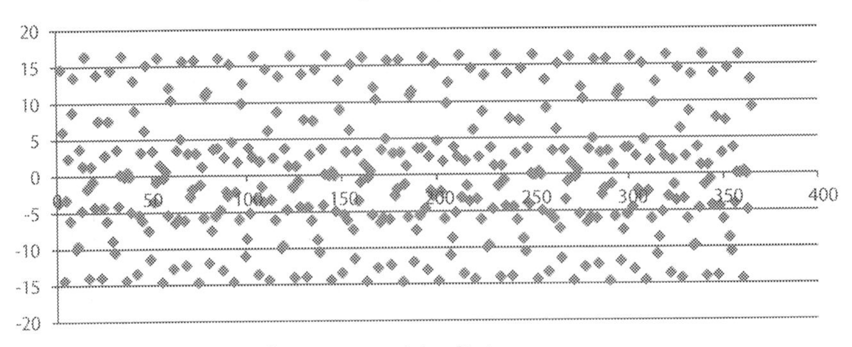

Figure 2.20 Assessment of extra-terrestrial radiation.

Table 2.24 Frequency distribution of equation of time of 364 days for a particular location.

Class interval	Frequency
−15 to −10	27
−10 to −5	34
−5 to 0	103
0 to 5	99
5 to 10	27
10 to 15	74

Table 2.25 Class midpoints, relative frequencies, and cumulative frequency.

Class interval	Frequency	Class midpoints	Relative frequencies	Cumulative frequencies
−15 to −10	27	−12.5	0.074176	27
−10 to −5	34	−7.5	0.093407	61
−5 to 0	103	−2.5	0.282967	164
0 to 5	99	2.5	0.271978	263
5 to 10	27	7.5	0.074176	290
10 to 15	74	12.5	0.203297	364

Table 2.26 Calculation of grouped mean.

Class interval	Frequency (F_i)	Class midpoints (M_i)	Cumulative frequencies	F_i M_i
−15 to −10	27	−12.5	27	−337.5
−10 to −5	34	−7.5	61	−255
−5 to 0	103	−2.5	164	−257.5
0 to 5	99	2.5	263	247.5
5 to 10	27	7.5	290	202.5
10 to 15	74	12.5	364	925
Total	364			$\sum F_i M_i = 525$

The first step in calculating a grouped median is to determine the value of n/2, which is the location of the median term. Since there are 364 values (n), the value of n/2 is 364/2 =182. The median is the 182th term.

An examination of these cumulative frequencies reveals that the 182th term falls in the fourth class interval because there are only 164 values in first three class intervals. Thus the median value is in the fourth class interval between 0 and 5. The class interval containing the median value is referred to as the median class interval. Since the 182th value is between 0 and 5. The difference between the location of the median value, n/2 = 182 the cumulative frequencies up to but not including the median class interval, 164, the frequency of median class is 99. Width of the class interval is 5. Now we put the value in the formula of median

$$\text{Median}=\left(0+\frac{\frac{364}{2}-164}{99}\right)5=0+\left(\frac{182-164}{99}\right)5=0+\left(\frac{18}{99}\right)5=0.909$$

Median = 0.909

The mode for grouped data is the class midpoint of the modal class. The modal class is the class interval with the greatest frequency. Modal class is 0 under 5, and then class midpoint is 2.5.

Mode = 2.5

2.6.2.2 Measures of variability

Table 2.27 shows calculating grouped variance and standard deviation with the original formula. Table 2.28 shows calculating grouped variance and standard deviation with the computational formula.

"Original Formulas" for Population Variance and Standard Deviation of Grouped Data:

$$\sigma^2=\text{Variance}=\frac{\sum F_i(M_i-\mu)^2}{n}=\frac{19717.7904}{364}=54.169$$

$$\text{Standard Deviation}=\sigma=\sqrt{\sigma^2}=\sqrt{54.169}=7.3600$$

Formula for "computational version" is

$$\sigma^2=\frac{\sum F_iM_i^2-\frac{\left(\sum F_iM_i\right)^2}{n}}{n}=\frac{20475-\frac{(525)^2}{364}}{365}=\frac{20475-757.211}{365}=54$$

$$\text{Standard deviation}=\sigma=\sqrt{\sigma^2}=\sqrt{54}=7.3600$$

Table 2.27 Calculating grouped variance and standard deviation with the original formula.

Class interval	Frequency (F_i)	Class midpoints (M_i)	$F_i M_i$	$(M_i - \mu)$ $\mu = 1.44$	$(M_i - \mu)^2$	$F_i (M_i - \mu)^2$
−15 to −10	27	−12.5	−337.5	−13.94	194.3236	5246.737
−10 to −5	34	−7.5	−255	−8.94	79.9236	2717.402
−5 to 0	103	−2.5	−257.5	−3.94	15.5236	1598.931
0 to 5	99	2.5	247.5	1.06	1.1236	111.2364
5 to 10	27	7.5	202.5	6.06	36.7236	991.5372
10 to 15	74	12.5	925	11.06	122.3236	9051.946
Total	364		$\sum F_i M_i = 525$		$\Sigma(M_i - \mu)^2 =$	$\Sigma \text{Fi}(M_i - \mu)^2 = 19717.7904$

Table 2.28 Calculating grouped variance and standard deviation with the computational formula.

Class interval	Frequency (F_i)	Class midpoints (M_i)	F_i M_i	F_i M_i^2
−15 to −10	27	−12.5	−337.5	4218.75
−10 to −5	34	−7.5	−255	1912.5
−5 to 0	103	−2.5	−257.5	643.75
0 to 5	99	2.5	247.5	618.75
5 to 10	27	7.5	202.5	1518.75
10 to 15	74	12.5	925	11562.5
Total	364		$\Sigma F_i M_i = 525$	$\sum F_i M_i^2 = 20475$

2.7 Conclusion

This chapter is the assessment of solar energy system through the data visualization and descriptive statistics. There are number of technical parameters in the solar energy system, but it is necessary to take perfect decision from cradle to grave process of solar power plant. From the mean, mode and median you can assess ideal value of solar radiation and load demand. All the data of solar energy system related parameters should not be analyzed in the same way statistically because the entities represented by the numbers are different. For this reason, it is necessary to apply the concept of data analysis and operational management in the field of solar power plant or solar energy system.

2.8 Exercise/question

1. Explain the concept of data visualization and descriptive statistics.
2. What is the significance of data visualization in the field of solar energy system?
3. What is the significance of descriptive statistics in the field of solar energy system?
4. What is the 3M concept of data analysis and how it can helpful in the field of solar energy system?
5. What is R language?
6. Explain with the help of an example percentile and quartiles measurement of solar energy system data.

7. What is the significance of variance and standard deviation in the solar energy system data analysis?
8. Write short note on the following:
   (i) Qualitative analysis of solar radiation data
   (ii) Quantitative analysis of solar radiation data
9. Write short note on the following:
   (iii) Cross tabulation of solar radiation data
   (iv) Scatter plot of solar radiation data
10. What is the significance of Z-Score in the data analysis of solar energy system?
11. Find mean, mode, and median of following solar radiation data.
    Table: Month-wise average solar radiation (Kwh/m^2/day)

Month	Location 1	Location 2	Location 3
January	5.5	5.3	4.1
February	5.1	7.1	4.2
March	6.3	8.4	4.7
April	7.5	7.6	5.9
May	6.8	8.8	5.3
June	6.8	9.0	6.1
July	7.3	7.6	5.8
August	6.1	7.1	4.9
September	5.2	4.5	4.9
October	5.4	6.6	4.7
November	5.7	5.3	5.3
December	6.0	5.3	7.7

12. Find the percentile and quartile of following solar radiation data.

Month	Location 2
January	7.3
February	7.1
March	8.4
April	7.6
May	8.8
June	9.0
July	7.6
August	7.1
September	6.5
October	5.6
November	4.3
December	7.3

13. Find the variance and standard deviation of following solar radiation data.

Month	Location 2
January	7.3
February	7.1
March	8.4
April	7.6
May	8.8
June	9.0
July	7.6
August	7.1
September	6.5
October	5.6
November	4.3
December	7.3

14. What are the prominent parameters to identify the shape of the solar energy system data?

15. Find real time application of data analysis in the field of solar energy system.

References

[1] D. Kumar, Satellite-based solar energy potential analysis for southern states of India, Energy Rep, Volume 6, Pages 1487–1500.

[2] J.M. Yeom, R.C. Deo, Exploring solar and wind energy resources in North Korea with COMS MI geostationary satellite data coupled with numerical weather prediction reanalysis variables, Renew. Sustain. Energy Rev. Volume 119, 109570.

[3] M. Noussan, M. Jarre, Data Analysis of the Energy Performance of Large Scale Solar Collectors for District Heating, in: Energy Procedia, 9th International Conference on Sustainability in Energy and Buildings, SEB-17, Chania, GREECE, 2017 5-7 July.

[4] S. Pashiardis, Statistical analysis for the characterization of solar energy utilization and inter-comparison of solar radiation at two sites in Cyprus, Appl. Energy, Volume 190, Pages 1138–1158.

[5] G.F. Viscondi, N.A. Solange, A Systematic Literature Review on *big data* for *solar* photovoltaic electricity generation forecasting, Sustain. Energy Technol. Asses. Volume 31, 54–63.

[6] Y.F. Huang, C. Shun-H., Mining optimum models of generating *solar* power based on *big data* analysis, Sol. Energy, Volume 15, 224–232.

[7] F.P. Pei, L. Dong-B., Double-layered *big data*analytics architecture for *solar* cells series welding machine, Comput. Ind, Volume 97, 17–23.

[8] D. Yang, D. Zibo, Analyzing *big* time series *data*in *solar* engineering using features and PCA *Solar Energy*, Volume 153, 317–328.

[9] Y. Xu, L. Hui, A distributed computing framework for *wind* speed *big data* forecasting on Apache Spark, Sustain. Energy Technol. Assess. Volume 37.

[10] X. Yin, Z. Xiaowei, *Big data* driven multi-objective predictions for offshore *wind*farm based on machine learning algorithms, Energy, Volume 186.

[11] B.K. Sovacool, A qualitative factor analysis of renewable energy and sustainable energy for all in the Asia-Pacific, Energy Policy 59 (2013) 393–403.
[12] F.G. Akinboro, I.A. Adejumobi, V. Makinde, Solar energy installation in Nigeria: observations, prospects, problems and solutions, Trans. J. Sci. Technol. 2 (4) (2013).

CHAPTER THREE

Facilities location and plant layout of solar energy system

Learning Objective

- Understand the basics of Facilities location and Plant layout.
- Know how different factor effect the choice of suitable location for solar power plant.
- Learn process product matrix of solar energy system.
- Understand the performance measures of power plant layout design.
- Discuss about the group technology based solar power plant layout design.

3.1 Introduction

Location decision pertain to the choice of appropriate geographical sites for locating various manufacturing and/or service facilities of an organizations. Locating facilities in regions that offer attractive cost advantages is one aspect of the trade-off. It may be expensive to set up a distribution system that is both efficient and responsive to an increase in transportation costs and the costs of coordinating and communicating with the supplier about products and availability. Location issues have more prominent in recent years due to the increased pace of economic reforms in several countries and the consequent globalization of markets.

Fig. 3.1 shows the affecting parameters for location decision in an organization. The most significant factor that drives globalization is the on-going financial and regulatory reforms in several developed and developing countries. Beginning in 1991, we embarked on a set of regulatory changes that made India much more attractive in terms of locating a manufacturing facility. Two events have been broadly responsible for this. First is the reduction in customs and excise tariffs and a move forward the single point value added tax regime. The other is the delicensing of several sectors of industry and the progressive removal of the cap on foreign direct investment. Another relevant point for location planning with respect to regulatory issue is the emergence of regional trading blocs. A trading bloc is essentially a group of geographically separated nations that provide advantageous access to markets,

Decision Science and Operations Management of Solar Energy Systems.
DOI: https://doi.org/10.1016/B978-0-323-85761-1.00007-X

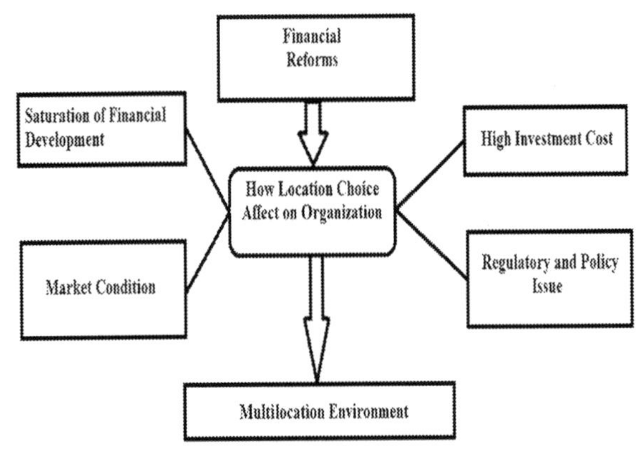

Figure 3.1 Affecting parameters for location decision on organization.

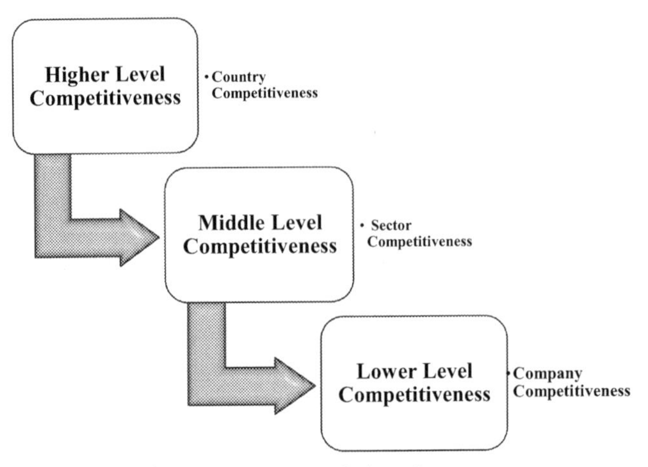

Figure 3.2 Assessment of competitiveness of a location.

manufacturing facilities, and technologies within the member countries. The factors that drive the globalization of operations are aptly summarized by the study on world competitiveness rating by Geneva-based World Economic Forum. Fig. 3.2 shows the assessment of competitiveness of a location. The model proposes three way of competitiveness, higher-level competitiveness, middle level competitiveness, and lower-level competitiveness. Higher-level competitiveness is the part of country competitiveness which include rules, regulation, and policy of the government toward the different industrial sector according to the business law. Middle level competitiveness is related to sector competitiveness, which include present market condition and cost of labor. Lower level competitiveness includes individual organization based

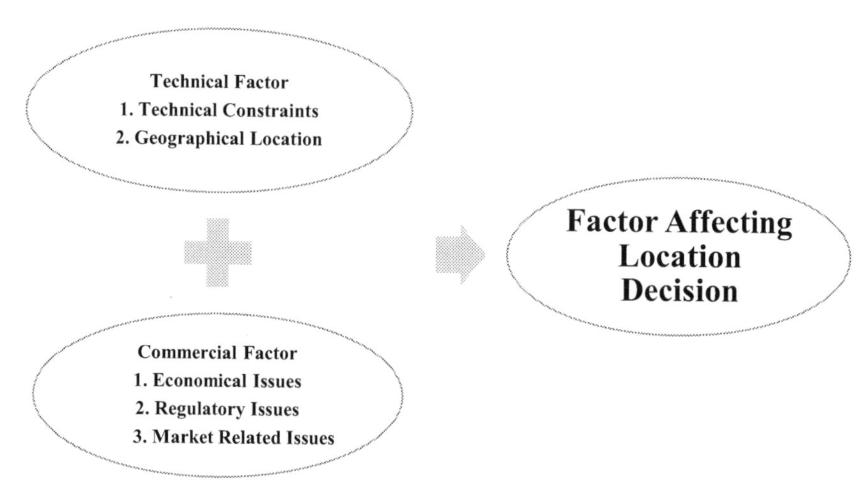

Figure 3.3 Factor affecting the location decision.

competitiveness, which include the ability to design, produce, and market products superior to competitors.

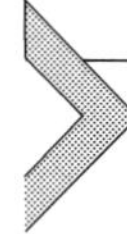

3.2 Factor affecting location decision of solar power plant

Solar energy system is one of the prominent sources of energy and it also reduces the greenhouse gas emission from the environment. When any organization or solar energy company or the government of any country want to install a solar power plant, then two points are arising "where" and "what." The meaning of "What" is the estimated capacity of solar power plant in KW, MW, and GW and after that, the most important point is "where," the meaning, where you want to install the solar power plant. Fig. 3.3 shows the factor affecting the location decision. Location decision is a very important step when your start the solar power plant project. Location planning is completed in three forms: identify the number of factors that could affect the location decision of solar energy system, create relative importance of these measures, and develop a framework to analyze the impact of these factors. There is different factor, which effect the location of solar power plant. The factors are classified into two categories Technical factor and Commercial factor. Technical factor is related to the technical constraints and geographical location, on the other hand different economical, regulatory and market related issues are the part of commercial factor.

In the solar power plant installation, all the factor is related to the prefeasibility analysis of the given system. Most important parameter is location decision is technical factor of given location, which include amount of solar radiation, wind velocity, and day wise temperature of a particular site. Another important measurement parameter is the longitude and latitude of a particular site.

The latitude is a geographic coordinate that indicates the North-South position of a particular point on the surface of the earth. Latitudes are angles that range from 0° at the equator to 90° North or 90° South at the poles. All latitude lines are parallel, and they run from East to West as circles, which are parallel to the equator. Together with the longitudes, latitudes are used to indicate a precise location of any feature on the earth's surface. Temperature is inversely related to latitude. As latitude increases, the temperature falls, and vice versa. Generally, around the world, it gets warmer toward the equator and cooler toward the poles.

3.2.1 How latitudes affect temperature & solar radiation?

Latitude is one of the primary factors that affect the temperature and temperature affects the amount of solar radiation. As one moves further away from the equator, the temperature falls because regions receive less sunlight. The reason behind this is the shape of the earth. The shape of the earth is an oblate spheroid. Thus, not all locations receive the same amount of sunlight heat or insolation (incoming solar radiation). Another reason for the difference in temperatures varying with latitude is the angle of solar incidence. The rays from the sun strike the surface of the earth at different angles. At the equator, the incidence of the sun's rays is at a right angle, and this translates to more heat because they are concentrated over a small area. It also implies that less heat is lost in the atmosphere because they travel a short distance in the atmosphere. On the other hand, at the poles, the Sun rays strike the surface of the earth at an acute angle. The rays from the sun are dispersed over a large area. This also implies that more heat is lost in the atmosphere because they travel long distances over the atmosphere before it hits the earth's surface. The existence of a market for the solar energy related products and their services significantly influences the location decision. It is very necessary to install the solar power plant close to the either the market or the source of raw material and other critical inputs for the system. Availability of raw material of solar panel and other equipment, if the company want to manufacturing at its own level. It is also identifying, what is the consumer demand at a particular location. Cost-related issues capture the desirability of competing locations on the basis of the cost of

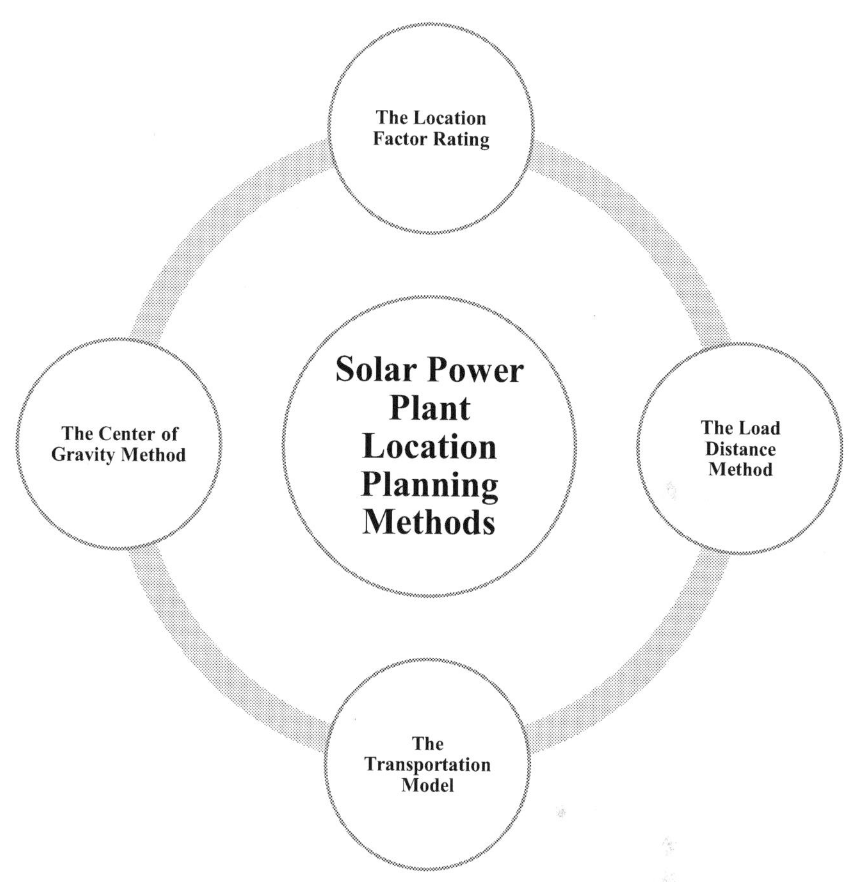

Figure 3.4 Solar power plant location planning method.

operating the solar energy system in alternative locations. Typically, logistics and distribution costs are often considered for the analysis simply because these costs are direct, tangible, and easy to measure and analyze.

3.3 Location planning method of solar power plant

In solar energy system location, finding is done in two ways: one input to multiple output and multiple input to multiple output. Here we discuss different location planning method to install the solar power plant, which works in efficient manner. Fig. 3.4 shows the solar power plant location planning method. Following are the location planning method:

1. **Location Factor Rating:** Factor rating is a simple methodology to assess the attractiveness of each potential location for solar power plant. This method involves four steps in which the relevant parameters are

Table 3.1 Factor rating.

Factors	**Rating (1–100)**
Amount of solar radiation	80
Consumer demand in KW	75
Shaded area near the location	70
Skilled technical workers	60
Government policy toward solar energy at a particular location	65
Subsidy policy toward solar energy at a particular location	65

identified, their relative importance established, the performance of each location in each factor assessed, and finally, all this information is combined to rank the locations.

The four step process of identifying an appropriate location is as follows:

a. Identify and list all the relevant parameters for the location decision for solar energy system.

b. Establish the relative importance of each parameter in the final decision for location of solar energy system.

c. Rate the performance of each candidate location using a rating mechanism.

Example 3.1. One of the solar energy companies is actively considering five alternative locations to install the solar power plant. Based on the survey related to solar power plant, company has six factors to be considered for final site selection for solar power plant. The ratings of each factor on a scale of 1–100 provide this information of location of solar energy system, as shown in Table 3.1.

The first step in the solution is to develop the relative importance of each parameter using a normalization factor. The sum of all the parameters related to solar energy system is 415. Therefore, by dividing each parameter rating by 415 one can obtain the relative weight of the parameters, which is shown in Tables 3.2 and 3.3.

Since the normalized weight of each parameter, one can compute how each location fares by weighing the rating of the location against each factor with the weight for the factor. The computation for location 1 is as follows:

$$(45 \times 0.192771084) + (30 \times 0.180722892) + (35 \times 0.168674699)$$
$$+(30 \times 0.144578313) + (40 \times 0.156626506) + (35 \times 0.156626506)$$
$$= 36.084$$

Table 3.2 Factor rating.

Factors	Rating (1–100)	Relative weight
Amount of solar radiation	80	0.192771084
Consumer demand in KW	75	0.180722892
Shaded area near the location	70	0.168674699
Skilled technical workers	60	0.144578313
Government policy toward solar energy at a particular location	65	0.156626506
Subsidy policy toward solar energy at a particular location	65	0.156626506
Sum of all factor rating	**415**	**1**

Similarly, factor rating of locations 2, 3, 4, and 5 is 46.433, ***63.698***, 45.337, and 48.216, respectively. So according to the above assessment location 3 is more feasible location for solar energy system. Factor-rating systems are the most widely used location techniques as they combine diverse factors in an easy-to-understand format. Lot of factors are taken into consideration which other processes don't take into consideration. Unforeseen factors which effect the establishment of plant.

2. **The Centre of Gravity Method:** In the location factor rating method, we utilize a simple weighted scheme to assess the attractiveness of each location for solar power plant. In the center of gravity method, all the demand points are represented in a Cartesian coordinate system. Each demand point will also have weights, which indicate the quantum of the demand per unit time. Now you can understand the basic concept of the center of gravity method with the help of the following example.

Example 3.2. One of the solar panel manufacturing company, developed four plants at different locations. Now owner of that company want to develop a data center, which should be easily connected with all the 4 locations of that company. Apply the center of gravity method to find out the suitable location for that data center. *X–Y* coordinate and quantity of product of all the four locations are given in Table 3.4. Fig. 3.5 shows the statistics of solar plant locations.

Let us represent the given information using the following notations:
Number of existing point $= n = 4 = w, x, y, z$
Coordinate of location i in the map $= (Xi, Yi)$
Quantity of Product with different points $i = Pi$

Table 3.3 Rating of each location of solar energy system against each factor.

Factors	Relative weight	Location 1	Location 2	Location 3	Location 4	Location 5
Amount of solar radiation	0.192771084	45	60	75	62	42
Consumer demand in KW	0.180722892	30	58	70	56	55
Shaded area near the location	0.168674699	35	41	63	45	47
Skilled technical workers	0.144578313	30	45	55	25	51
Government policy toward solar energy at a particular location	0.156626506	40	40	60	42	50
Subsidy policy toward solar energy at a particular location	0.156626506	35	30	55	35	45
Sum of all factor rating		**36.084**	**46.433**	**63.698**	**45.337**	**48.216**
Ranking of the Locations		**5**	**3**	**1**	**4**	**2**

Table 3.4 Data of four locations.

Solar plant location	X-coordinates	Y-coordinates	Quantity of solar panel produced per day
W	135	120	1200
X	260	85	1600
Y	500	350	1100
Z	250	400	2100

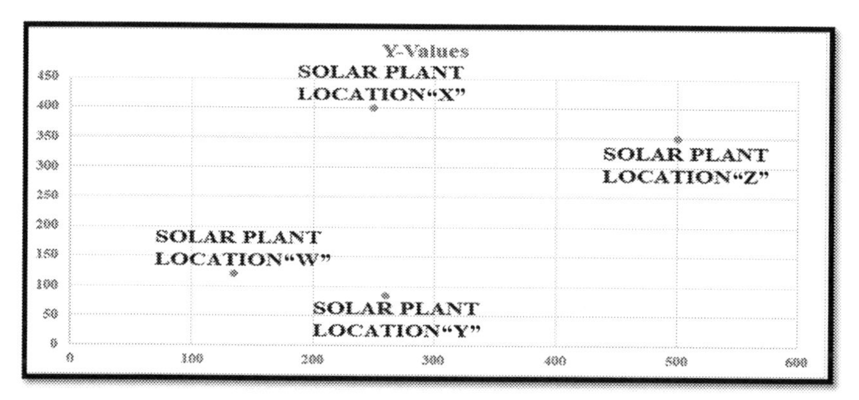

Figure 3.5 Statistics of solar plant location.

So the coordinates of the center of gravity are given by:

$$X_C = \frac{\sum_{i=1}^{n} (x_i) \times P_i}{\sum_{i=1}^{n} P_i} \text{ And } Y_C = \frac{\sum_{i=1}^{n} (y_i) \times P_i}{\sum_{i=1}^{n} P_i}$$

$$X_C = \frac{(135 \times 1200) + (260 \times 1600) + (500 \times 1100) + (250 \times 2100)}{1200 + 1600 + 1100 + 2100}$$

$$= \frac{165,3000}{6000} = 275.5$$

$$Y_C = \frac{(120 \times 1200) + (85 \times 1600) + (350 \times 1100) + (400 \times 2100)}{1200 + 1600 + 1100 + 2100}$$

$$= \frac{150,5000}{6000} = 250.83$$

Fig. 3.6 shows the data center of four solar power plant locations. This method is purely location-based because the location is such a significant cost and revenue driver, location often has the power to make or break a company's business strategy and obtains best geographical location by considering the distance between each facility.

3. **Load Distance Method:** The load-distance technique is a mathematical model for evaluating sites based on their proximity to one another. The

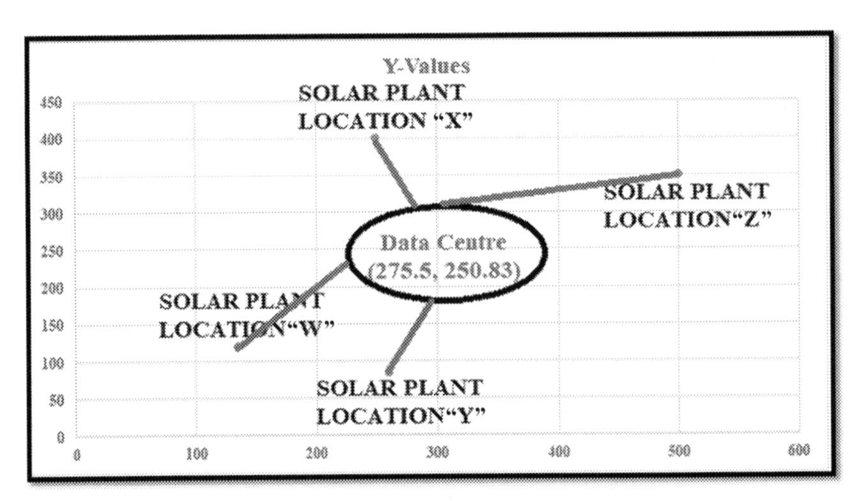

Figure 3.6 Data center of four solar power plant locations.

goal is to choose a site that reduces the total weighted loads entering and leaving the facility. When two points are assigned to grid coordinates on a map, the distance between them is stated. Instead of using distance, time can be used as an alternative.

Suppose that a new warehouse of one of the solar panel manufacturing company is to be located to serve Mumbai. It will receive inbound shipments from several suppliers, including one in Pune. If the new warehouse of solar panel were located in Hyderabad, what would be the distance between the two facilities? If shipments travel by truck, the distance depends on the highway system and the specific route taken. The exact mileage between any two places in the same county can be calculated using computer software. However, a rough estimate that is either Euclidean or rectilinear distance measure can be employed for the load-distance approach. The straight-line distance, or shortest possible path, between two places is known as Euclidean distance.

The point A on the grid represents the solar panel supplier's location in Pune, and the point B represents the possible warehouse location at Hyderabad. The distance between points A and B is the length of the hypotenuse of a right triangle or $d_{AB} = \text{Sqrt}((X_A - X_B)^2 + (Y_A - Y_B)^2)$

where d_{AB} = distance between points A and B

X_A = x-coordinate of point A

Y_A = y-coordinate of point A

X_B = x-coordinate of point B

Y_B = y-coordinate of point B

Table 3.5 Data of solar panel manufacturing service centers.

S.N.	Solar panel manufacturing company service centers	(x,y)	Consumers
1	M	(3.5, 5.5)	300
2	N	(3.5, 3.5)	600
3	O	(6.5, 5.5)	1000
4	P	(6, 3)	700
5	Q	(9, 6)	1000
6	R	(8, 3)	2000
7	S	(10, 3.5)	1400

Rectilinear distance measures the distance between two points with a series of 90° turns as city blocks. Essentially, this distance is the sum of the two dashed lines representing the base and side of the triangle. The distance traveled in the x-direction is the absolute value of the difference in x-coordinates. Adding this result to the absolute value of the difference in the y-coordinates gives

$$D_{AB} = |X_A - X_B| + |Y_A - Y_B|$$

Assume that a company planning a new location wishes to choose a location that minimizes the distances that loads, especially larger ones, must travel to and from the location. A load can be shipments from suppliers, between factories, or to customers, or it can be customers or staff traveling to or from the facility, depending on the industry.

Example 3.3. A solar panel manufacturing company is targeted to develop seven service centers in Mumbai. Table 3.5 shows the coordinates for the center, along with the projected electricity consumer. Consumer will travel from the seven service centers to the new facility, when they need services related to the solar energy system. Two locations being considered for the new service center are at (6.5, 5.5) and (8, 3), which are the service centers; coordinate distances along with the population for each service center are given below. If we use electricity consumer as a load and use rectilinear distance which location is better in terms of its total load distance score.

Solution: Calculate the load-distance score for each location in Table 3.6. Using the coordinates from the above table calculate the load-distance score for each tract.

Using the formula $D_{AB} = |X_A - X_B| \, |Y_A - Y_B|$

Table 3.6 Calculation of load distance score.

S.N.	Solar panel manufacturing company service centers	(x,y)	Consumers	Locate at (6.5, 5.5)		Locate at (8, 3)	
1	M	(3.5, 5.5)	300	3 + 0 = 3	900	4.5 + 2.5 = 7	2100
2	N	(3.5, 3.5)	600	3 + 2 = 5	3000	4.5 + 0.5 = 5	3000
3	O	(6.5, 5.5)	1000	0 + 0 =0	0	1.5 + 2.5 = 4	4000
4	P	(6, 3)	700	0.5 + 2.5 =3	2100	2 + 0 = 2	1400
5	Q	(9, 6)	1000	2.5 + 0.5 =3	3000	1 + 3 = 4	4000
6	R	(8, 3)	2000	1.5 + 2.5 = 4	8000	0 + 0 = 0	0
7	S	(10, 3.5)	1400	3.5 + 2 = 5.5	7700	2 + 0.5 = 2.5	3500
				Total	24700	Total	18000

Summing the scores for all service centers gives a total load-distance score of 24,700 when the facility is located at versus a load-distance score of 18,000 at the location (8, 3). Therefore, the location in solar panel manufacturing company service center F is a better location. To calculate a load-distance for any potential location, we use either of the distance measures and simply multiply the loads flowing to and from the facility by the distances traveled.

4. **The Transportation Model:** The transportation model is an interesting variation of the basic linear programming model, which could be used to optimally identify a subset of supply points from a potential list that can satisfy the demand at various demand points. For example, there are four solar panel manufacturing company, that could supply solar panel in the five different locations. The last column in the table shows the capacity available in each manufacturing company. The last row shows the demand for the product in each of the five different locations. You can better understand the concept of the transportation model by following the example of Table 3.7:

Table 3.7 Data of five different locations.

	Location 1	Location 2	Location 3	Location 4	Location 5	Supply
Company A	1000	600	500	300	400	7000
Company B	400	900	400	1200	500	5000
Company C	800	200	700	400	300	4000
Company D	300	400	900	900	700	4000
Demand	3000	5000	3000	4000	5000	20000

Table 3.8 Data of different locations.

	Location 1	Location 2	Location 3	Location 4	Location 5
Company A	1000	600	500	300	400
Company B	400	900	400	1200	500
Company C	800	200	700	400	300
Company D	300	400	900	900	700

Table 3.9 Calculation of supply of solar panel.

	Location 1	Location 2	Location 3	Location 4	Location 5	Supply of solar panel
Company A	1000	600	500	300	400	9000
Company B	400	900	400	1200	500	7000
Company C	800	200	700	400	300	13500
Company D	300	400	900	900	700	15500
Demand of Solar Panel	12000	10000	8000	7000	8000	45000

Example 3.4. A manufacturer of solar panel find five best location to deliver the product in central part of India, where the atmospheric condition is suitable for solar power plant. Based on the forecasting of solar company, it has been found that the average monthly demand of solar panel is 8000, 6000, 12,000, 14,000, and 4000 in each of the location segments. Based on the forecast and other parameters, it has been decided to build four storage centers with a capacity to handle monthly requirement to the extent of 9000, 7000, 13,500, 15,500 units, respectively, in Table 3.8.

Solution: Constructing the transportation Table 3.9 for the calculation of supply of solar panel, there are several heuristic methods to solve the transportation problem. Optimal methods are also available to solve out such types of problems. The transportation model addresses the concept of moving a thing from one place to another without change. It assumes that any damage en route has negative consequences, and so it is used to analyze transportation systems and find the most efficient route for resource allocation.

3.3.1 Location and performance assessment of solar system by PVSYST 7.1

PVsyst 7.1 is a PC software package for the study, sizing, and data analysis of complete PV systems. It deals with grid-connected, stand-alone, pumping, and DC-grid (public transportation) PV systems, and includes extensive Mateo and PV systems components databases, as well as general solar energy

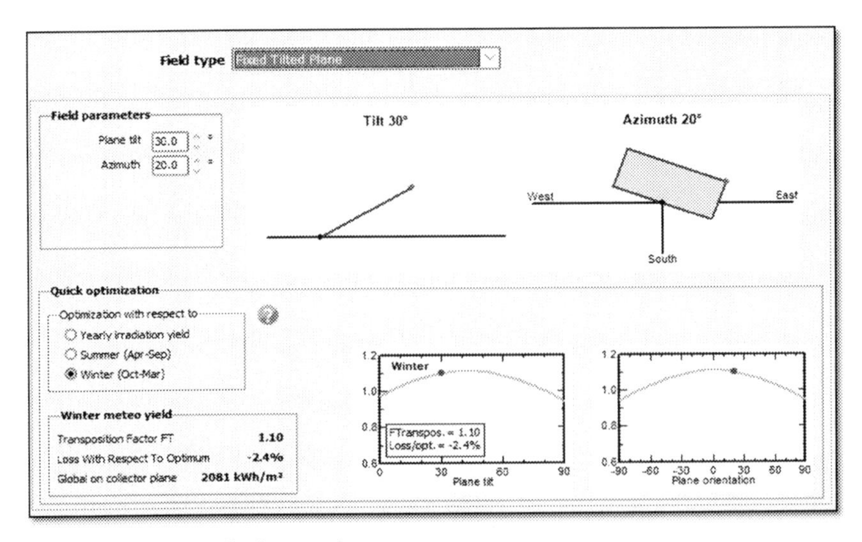

Figure 3.7 Position of solar panel.

tools. This software is geared to the needs of architects, engineers, researchers. It is also very helpful for educational training [1].

Here we are develop 1300 W solar energy system through PVSYST 7.1 at the location of Ahmadabad, state of Gujarat, India. Such installations are using one "solar chargers," equipped with MPPT converters, and DC/AC battery inverters. The controller is not a single device anymore: the charging control is ensured by the solar charger, and the discharge control by the inverter. In PVsyst, consider the solar charger as the controller for standalone component. For historical reasons, this PVsyst component has also to control the discharge and the back-up generator. But as the parameters for these functions are not defined within the "solar charger" physical devices, the internal parameters for these controls will remain in the dialog of this component, but their values will be defined within the system. Their initial values will be adapted from the system configuration and will be modifiable according to the real Inverter used, or other control devices. Fig. 3.7 shows that the photovoltaic (PV) panel is connected in the fixed tilted plane with 30° and in that case azimuth angle is 20°. We can optimize the data year-wise and according to the summer and winter season. The system is installed at the Ahmadabad, state of Gujarat, India, where latitude and longitude become 23.07°N and 72.63°E, respectively, at 55 m altitude and latitude, longitude, tilt angle, and the Azimuth angle is the parameters, which effect the decision to find out the best location for the solar power plant. To find out the optimal value of the location, it is necessary to optimize the system

Table 3.10 Month wise global horizontal irradiation.

Month	Global horizontal irradiation (kWh/m^2)	Effective global corr. for IAM and shadings	Available solar energy kWh
Jan.	144	194.2	213.2
Feb.	150.9	184.6	199.8
Mar.	193.5	209.2	221.5
Apr.	201.8	193.7	204.1
May	208.7	183.2	194.0
Jun.	173	147.2	158.3
Jul.	126.7	110.7	120.6
Aug.	126.9	115.9	126.1
Sep.	150.3	154.2	167.1
Oct.	158.2	182.4	196.0
Nov.	136.5	177.3	192.4
Dec.	123.5	168.3	185.6

separately according to the summer season, winter season, and whole year wise. Table 3.10 shows that month-wise global horizontal solar radiation, effective global correlation, and available solar energy kWh of a particular location and these are other three parameters that effect the choice of select the suitable location.

When choosing the (fixed) plane orientation, an information panel indicates the corresponding transposition factor, the difference (loss) with respect to the optimum orientation, and the available irradiation on this tilted plane.

In the second step, you are asked to define the user's needs. By default for small systems, this is proposed as a list of domestic appliances and their consumption (may be seasonal or monthly). For industrial or bigger systems, you have many possibilities of defining a load profile (including by a list of hourly values). Fig. 3.8 shows those daily household consumptions for the year. Total daily energy consumed is 4874 Wh/day and monthly energy consumption is 146.2 kWh/mth [2].

When you click on the "hourly distribution" tab, the respective watch dials for your devices will be empty. Only the dial for the refrigeration systems will be completed beforehand. Each dial is made up of 48 sections. Each section represents 30 min of the day. To define a schedule or a time range you can do so using the left mouse button. To delete a schedule or a time range you can do this by using the right click of your mouse. Fig. 3.8 shows the hourly distribution of daily household consumptions for the year, where the attributes are Lamps, TV, domestic appliances, dish &

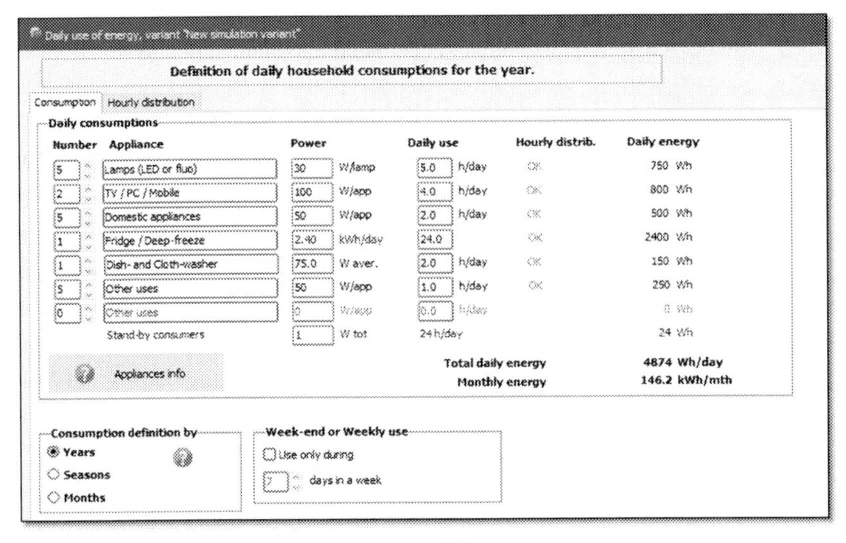

Figure 3.8 Daily household consumptions of the year.

cloth washer, and fridge. Graph also shows the hourly consumptions in a watt in throughout the day.

Once you have finished defining the system orientation and your user needs, then you must define:

- Characteristics of the battery pack.
- Characteristics of the PV field.
- Characteristics of the charge/discharge regulator.

In a stand-alone, PV system with direct coupling to the user (without an inverter), the battery voltage determines the distribution voltage. Nowadays many DC appliances can be found in 24 V as well as in 12 V, this choice should be made according to the system and/or appliance's power, as well as the extension of the planned distribution grid to minimize the ohmic wiring losses.

Fig. 3.9 shows the specification of battery set, of lithium-ion battery with 25.6 V, 3.1 Ah with stored energy of 20.7 kWh at the number of cycles of 2000 and total stored energy during the battery life 36,471 kWh. With the lithium-ion model, the capacity correction according to the discharge rate is handled through the peukert coefficient. The capacity correction according to the temperature is handled similarly as the lead-acid model, with experimental correction coefficients which are fit to catch the behavior as best as possible.

Fig. 3.10 shows the basic data of battery system in the terms of basic parameters, behaviors at limits and full battery indicators. In the overall

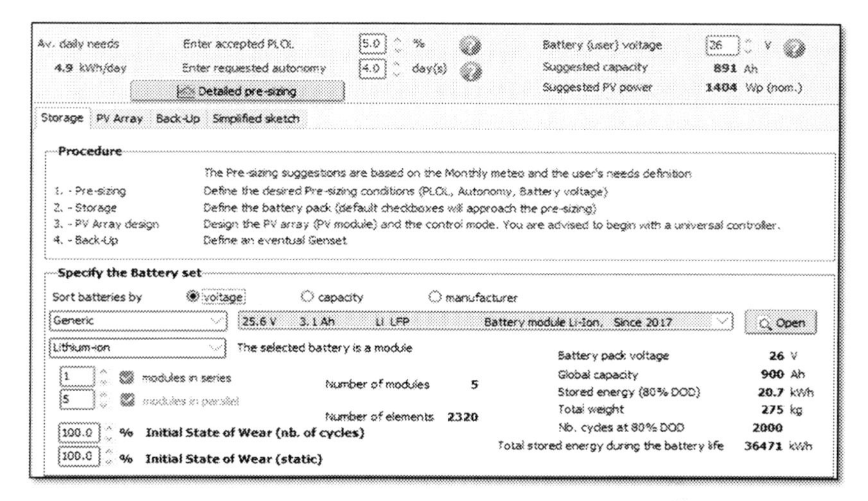

Figure 3.9 Specification of battery.

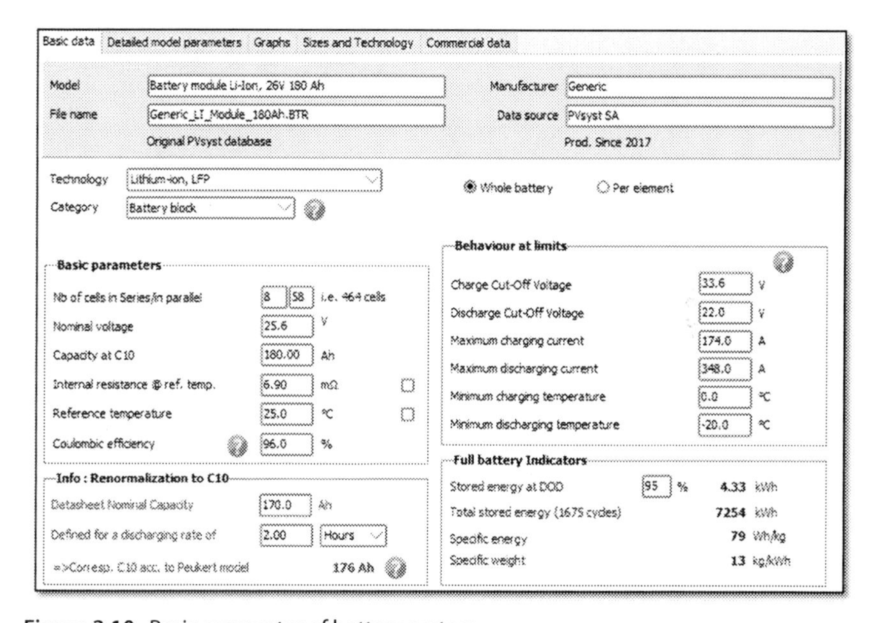

Figure 3.10 Basic parameter of battery system.

battery system 464 cells are used, where 8 are connected in series and 58 are connected in parallel. Charge and discharge cut-off voltage of battery are 33.6 V and 22 V, respectively. Maximum charging and discharging current of battery is 174 A and 348 A, respectively. With lithium-ion batteries, the nominal capacity is usually specified as C10 or C100 [Ah]. This means the

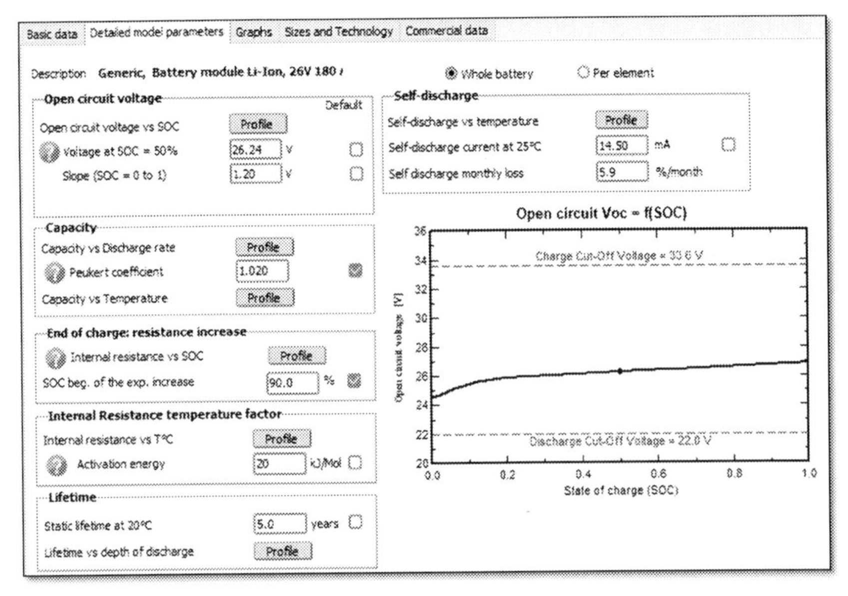

Figure 3.11 Detailed model parameters of battery system.

capacity when discharging in 10 h, or 100 h, that is, with a current of C10/10 or C100/100 [A]. In this case, the capacity at C-10 is 180 Ah [3].

Fig. 3.11 shows the detailed model parameters of battery system in terms of open-circuit voltage, capacity, and end of charge, internal resistance temperature factor, lifetime, and self-discharge rate. The ratio of the quantity of energy currently stored in the battery to the nominal rated capacity is known as the battery state of charge (BSOC or SOC). For a battery with a 500 Ah capacity and an 80% SOC, the energy stored in the battery is 400 Ah. The voltage of the battery is often measured and compared to the voltage of a fully charged battery to determine the BSOC. A common way of specifying battery capacity is to provide the battery capacity as a function of the time in which it takes to fully discharge the battery (note that in practice the battery often cannot be fully discharged). The notation to specify battery capacity in this way is written as Cx, where x is the time in hours that it takes to discharge the battery. C10 = Z (also written as C10 = xxx) means that the battery capacity is Z when the battery is discharged in 10 h. When the discharging rate is halved (and the time it takes to discharge the battery is doubled to 20 h), the battery capacity rises to Y. The discharge rate when discharging the battery in 10 h is found by dividing the capacity by the time. Therefore, C/10 is the charge rate. This may also be written as 0.1C. Consequently, a specification of C20/10 (also written as 0.1C20) is the charge rate obtained when the battery capacity (measured when the battery

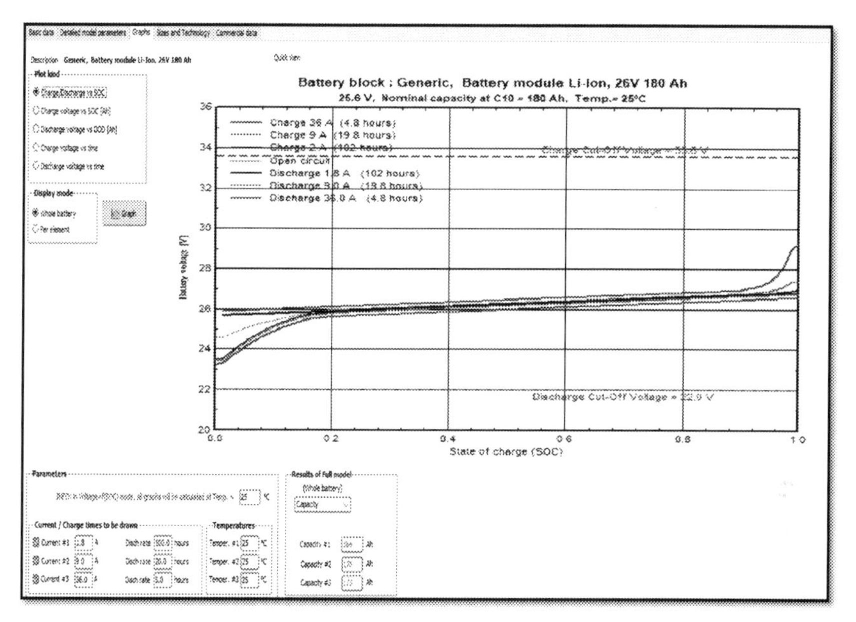

Figure 3.12 State of charge of battery.

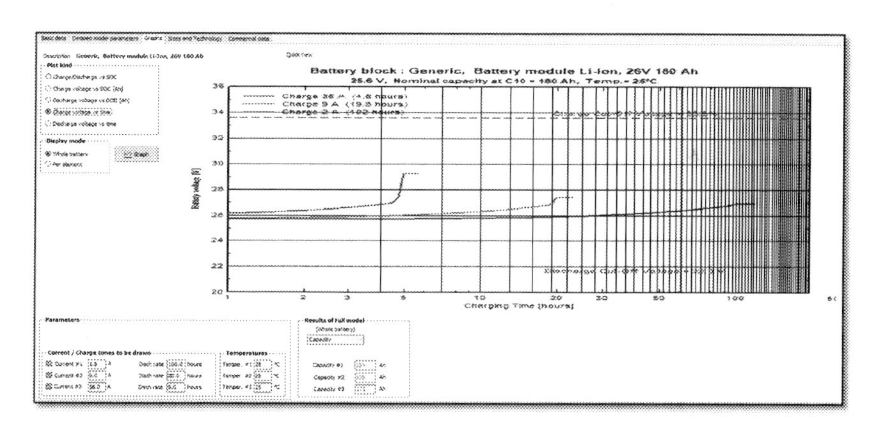

Figure 3.13 Graphical representation between charging time and battery voltage.

is discharged in 20 h) is discharged in 10 h. Such relatively complicated notations may result when higher or lower charging rates are used for short periods of time.

Fig. 3.12 shows the representation between the state of charge and battery voltage w.r.t the charge rate at 36 A, 9 A, and 2 A and discharge rate of 1.8 A, 9 A, and 36 A, respectively.

Fig. 3.13 shows the graphical representation between charging time and battery voltage for the charge rate of 36 A, 9 A, and 2 A at the temperature rate of 25°C [4].

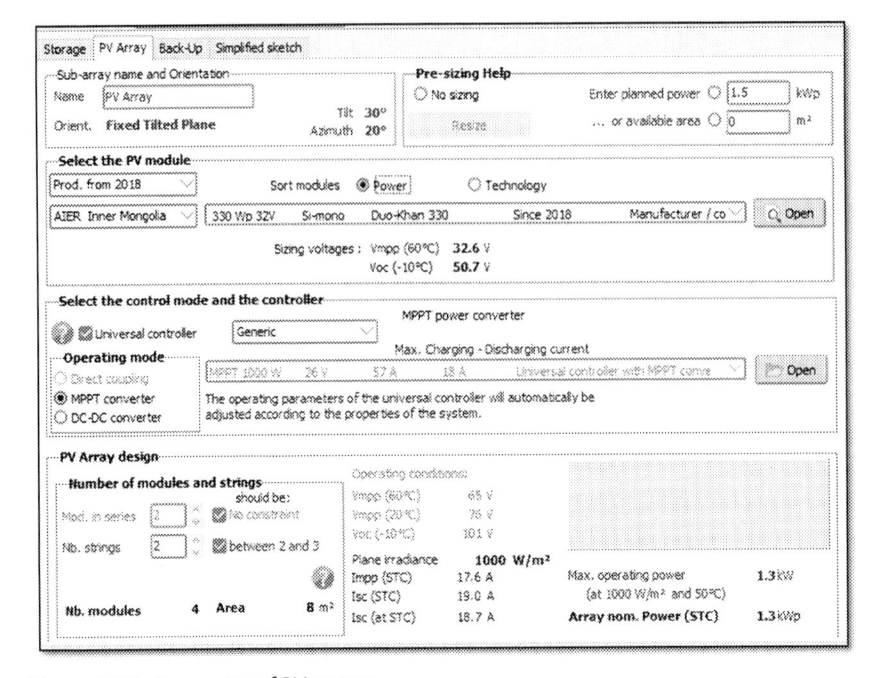

Figure 3.14 Parameter of PV system.

In the next stage, define the array configuration and control strategy

- Acknowledge the presizing propositions (planned power or available area).
- Choose a PV module model in the database.
- Choose the control strategy (direct coupling, MPPT or DCDC converter).

In a first step, you are advised to choose the "Universal controller," to get rid of the specific control conditions. The program determines the number of modules in series and in parallel, according to the battery voltage or MPPT conditions and required PV power.

Fig. 3.14 shows the parameters of PV system of 330 W with 32 V voltage supply, where open circuit voltage is varied from 65 V to 101 V with short circuit current of 17.6–19 A. The maximum operating voltage of PV panel is 1300 W. The operating conditions at 60°C, 20°C, and -10°C are 65 V, 76 V, and 101 V, respectively.

Fig. 3.15 shows the technical parameters of PV panel with manufacturer specification. The short circuit and maximum current of PV panel is 9.37 and 8.82 A, respectively. The maximum value of series and shunt resistance is 0.35Ω and 350Ω, respectively. In the next step, assess the parameters of

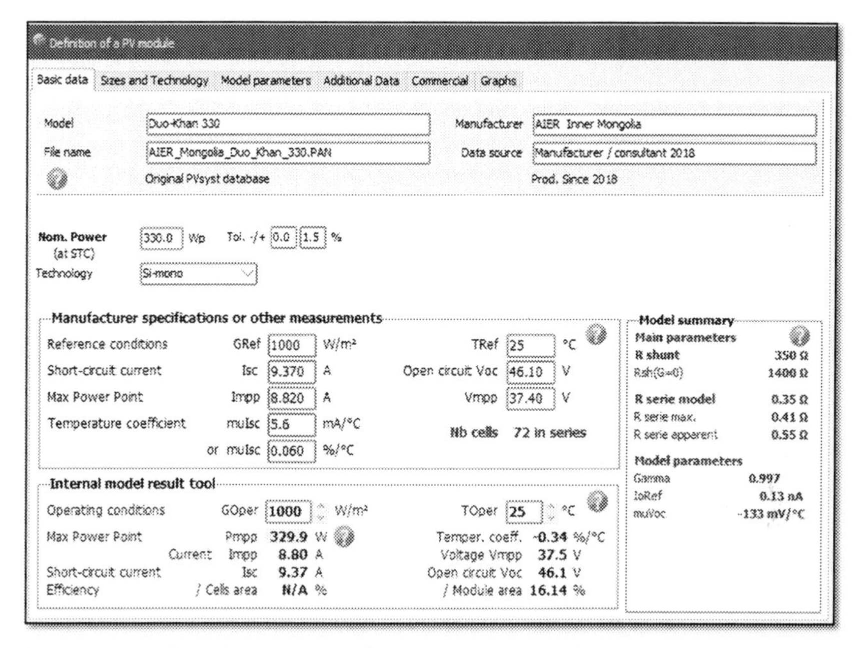

Figure 3.15 Technical parameters of PV panel with manufacturer specification.

universal controllers. During the early stage of a stand-alone system study, the main issues are the global system sizing, that is, the determination of the battery pack, the PV array power, as function of the user's needs and the meteorological conditions. After the quick presizing evaluation, the results must be assessed with a detailed hourly simulation. The exact regulator strategy does not matter. To get rid of the control constraints, PVsyst introduces a general purpose "generic" universal controller, for the three different strategies: Direct coupling, MPPT converter or DCDC converter. During the sizing process (specification of the battery pack and PV array), these special devices will adapt their parameters to the system, in order to always stay compatible with a normal behavior without control losses during the hourly simulation.

Fig. 3.16 shows charging and discharging controller parameters with maximum charging and discharging current of 56.7 A and 17.6 A, respectively. Controller is used for load management and back-up generator controller. Maximum charging and discharging current, maximum backup current, converter nominal power is the main electrical characteristics of charged controller. During the early stage of a stand-alone system study, the main issues are the global system sizing, that is, the determination of the Battery pack, the PV array power, as function of the user's needs and the

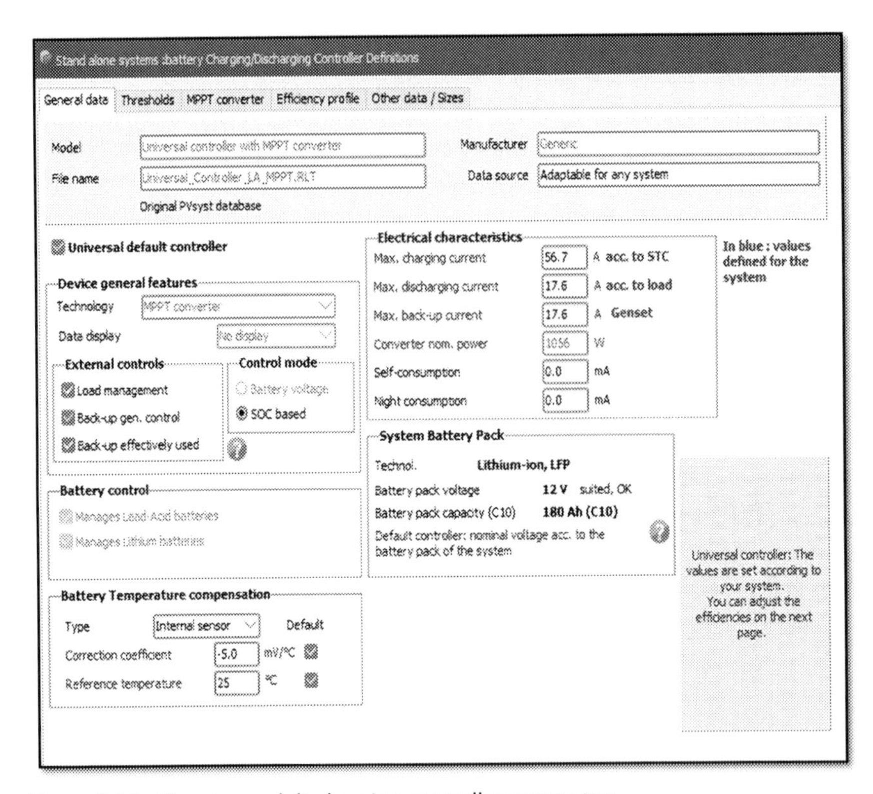

Figure 3.16 Charging and discharging controller parameters.

meteo conditions. During the sizing process (specification of the battery pack and PV array), these special devices will adapt their parameters to the system, in order to always stay compatible with a normal behavior without control losses during the hourly simulation.

Fig. 3.17 shows the threshold value of the universal controller w.r.t. the charging, discharging and recharging threshold. A wide variety of battery charging and discharging controller options are available to you, such as 1 × usb, 2 × usb. You can also choose from qi, qc4.0, and qc1.0 battery charging and discharging controller, As well as from micro usb, dc, and single usb. And whether battery charging and discharging controller is toys, electric power systems, or power tools. There are 2617 battery charging and discharging controller suppliers, mainly located in Asia. The top supplying countries or regions are battery charging and discharging controller, China, and 100%, which supply {3}%, {4}%, and {5}% of {6}, respectively.

Fig. 3.18 shows the parameters of the backup generators, which is the capacity of 1.5 KW and charging current is varied from 39.1 A to 58.6 A.

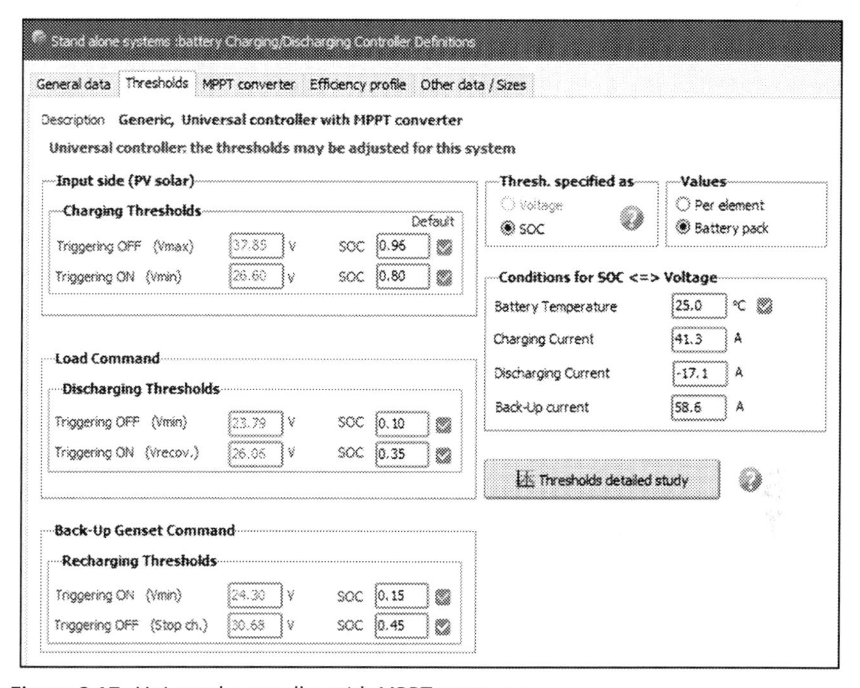

Figure 3.17 Universal controller with MPPT converter.

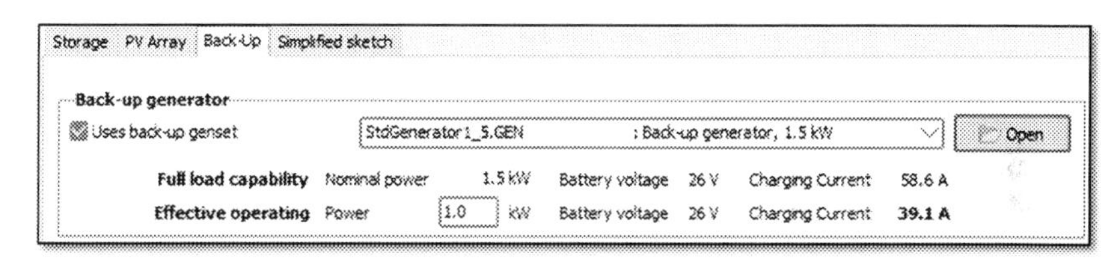

Figure 3.18 Parameters of backup generators.

Fig. 3.19 shows the typical layout design of solar power plant. Fig. 3.19 shows PV array, back-up generator, batteries are connected through the regulator. Stand-alone PV systems are ideal for remote rural areas and applications where other power sources are either impractical or are unavailable to provide power for lighting, appliances, and other uses. In these cases, it is more cost effective to install a single stand-alone PV system than pay the costs of having the local electricity company extend their power lines and cables directly to the home. A stand-alone PV system is an electrical system consisting of and array of one or more PV modules, conductors, electrical components, and one or more loads. But a small-scale PV system does not have to be attached to a roof top or building structures for domestic applications, they can be used for camper vans, RV's, boats, tents, camping

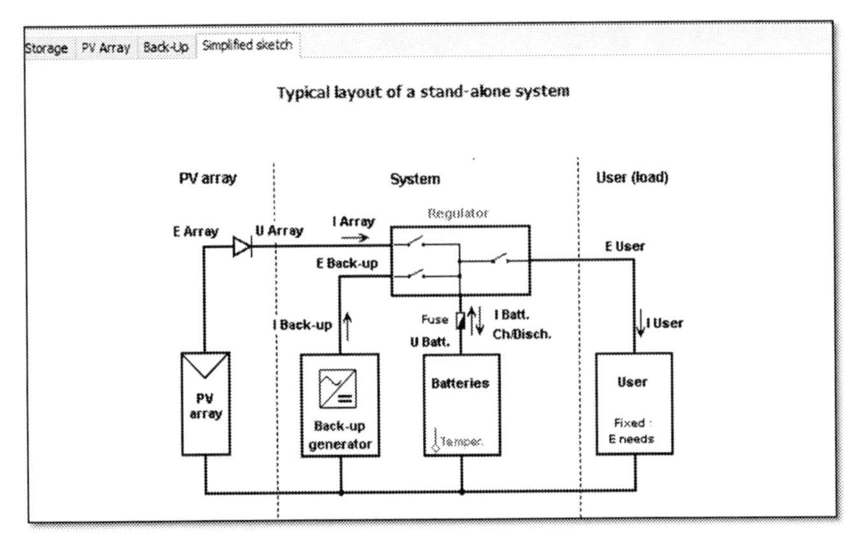

Figure 3.19 Typical layout design of solar power plant.

System summary

Stand alone system		**Stand alone with back-up generator**	
PV Field Orientation		**User's needs**	
Fixed plane		Daily household consumers	
Tilt/Azimuth	30 / 20 °	Constant over the year	
		Average	4.9 kWh/Day

System information			
PV Array		Battery pack	
Nb. of modules	4 units	Technology	Lithium-ion, LFP
Pnom total	1320 Wp	Nb. of units	5 units
		Voltage	26 V
		Capacity	900 Ah

PVsyst TRIAL

Results summary

Available Energy	2179 kWh/year	Specific production	1651 kWh/kWp/year	Perf. Ratio PR	63.28 %
Used Energy	1779 kWh/year			Solar Fraction SF	96.66 %

Figure 3.20 System and result summary of solar power plant.

and any other remote location. Many companies now offer portable solar kits that allow you to provide your own reliable and free solar electricity anywhere you go even in hard-to-reach locations.

Fig. 3.20 shows the system and result summary of the solar system project. The available energy and used energy is 2179 kWh/year and 1779 kWh/year, respectively, with solar fraction of 96.66%.

Fig. 3.21 shows the overall losses representation of overall solar project, where losses are classified into losses through the solar system, battery system, and converter. Generally, array losses can be defined as all events which penalize the available array output energy with respect to the PV-module

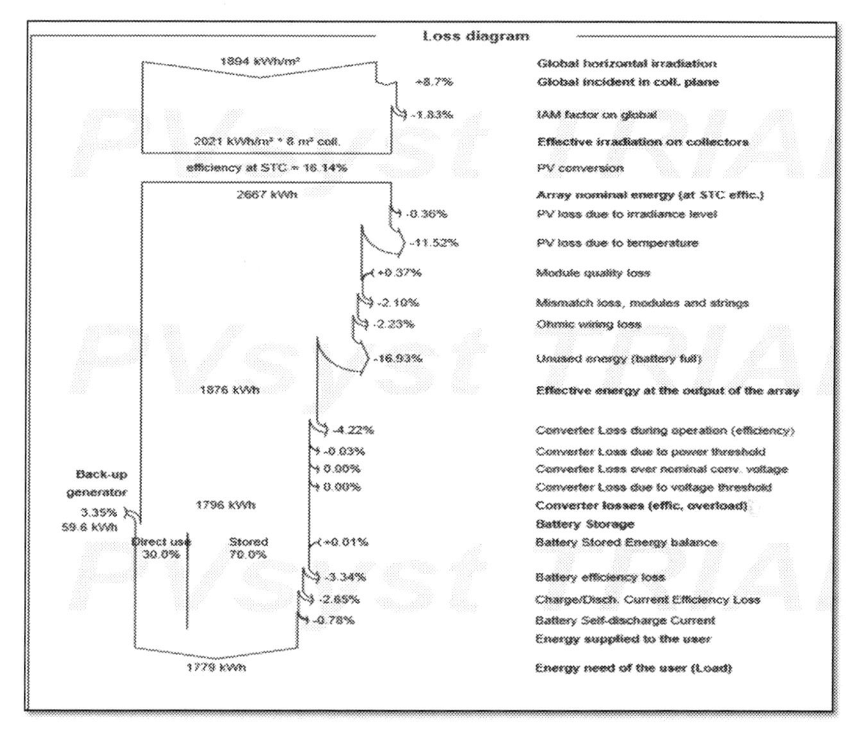

Figure 3.21 Overall losses representation of solar project.

nominal power as quoted by the manufacturer for STC conditions. This is the philosophy stated by the JRC/Ispra European Center recommendations, through the normalized performance index and the performance ratio. Several of these loss sources are not directly measurable. Starting with incident irradiation in the collector plane, one can imagine that an ideal PV-array should yield 1 kW/kWp under an irradiance of 1 kW. That is, assuming a linear response according to Ginc, the ideal array will produce one kWh energy under one kWh irradiance for each installed kWp. Incidence angle modifier (IAM) is an optical effect (reflection loss) corresponding to the weakening of the irradiation really reaching the PV cells surface, with respect to irradiation under normal incidence. Irradiance loss is the nominal efficiency specified for the STC (1000 W/m^2), but is decreasing with irradiance according to the PV standard model.

Fig. 3.22 shows the month-wise normalized energy generation through the solar power plant. The unused energy, when the battery is full is 0.79 kWh/kW/day, PV array loss is 0.95 kWh/kW/day, system losses and

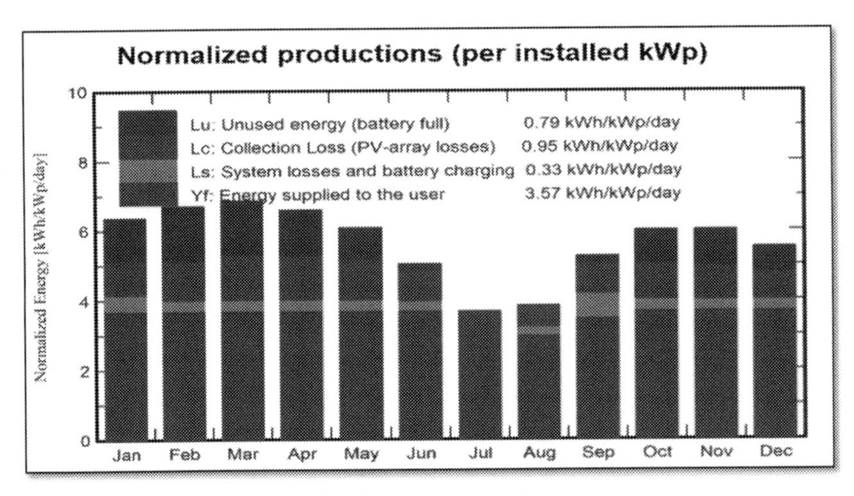

Figure 3.22 Month-wise normalized energy generation.

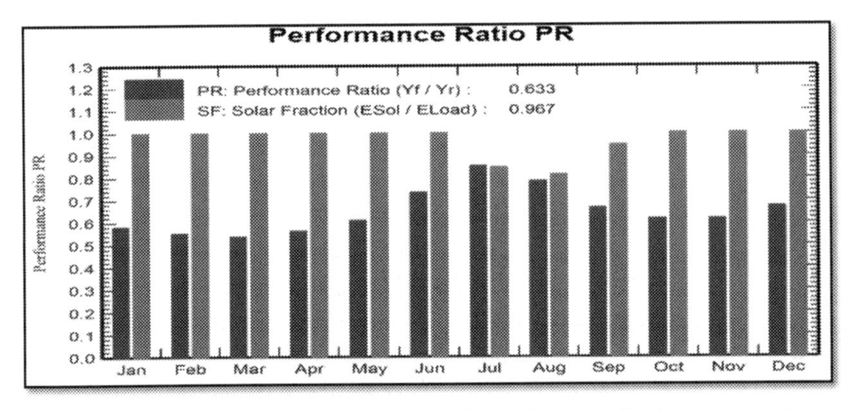

Figure 3.23 Month wise performance ratio and solar fraction of solar energy system.

battery charging is 0.33 kWh/kW/day and energy supplied to the users is 3.57 kWh/kW/day.

Fig. 3.23 shows the month-wise performance ratio and solar fraction of solar energy system. The value of the performance ratio and solar fraction is 0.633 and 0.967, respectively. The performance ratio is the ratio of the energy effectively produced (used), with respect to the energy which would be produced if the system was continuously working at its nominal STC efficiency. The PR is defined in the norm IEC EN 61,724. Table 3.11 shows the month-wise parameter of solar energy system. According to Table 3.11, maximum and minimum value of unused energy is in the month of Mar. and Jul., and Aug., respectively. Solar fraction is lies between 0.846 and 1.

Table 3.11 Month-wise parameter of solar energy system.

Month	Unused energy (battery full)	Energy supplied to the user	Energy need of the user (load)	Solar fraction
Jan.	49.37	151.1	151.1	1
Feb.	57.67	136.5	136.5	1
Mar.	63.85	151.1	151.1	1
Apr.	51.56	146.2	146.2	1
May	37.00	151.1	151.1	1
Jun.	7.59	146.2	146.2	1
Jul.	0.00	151.1	151.1	0.846
Aug.	0.00	151.1	151.1	0.814
Sep.	8.27	146.2	146.2	0.945
Oct.	38.26	151.1	151.1	1
Nov.	40.35	146.2	146.2	1
Dec.	28.37	151.1	151.1	1
				0.967

The objective of this section is to closely compare on-site measured data with simulated values, either in hourly or in daily values. It has a two-sided function.

On the one hand, it has helped us for the validation the software by comparing its results with carefully measured data in different installations. On the other hand, it constitutes a powerful tool for the analysis of the operation of PV systems in use, allowing for the detection and identification of even the smallest malfunctioning. This involves a much more complex process than the simple system simulation, which includes the following steps:

- **Importing the measured data:** This is done by a programmable data interpreter, which accepts almost any ASCII file, provided that it holds records of hourly or subhourly steps, each one on a single ASCII line. It allows choosing, among the measured variables, those which suit the simulation variables.
- **Checking the imported data:** In order to verify the validity of the imported data file, a number of tables and graphs in hourly, daily, or monthly values, may be drawn. Further, some specific graphs often used in PV data analysis are also available, allowing, at this stage, for using PVsyst as a complete tool for the presentation of measured data.
- **Defining the system parameters:** You have to define a project and variant parameters, exactly in the same way as for the usual simulation. At this stage, you should carefully introduce the real properties of your system.

- **Comparisons between measured and simulated values:** After performing the simulation, you will obtain close comparison distributions for any measured variable. According to the observed discrepancies, you probably will analyses their cause and modify the input parameters accordingly. This offers a powerful way to exactly determine the real system parameters, as well as temporarily misrunning's.
- **Elimination of break-down events:** Most of time measured data hold undesired records (break-down of the system or the measurement equipment), easily identifiable on the graphs. These can be selectively eliminated in order to obtain clean statistical indicators mean bias and standard deviation corresponding to the normal running of the system.

3.4 Process–product matrix of solar power plant

The Hayes–Wheelwright Matrix, also known as the product-process matrix, is a tool to analyze the fit between a chosen product positioning and manufacturing process. The first dimension of the matrix, the product lifecycle, is a measure of the maturity of the product or market. The matrix itself consists of two dimensions, product structure/product life cycle and process structure/process life cycle. The production process used to manufacture a product moves through a series of stages, much like the stages of products and markets, which begins with a highly flexible, high-cost process, and progresses toward increasing standardization, mechanization, and automation, culminating in an inflexible but cost-effective process. The process structure/process life cycle dimension describes the process choice and process structure while the product structure/product life cycle describes the four stages of the product life cycle (low volume to high volume) and product structure (low to high standardization). When this concept apply on the solar energy system, so in that case you can identify structure and life cycle of solar panel, solar inverter, and battery through the product structure and product life cycle concept. In the process structure and process cycle, you can identify the cradle to grave process of solar energy system which is shown in Table 3.12.

Solar energy product manufacturing firms utilizing batch processes provide similar items such as 300 W solar panel on a repeat basis, usually in larger volumes than that associated with job shops. Solar energy products are sometimes accumulated until a lot can be processed together. When the most effective manufacturing process of solar energy product has been

Table 3.12 Process–product matrix of solar energy system.

Process structure & life cycle stage of solar energy system	Product structure and life cycle stage of solar energy system	Low volume unique solar energy product	Low volume multiple solar energy product	Higher volume standardized solar energy product	Very high volume commodity solar energy product
Jumbled flow		Job shop			
Disconnected line flow			Batch		
Connected line flow				Assembly line	
Continuous flow					Continuous

determined, the higher volume and repetition of requirements can make more efficient use of capacity and result in significantly lower costs of solar energy products. When product demand is high enough, such as requirement of 300 W solar panel and 1 KW solar inverter and in that case the appropriate process is the assembly line. Often, this process is referred to as mass production. The assembly line treats all outputs as basically the same. Firms characterized by this process are generally heavily automated, utilizing special-purpose equipment. Frequently, some form of conveyor system connects the various pieces of equipment used. There is usually a fixed set of inputs and outputs, constant throughput time, and a relatively continuous flow of work. Because the product is standardized, the process can be also, following the same path from one operation to the next. Routing, scheduling, and control are facilitated since each individual unit of output does not have to be monitored and controlled. This also means that the manager's span of control can increase and less-skilled workers can be utilized.

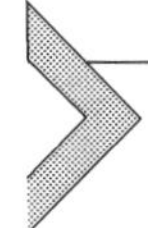

3.5 Performance measures of solar power plant layout design

Layout planning in manufacturing and service organizations involves the physical arrangement of various resources available in the system to improve the performance of the operating system, thereby providing better customer service. Layout planning of solar energy system provides a set of tools and techniques that help an operations manager decide where to locate resources and also to assess the impact of the alternative choices that he/she may have for locating the resources.

$$\text{Actual (desired) cycle time} = \frac{\text{Available Time}}{\text{Actual (Desired)Production}}$$

$$\text{Minimum no. of required workstations required} = \frac{\text{Sum of all task times}}{\text{Cycle Time}}$$

$$\text{Average resource utilization} = \frac{\text{Sum of all task times}}{\text{Number of workstation } \times \text{Cycle time}}$$

You can understand the basic concept of performance measures of solar power plant layout design by the following example.

Example 3.5. A solar manufacturing company working in two shifts, each of 9 h produces 20,000 solar panel each of 300 W using a set of workstations. Using this information, compute the actual cycle time of the solar panel manufacturing plant operation. There are 18 tasks required to manufacturer the solar panel. The sum of all the task times is 10,000 s. How many workstations are required to maintain this level of production assuming that combining of 18 tasks into those workstations is a feasible solutions?

Solution:
Available time $= 2 \times 9 \times 60 \times 60 = 64{,}800$ s

Actual production $= 20{,}000$ Solar Panel

Now we compute the cycle time for each bulb as

$$\text{Actual (desired) cycle time} = \frac{\text{Available Time}}{\text{Actual (Desired)Production}} = \frac{64,800}{20,000}$$

$$= 3.24\text{s}$$

This means the company is producing 1 solar panel every 3.24 s

Number of workstation required $= 10{,}000/3.24 = 3086.42 = 3087$

Therefore the task are to be split among the five stations such that each workstation will have 3.24 s as the sum of its task time.

Example 3.6. A solar manufacturing company working in two shifts, each of 8 h produces 15,000 solar panel each of 250 W using a set of workstations. Using this information, compute the actual cycle time of the solar panel manufacturing plant operation. There are 16 tasks required to

Table 3.13 The tasks and their duration.

Task	Description	Duration (s)
A	Cell cutting and stringing process	1000
B	Solar glass and visual inspection	800
C	Taping and connection	1200
D	Mirror observation and EI testing	600
E	Lamination process and trimming backsheet	1000
F	Frame cutting and punching	1400
G	Sealant Filling/Framing and fixing junction box	1500
H	Module testing and packing	1000

manufacturer the solar panel. The sum of all the task time is 8000 s. How many workstations are required to maintain this level of production assuming that combining of 16 tasks into those workstations is a feasible solution?

Solution:

Available time $= 2 \times 8 \times 60 \times 60 = 57{,}600$ s

Actual production = 15,000 Solar Panel

Now we compute the cycle time for each bulb as

$$\text{Actual (desired) cycle time} = \frac{\text{Available Time}}{\text{Actual (Desired)Production}} = \frac{57{,}600}{15{,}000}$$

$$= 3.84\ \text{s}$$

This means the company is producing 1 solar panel every 3.84 s

Number of workstation required $= 8000/3.84 = 2083.33 = 2084$

Therefore the task is to be split among the five stations such that each workstation will have 3.84 s as the sum of its task time.

Example 3.7. A solar panel manufacturer company needs to design a solar panel. The company currently works for two shifts of 9 h. The tasks and their durations are given in Table 3.13:

(a) If the cycle time is 110 s what will the daily production of solar panel?

(b) If the desired production rate is 400 solar panel each of 300 W per day, what is maximum permissible cycle time?

(c) What is the maximum and minimum number of workstations required to maintain this daily production rate?

(d) Find the average utilization.

Solution:

(a) Total available time per day $= 2 \times 9 \times 60 \times 60 = 64{,}800$ s.

If the cycle time is 110 s, then the daily production of solar panel can be obtained by the following equation:

Daily production rate = Total available time/cycle time= 64,800/ 110 = 590

(b) Since the desired production rate is only 400 solar panel, one can obtain the maximum permissible cycle time for the assembly stations.

Maximum cycle time = Total available time/desired production rate = 64,800/400 = 162 s

The computation shows that while a cycle time of 110 s may yield may yield a much higher production rate, the workstations can be designed for a cycle time of up to 162 s without short of the desired daily production.

(c) The minimum number of workstation is given by the equation

Min. number of workstation required = Sum of all task times/cycle time = 8500/162 = 52.46 = 53

(d) The average utilization of the resources can be computed as follows:

Average utilization = Sum of all task times/(no. of workstation $\times$ cycle time) = $8500/(53 \times 162) = 98.99\%$

3.6 Design of group technology solar plant layout

"Group technology (GT) is the realization that many problems are similar and that, by grouping similar problems, a single solution can be found to a set of problems, thus saving time and effort." GT is a concept that currently is attracting a lot of attention from the manufacturing community. The essence of GT is to capitalize on similarities in recurring tasks in three ways:

- By performing similar activities together, thereby avoiding wasteful time in changing from one unrelated activity to the next.
- By standardizing closely related activities, thereby focusing only on distinct differences and avoiding unnecessary duplication of effort.
- By efficiently storing and retrieving information related to recurring problems, thereby reducing the search time for the information and eliminating the need to solve the problem again.

GT layout is constructed with the purpose of dividing a process into a subgroups. GT layout design of solar energy system is done with the systematic assessment of a machine component incident matrix. Consider an example of 20 subprocess to design the solar panel, which is done

Table 3.14 Incident matrix for hypothetical example.

	1	2	3	4	5	6	7	8	9	10	11	12	13	14	15	16	17	18	19	20
M	*			*			*													
N			*	*	*			*												
O		*	*		*			*												
P	*			*			*													
Q													*	*	*		*	*		*
R	*			*			*													
S						*			*		*	*				*			*	
T						*			*		*	*				*			*	
U													*	*	*		*	*		*
V						*			*		*	*				*			*	

Table 3.15 Arrangement of rows and columns.

	2	3	5	8	10	1	4	7	20	18	17	15	14	13	6	9	11	12	16	19
N	*	*	*	*	*															
O	*	*	*	*	*															
P						*	*	*												
M						*	*	*												
R						*	*	*												
Q									*	*	*	*	*	*						
U									*	*	*	*	*	*						
S															*	*	*	*	*	*
T															*	*	*	*	*	*
V															*	*	*	*	*	*

Table 3.16 Four cells with process and machine groups of solar panel manufacturing.

	Process families	Machine group
Cell 1	2,3,5,8,10	N,O
Cell 2	1,4,7	M,P,R
Cell 3	13,14,15,17,18,20	Q,U
Cell 4	6,9,11,12,16,19	S,T,V

through the 10 machines. Let us denote the process using numbers and the machine using alphabets. A machine and process incident matrix indicates the relationship between the process and machines using a "*" representation. In this process, if the subgroup is nonoverlapping, then the resulting design has independent resources and process. Each subgroup is also referred as a cell. The structure of the matrix is known as a block diagonalization structure shown in Tables 3.14–3.16.

3.7 Conclusion

Nowadays, it is very necessary to identify the suitable location for the installation of solar power plant. Based on the above discussion, it is find out factor rating method, load distance, center of gravity, and transportation model is a suitable method for location finding. PV syst software includes a detailed contextual help menu that explains the procedures and models that are used and offers a user-friendly approach with a guide to develop a project. PVsyst is able to import meteo data, as well as personal data from many different sources. This chapter also discussed matrix facilitates broader thinking about organizational competence and competitive advantage by including stages of the product lifecycle and its choice of the production processes for different products into its strategic planning process. At the end GT provides futuristic insights for the solar energy system.

3.8 Exercise/ Question

1. Define the term "Facility Location" and "Plant Layout" w.r.t. the solar power plant.
2. Explain different parameter of location decision for an organization.
3. Explain the concept of higher level, middle level and lower level competitiveness of a location.
4. How latitude affect the temperature and solar radiation.
5. Write the name of different methods of location planning for solar power plant.
6. Explain the concept of location factor rating method for location finding for solar power plant.
7. Explain the concept of center of gravity method for location finding for solar power plant.
8. Explain the concept of load distance method for location finding for solar power plant.
9. Explain the concept of transportation model for location finding for solar power plant.
10. How we assess location and performance of solar energy system by PVSYST 7.1.
11. Explain the concept of process-product matrix w.r.t. the solar power plant.
12. What are the different parameters of solar power plant layout design?

13. A solar manufacturing company working in two shifts, each of 7 h produces 30,000 solar panel each of 200 W using a set of workstations. Using this information, compute the actual cycle time of the solar panel manufacturing plant operation. There are 16 tasks required to manufacturer the solar panel. The sum of all the task time is 10,000 s. How many workstations are required to maintain this level of production assuming that combining of 18 tasks into those workstations is a feasible solution?
14. A solar panel manufacturer company needs to design a solar panel. The company currently works for two shifts of 6 h. The tasks and their durations are given in the Table:
    (a) If the cycle time is 120 s what will the daily production of solar panel?
    (b) If the desired production rate is 300 solar panel each of 300 W per day, what is maximum permissible cycle time?
    (c) What is the maximum and minimum number of workstations required to maintain this daily production rate?
    (d) Find the average utilization.

The Tasks and Their Duration

Task	Description	Duration (s)
A	Cell cutting and stringing process	900
B	Solar glass and visual inspection	900
C	Taping and connection	1100
D	Mirror observation and EI testing	600
E	Lamination process and trimming back sheet	900
F	Frame cutting and punching	1200
G	Sealant filling/Framing and fixing junction box	1500
H	Module testing and packing	1100

References

[1] R. Ruparathna, Developing a level of service (LOS) index for operational management of public buildings, Sustain. Cities Society 34 (2017) 159–173.
[2] P. Wide, Improving decisions support for operational disruption management in freight transport, Res. Transportation Business Manage. 37 (2020) 100540.
[3] W. Dong, Adaptive optimal fuzzy logic based energy management in multi-energy microgrid considering operational uncertainties, Appl. Soft Comput. 98 (2021) 106882.
[4] D. Prajogo, The relationships between information management, process management and operational performance: Internal and external contexts, Int. J. Production Economics 199 (2018) 95–103.

CHAPTER FOUR

Productivity and manufacturing economics of solar energy system

Learning objective

- Understand the basics of aggregate operation planning in the field of solar energy system.
- Discuss the concept of level, chase and mixed strategy w.r.t. the solar energy system.
- Learn master operation scheduling of solar industry and solar power plant.
- Understand the concept of manufacturing resource planning.
- Discuss about the enterprise resource planning w.r.t. the solar energy system.

4.1 Introduction

The world's primary energy sources are fossil fuels such as crude oil, natural gas, and coal. Despite being a nonrenewable resource, fossil fuels continue to be in great demand due to their affordability and dependability. Fossil fuels play an important part in energy production and the global economy, from heating and lighting homes to fueling automobiles. Even with tremendous technological advancements, sustainable energy has unable to supplant traditional fossil fuels. Governments have imposed tax credits for solar energy, which were formerly significantly more expensive than the status quo, in order to encourage the use of renewable energy. Consumers' direct costs of solar energy have reduced as a result of increased production, government subsidies, and mounting environmental concerns. In fact, some markets provide renewable energy for consumers at a lower cost than fossil fuels. Solar energy has both business and residential applications, but wind energy, such as wind farms, is mostly employed for commercial purposes [1,2].

Now a days rather than technical and financial aspects of, it is necessary to identify the productive and manufacturing economics of solar energy system. Aggregate operation planning (AOP), master operation scheduling (MOS), dependent demand attributes, manufacturing resource planning (MRP) and enterprise resource planning is the main attributes of productive

Decision Science and Operations Management of Solar Energy Systems.
DOI: https://doi.org/10.1016/B978-0-323-85761-1.00004-4

and manufacturing economics of solar energy system and also of different types of solar industry. The aggregate operation planning is done using the aggregate data of different operation at different levels and such type of planning provides comprehensive analysis of overall system. It deals with the number of resources such as effective solar radiation for electricity generation and resources of raw material of solar energy product to be committed, the cost at which solar product and services need to be produced during a given time period and inventory to be carried out from one period to the another [1]. Master operations scheduling is the process of linking between planning and execution of solar industry. It is the next stage in the operations planning process in any solar industry. MOS makes use of actual electricity consumer orders for the purpose of capacity planning and resource allocation to specific consumer orders. Manufacturing resource planning is a way for effectively planning all of the manufacturing company's resources. It is an extension of closed-loop MRP and addresses operational planning in units, financial planning, and has a simulation capability to answer "what-if" concerns. ERP stands for "Enterprise Resource Planning," and it refers to the software and tools that a business uses to plan and manage all of its essential supply chain, manufacturing, services, financial, and other processes [2]. Individual activities within a corporation or organisation, such as accounting and procurement, project management, customer relationship management, risk management, compliance, and supply chain operations, can all be automated and simplified with Enterprise Resource Planning software.

This chapter is categorized into seven section. Section 4.1 explained the introductory part of this chapter. Section 4.2 elaborate the aggregate operation planning of solar energy system. Level, chase and mixed strategy is represented in section 4.3. Sections 4.4 and 4.5 described master operation scheduling and dependent demand attributes of solar energy system. Manufacturing resource planning and enterprise resource planning is explained in sections 4.6 and 4.7. Chapter is concluded with section 4.8.

4.2 Aggregate operations planning of solar energy system

Operating systems in a solar business play the very important role of providing a continuous flow of electricity as well solar system related devices and their services to the electricity consumer. In the different circumstances

Table 4.1 Aggregate units for operations planning of solar energy system.

Parameter	Aggregate unit of capacity
Solar radiation	kWh/meter^2/day
Wind velocity	Meter/second
Temperature	°C
Solar panel	Watt
Load demand	KW
Battery	KWh
Inverter	KW
Converter	KW

solar panel manufacturer may be producing at a daily rate of 20,000 solar panel, 5000 solar inverter, and 2000 solar battery in its factory. The operating system of solar energy system is characterized by the complex relationship between, electricity consumer, number of resources such as solar radiation, temperature, humidity and wind velocity, raw material for solar energy product and different parameters, which effect the overall process of solar energy system. Therefore, it is necessary to proper planning and different strategies for dealing with real time changes play a very important role in managing operating systems and ensuring a continuous flow of goods and services to the solar energy market. Table 4.1 shows the aggregate units for operations planning of solar energy system [3].

In the aggregation planning of the solar energy system, overall planning is develop through the different parameters, which is the composition of technical, financial and managerial factors. Usually, the decision involve the number of prefeasibility analysis parameters, the rate at which solar energy related product and their services need to be produced during a winter, summer and rainy season and the solar energy system inventory to be carried forward from one season to another season. In the planning stage one point keep in mind, the sale of the solar energy product, their services and inventory are changed according to the different circumstances and seasons [4].

At the end of an aggregate operation planning, a solar energy product manufacturer and service provider may arrive at the following plan:

- Produce 5000 solar panel every day during the period of February to June, decrease the rate 2000 solar panel every day during the period of July to September and further increased the production rate up to 3500 solar panel per day during the season of October to January. Therefore, in this example critical decision has to be made because the sale of solar panel is decreased or increased according to the seasons. If the solar energy product manufacturing company distributes the solar product in all over the world, so in that case company can manufacture product at the constant rate, because environmental condition of different country is different and in every country, the market of solar energy product is increased during the summer seasons [5].

- Work on one shift basis throughout the year with 30 percent overtime during February to June.

➢ **The need for aggregate operations planning:**

There are several reasons for every solar energy company, whether engaged in solar energy related product manufacturing and service delivery for electricity consumer.

- **Load demand fluctuations:** Consumer electricity demand always varied according to the different circumstances. The load of the consumer, who receives the electricity from the solar energy system, is different in peak hours and non-peak hours and varied in summer, winter and rainy season.
- **Capacity fluctuations:** In terms of sustaining profitability, projecting revenue returns, and assuring customer quality of service, solar PV (photovoltaic) plant operators need to be able to precisely predict and prevent power fluctuations. Solar irradiance changes can induce fast swings in power generation, lowering the quality and reliability of power provided by large grid-connected PV systems. A number of factors, including shadowing because of cloud cover and dust gathering on PV panels, can cause such inconsistency in solar irradiance. For intervals of less than ten minutes, these fluctuations are directly absorbed by PV electricity systems, which results in variations in power frequency. Utility operators are powerless to correct these imbalances, which can ultimately result in electrical power systems failure.

It is evident that for several reasons, electricity supply through the solar energy system and consumer demand may not match exactly on a season by season at the particular location. Therefore, aggregate operations planning is done in solar energy system to match the consumer electricity demand with the electricity supply through the energy system basis in a cost effective manner [6]. Matching the demand with the supply can be done in any one of the following ways:

i. For the electricity supply capacity of solar energy system has during a period, ways by which the load demand of the consumer could be suitably modified are identified so that both are matched.

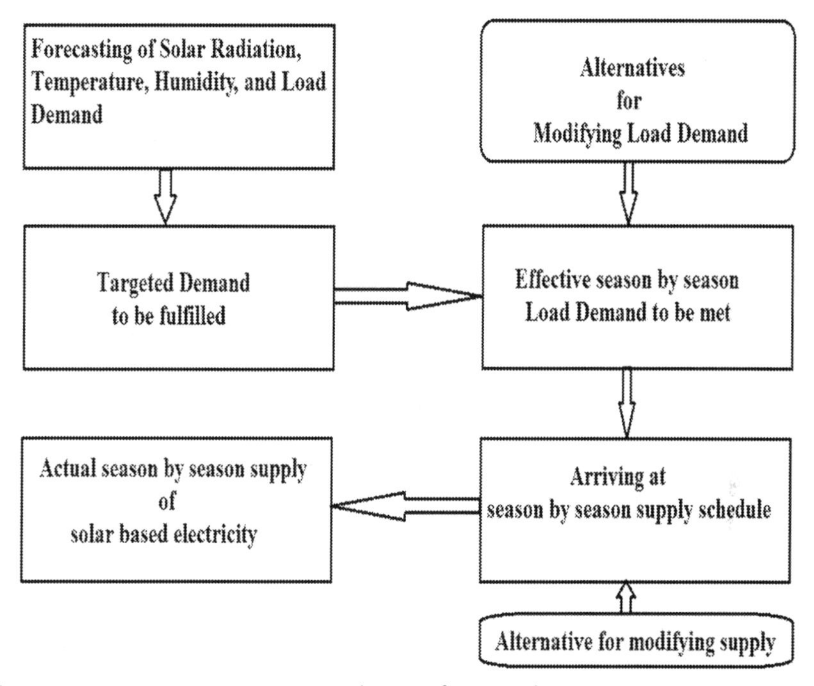

Figure 4.1 An aggregate operations planning framework.

ii. For the given load demand of the consumer to be fulfilled in every period, ways by which the electricity supply through the solar energy system is altered are identified so that both are matched.
iii. During every season, both the load demand and the electrical energy are adjusted in such a way that both are matched.

The aggregate operations planning process starts with a look at several approaches to adjust demand and then calculating period-by-period net demand throughout the planning horizon [7]. Fig. 4.1 shows an aggregate operations planning framework.

4.2.1 Alternatives for managing demand

In the solar energy system, there are two types of managing demand, one demand is related to the retailers of solar energy product and their demand through the manufacturer company of solar energy product and another type of demand is load demand of the electricity consumer from the solar power plant. The aggregate operations planning exercise provides a few alternatives to achieve this goals.

Reservation of capacity: To manage the demand of retailers, it is necessary to Manufacturer Company; create the reservation of capacity of solar energy products. In the case of load demand, it is necessary to develop flexible solar energy system, which have a capability of increase or decrease the capacity of plant. At the initial level of power plant, it is necessary to develop 30% more capacity w.r.t the actual load demand. Capacity reservation is a very popular method in service system [8].

Inventory-based alternatives: In the solar energy product development company, matching the demand of solar energy based product could be done in different ways using proper inventory mechanism. One is to build inventory during periods of lean demand and consume them during periods of high demand. Let us consider XYZ solar product manufacturer and suppose the forecast demand for solar panel during the month of February to June is 20,000 number of solar panel of each 200 W rating and in July to November is 30,000 number of solar panel. Then it is necessary to manufacturing company to produce 25,000 number of solar panel during the February to November and carry the inventory from the first five months to meet the demand during the next five months [9].

Another option is to back order the current periods demand in a future period. Backordering is a method of pushing an order to a future period because of insufficient inventory and/or capacity to supply during the current periods. For the example in the solar panel manufacturing company, if the demand for solar panel during the month of May is 25,000 and if the company has capacity to manufacturer only 15,000 solar panel and has on hand inventory of 5000 solar panel, then the balance 5000 solar panel is scheduled in the month of June, although the demand is for the month of May.

Another possibility is the stock-out. By stock-out, we mean a deliberate strategy of leaving a portion of the demand unfulfilled. It is also possible that the solar panel manufacturer can go its consumer and express its inability to supply 5000 solar panels.

The fourth variation is to backlog the orders. In this type of inventory process, instead of promising the customer a due date and later incurring a back order or a stock out, the solar energy company could specify a waiting time for supply. Thus during periods of high demand, the backlog of order could be large.

Demand capacity adjustment alternatives: In the solar product manufacturing company another possibility to meet the solar product demand during the planning horizon. In that case, increase the manufacturing

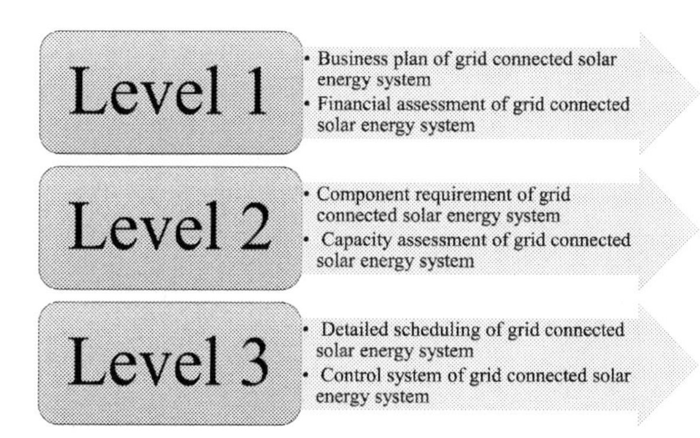

Figure 4.2 Aggregate operation planning of grid connected solar energy system.

capacity of solar product through the increase of number of working hours and number of employee. In the certain areas of manufacturing system, it is possible to increase or decrease capacity and operations rate by varying the number of workers. But in the solar product manufacturing company, it is not easy to increase the quantity of workers instantaneously, because in such type of company very skilled workers are required, who possesses the knowledge of technology of solar energy system and also aware about the basic concept of solar energy system.

4.2.2 Aggregate operation planning of grid connected solar energy system

The aggregate operation planning of grid connected solar energy system is more complicated rather than standalone solar energy system. The grid connected photo-voltaic system are worked parallel with the electricity utility grid. The aggregate operation planning of grid connected solar energy system is divided into three level which is shown in Fig. 4.2. In level 1 create business plan of grid connected solar energy system through the marketing and financial planning. In this level, it is necessary to develop proper financial assessment in terms of cost management, loan management, insurance management and warranty management. In the next level or in level 2 master operations scheduling of grid connected solar energy system. In this also assess the proper working or scheduling of grid connected solar energy system. The primary component of grid connected solar system is power conditioning unit (PCU), where the PCU converts the DC power produced by the PV array into AC power. The proper aggregate operation planning is very necessary to increase the power quality of solar energy system [10,11]. To develop the proper operation planning, a bi-directional interface is made between the PV system AC output circuits and the electric utility network, typically at an onsite distribution panel or service entrance. In level 2, it is also necessary to create component requirement plan and grid connected capacity requirement plan. The capacity of plant is depend on the consumer load demand, because at night and during other periods when the electrical loads are greater than the solar system output, the balance of power required by the loads is received from the electric utility. In the level 3 it is necessary to create detailed scheduling and real time control of grid connected solar energy system. The control system is required in all grid connected solar energy system, and ensure that the grid connected solar system will not continue to operate and feed back into the utility grid when the grid is down for service or repair.

4.3 Level, chase, and mixed strategy of solar energy system

A level strategy definition can be summarized as a detailed outline which incorporates a solar product manufacturing companies and solar energy generation power plant policies, goals, and actions with the focus on being how to deliver value to electricity consumer while maintaining a competitive advantage. The focus of the level strategy is not to disrupt the present operations of the solar product manufacturing company or the solar power plant in any way. This suggests that the system would employ a consistent workforce and/or keep working hours in the consistent way. Inventory is crucial in this method since it connects one period to the next. As a result, solar energy companies frequently use inventory-related solutions to address supply-demand imbalances.

On the other hand if any type of solar industry, if apply the chase strategy, so in that case may be minimum or no inventory is carried from one season to another season. In the solar product manufacturing company and solar power plant, such type of strategy is used in very rare case because supply demand mismatch is addressed in each period, because solar product demand is varies according to the whether condition of the particular location. On the other hand load demand in KW is also varies season to season, such as load demand in summer season is maximum compare to the winter season. Similarly, during periods of low demand, some workers are laid off, while others are allowed to continue on a part-time basis, and the number of

Table 4.2 Month-wise data set.

Month	Solar panel demand	No. of working days
January	900	12
February	930	13
March	1,000	15
April	1100	17
May	1,050	16
June	1,000	15
July	1,050	16
August	900	12
September	870	11
October	910	12
November	930	13
December	800	10

working hours is lowered by lowering the number of shifts or, in extreme cases, even the shift duration.

Two effective aggregate operations planning are the chase and level strategies. As a result, a mixed strategy in the solar energy business is characterized as one that uses a combination of the available options for aggregate operations planning. The following example illustrates this concept:

- Hire a number of skilled workers which is efficient in the field of solar energy at the beginning of the any type of the solar energy company.
- Do not hire or lay off workers, adjust the operations rate by varying the number of shifts, use inventory and subcontracting to match supply with demand [10].

Example 4.1. A manufacturer of solar panel is in the process of preparing the aggregate production plan for the next year. Table 4.2 shows the details pertaining to the force demand for the equivalent model of solar panel and the number of working days available during the planning horizon.

The following relevant details are also available:

1. The solar panel manufacturer currently works on a single-shift basis for 8 hours and employ 100 workers.
2. One unit of solar panel requires 10 hours of production time.
3. It is expected that at the beginning of the planning horizon, there will be a finished goods inventory of 1000, each 300W solar panel.
4. Inventory carrying is Rs. 300 per solar panel per month and unit shortage/backlogging costs are 150 percent of unit carrying cost.

Solution:

The first step in aggregate operation planning of solar panel manufacturing company is to compute the solar panel manufacturer firm works for 8 hours a day, employing 100 employee, then the capacity available in the company is 800 hours per day. Similarly each solar panel requires 10, we can compute the number of hours required during a period. Tables 4.3 and 4.4 show the relevant calculation:

In the above example cost of total inventory is Rs. 798000 and the cost of inventory is maximum in January and minimum in December.

Example 4.2. A solar panel of one of the Manufacturer Company, produces four varieties of solar panel 'A', 'B', 'C', 'D' and 'D' requires a large amount of customization than the other three. The production planning department has estimated the number of hours required for manufacturing one unit of solar panel of each variety. Type 'A' requires 25 hours; Type 'B' requires 35 hours, Type 'C' requires 40 hours, Type 'D' requires 50 hours. The forecast for the four varieties are given in Table 4.5:

The overtime premium is estimated to be Rs. 10 per hour, while under time cost is 40% of the overtime premium. The cost of carrying inventory is Rs. 5 per hour per month and the backordering/shortage cost are 150% of inventory carrying cost. The solar panel manufacturing company can subcontract capacity from a reputed vendor and the additional costs related to subcontracting is Rs. 15 per hour of capacity.

Evaluate the following alternative for AOP:

(a) Prepare a level strategy. Use inventory backlogging to absorb the demand-supply mismatch.

Solution: Since there are four types of solar panel, we use number of hours as the basis for aggregation and convert the demand in units of each variety to demand in hours shown in Table 4.6. Let us consider the month of September. Since the solar panel demand for category A, B, C and D is 15000 hours, 21000 hours, 32000 hours and 2500 hours respectively. Therefore, the aggregate demand during September is 70500 hours. All the values are calculated in a similar fashion.

Plan A: level strategy: A level strategy requires a constant production rate which is shown in Table 4.7. We find that the average requirement during the planning horizon to be 64000hours. Therefore every month, therefore production will be 64000 hours of solar panel. We then compute the demand-supply mismatch during each period.

Table 4.3 Month-wise computation of demand-supply mismatch.

Month	Solar panel demand	Hours required per unit solar panel production	Demand (hours)	No. of work-ing days	Working hours per day	No. of workers	Capacity available (hours) = no. of working days* no. of working hours* no. of workers	Solar panel demand supply mismatch = capacity available (hours) - demand (hours)
January	900	10	9,000	12	8	100	9,600	600
February	930	10	9,300	13	8	100	10,400	1,100
March	1,000	10	10,000	15	8	100	12,000	2,000
April	1,100	10	11,000	17	8	100	13,600	2,600
May	1,050	10	10,500	16	8	100	12,800	2,300
June	1,000	10	10,000	15	8	100	12,000	2,000
July	1,050	10	10,500	16	8	100	12,800	2,300
August	900	10	9,000	12	8	100	9,600	600
September	870	10	8,700	11	8	100	8,800	100
October	910	10	9,100	12	8	100	9,600	500
November	930	10	9,300	13	8	100	10,400	1,100
December	800	10	8,000	10	8	100	8,000	0

Table 4.4 Total cost of the level production strategy.

Month	Demand hours	Capacity available (hours) = no. of working days* no. of working hours*no. of workers	Solar panel demand supply mismatch = capacity available (hours) - demand (hours)	Supply-demand (units)	Opening inventory	Closing inventory	Average inventory	Cost of inventory
January	9,000	9,600	600	60	1,000	940	970	194,000
February	9,300	10,400	1,100	110	940	830	885	177,000
March	10,000	12,000	2,000	200	830	630	730	146,000
April	11,000	13,600	2,600	260	630	370	500	100,000
May	10,500	12,800	2,300	230	370	140	255	51,000
June	10,000	12,000	2,000	200	140	60	100	20,000
July	10,500	12,800	2,300	230	60	170	115	23,000
August	9,000	9,600	600	60	170	110	140	28,000
September	8,700	8,800	100	10	110	100	105	21,000
October	9,100	9,600	500	50	100	50	75	15,000
November	9,300	10,400	1,100	110	50	60	55	11,000
December	8,000	8,000	0	0	60	60	60	12,000
								798,000

Table 4.5 Forecast for the four varieties.

Month	Forecast of solar panel demand (units)			
	Type 'A'	Type 'B'	Type 'C'	Type 'D'
January	1,000	400	600	100
February	900	300	300	50
March	800	500	500	100
April	1,300	400	700	200
May	700	300	300	150
June	900	500	500	50
July	1,000	400	500	100
August	1,100	300	400	150
September	600	600	800	50
October	1,100	200	300	100
November	1,200	400	400	50
December	1,400	500	700	100

4.4 Master operations scheduling (MOS) of solar energy system

Master operations scheduling is the process of linking between planning and execution of solar industry. It is the next stage in the operations planning process in any solar industry. MOS makes use of actual electricity consumer orders for the purpose of capacity planning and resource allocation to specific consumer orders.

In this section Master Operations Scheduling of solar energy system is categorized into two category, first one MOS of solar product manufacturing company and other one is MOS of solar energy generation station.

4.4.1 MOS of solar product manufacturing industry

Master operation scheduling is the process by which disaggregation of varieties is done. The critical link between MOS and other solar energy system planning processes in solar industry are shown in Fig. 4.3. In the solar industry three important factor for proper scheduling is Market of solar product, skilled level who is efficient in the field of solar energy and Vendors. There are two stages involved in the MOS. The first step is to update the projected demand of solar energy product based on earlier forecast and current market information. The second step involves disaggregation of information and relating it to specific material and capacity requirements.

In the solar product manufacturing industry demand of solar energy product for March, April, May would have been based on some prediction

Table 4.6 Aggregate units of the forecast demand.

Month	Forecast of solar panel demand (units)				Forecast of solar panel demand (hours) = forecast of solar panel demand (units) x no. of hours A = 25 hours, B = 35 hours, C = 40 hours, D = 50 hours				
	Type 'A'	Type 'B'	Type 'C'	Type 'D'	Type 'A'	Type 'B'	Type 'C'	Type 'D'	Total
January	1000	400	600	100	25000	14000	24000	5000	68000
February	900	300	300	50	22500	10500	12000	2500	47500
March	800	500	500	100	20000	17500	20000	5000	62500
April	1300	400	700	200	32500	14000	28000	10000	84500
May	700	300	300	150	17500	10500	12000	7500	47500
June	900	500	500	50	22500	17500	20000	2500	62500
July	1000	400	500	100	25000	14000	20000	5000	64000
August	1100	300	400	150	27500	10500	16000	7500	61500
September	600	600	800	50	15000	21000	32000	2500	70500
October	1100	200	300	100	27500	7000	12000	5000	51500
November	1200	400	400	50	30000	14000	16000	2500	62500
December	1400	500	700	100	35000	17500	28000	5000	85500
Total	12000	4800	6000	1200	300000	168000	240000	60000	768000
Average	1000	400	500	100	25000	14000	20000	5000	64000

Table 4.7 Plan A - level strategy with a monthly production of 64,000 hours.

Month	Demand (hr.)	Production rate (hours)	Opening inventory	Production - demand (hours)	Closing inventory (hours)	Cost of inventory/backorder = (rs. 5 × (opening inventory + closing inventory)/2)
January	68000	64000	–	4000	4000	10000
February	47500	64000	4000	16500	12500	41250
March	62500	64000	12500	1500	14000	66250
April	84500	64000	14000	20500	6500	51250
May	47500	64000	6500	16500	10000	41250
June	62500	64000	10000	1500	11500	53750
July	64000	64000	11500	0	11500	57500
August	61500	64000	11500	2500	14000	63750
September	70500	64000	14000	6500	7500	53750
October	51500	64000	7500	12500	20000	68750
November	62500	64000	20000	1500	21500	103750
December	85500	64000	21500	21500		53750
Average 12 months	64000			Total cost of the plan		665000

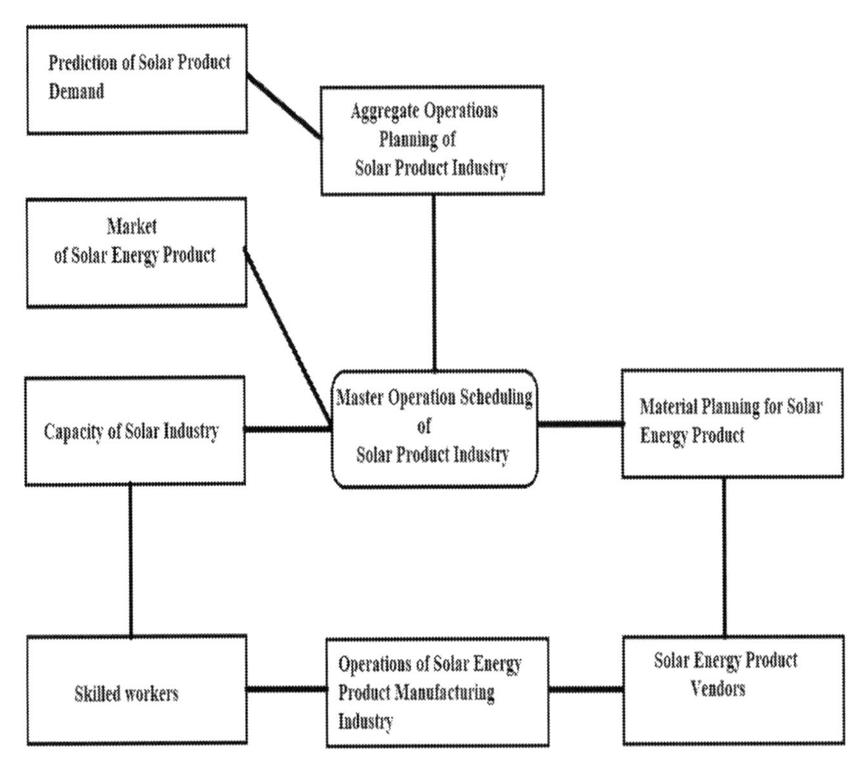

Figure 4.3 MOS of solar industry.

and in that case it is necessary to complete MOS exercise in the month of January, because it is possible that some solar product consumer would have already placed orders. Therefore while planning is done at that time it is important that solar product orders are also taken into consideration. If the solar product orders are more than the forecast quantity, then it is desirable to incorporate this information for the purpose of next level MOS planning.

4.4.2 MOS of solar energy generation system

The goal of this MOS is to reduce the cost of photovoltaic (PV) systems as well as reduction of operations and maintenance (O&M) of hybrid PV/energy storage systems while boosting their effectiveness. O&M issues should be considered in design, engineering, and construction in order to:

- Select low or no maintenance alternatives which is available for the effectiveness of overall system.
- Track the performance of solar energy system component and identify & specify the models with the lowest failure rates and the best reliable services, which possess higher level of reliability.

- Remote testing, software setups and/or updates, and remote resets can all be done with network-connected inverters.
- Provide required access to and clearance around equipment of solar energy system for maintenance.

Following are prerequisites to address in the design and installation of a system in hurricane-prone areas prior to O&M tenure, based on inspections of PV systems damaged in hurricanes:

- Self-tapping screws are favored over through-bolting because self-tapping screws are more likely to pull out, and corrosion weakens them over time due to the hardened threads and absence of galvanised coating on the threads.
- In rack hardware, through-bolting is favoured over clamps. When forces are applied in the appropriate direction, clamps may have the necessary strength, but relative movement, rotation, or vibration of rack pieces in a storm can cause them to release. If one module fails, an entire row of modules can fail, and they are kept together by T-shaped clamps that hold two modules at once.
- Stainless-steel boxes with thick rubber gaskets and multiple closure attachments stay intact and exclude moisture more effectively than those with only handle-actuated attachments (specify NEMA 6 rating).
- Mastheads can break off or leak in a storm and must be sealed with foam packing provided for that purpose to avoid water infiltration and keep insects out.
- Drainage from boxes and conduit runs must be provided in the case of water infiltration.

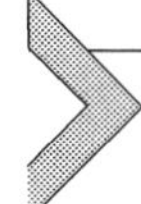

4.5 Dependent demand attributes of solar energy system

Dependent demand describes how the demand for one product is influenced by the demand for another. Both categories are counted as products with dependent demand since demand for both products can move in tandem or in the opposite direction. Estimating for a product with dependent demand is more difficult because the company also consider how demand for its counterparts effects other demand, as well as how external economic forces are expected to affect both items. In the solar energy system demand of battery and capacity of battery is directly depend on the demand of solar panel and capacity of solar power plant. For the

Table 4.8 Attribute and their dependent demand.

Attribute	Dependent demand
Nature of demand of solar product	Parent-child relationship between solar panel and battery or solar panel and solar inverter
Service level of solar product manufacturing company	100% service level a necessity, feasible to achieve
Demand occurrence	Often lumpy
Estimation of product and load demand	By electricity generation planning
How much to order the solar product	Known with uncertainty
When to order the solar product	Very critical, can be estimated

example 20 KW power plant require large capacity of battery compare to the 10KW power plant. In the solar energy system demand of solar inverter and capacity of solar inverter is also depend on the capacity of solar power plant. Table 4.8 shows the attribute and dependent demand of solar energy system.

Before we unfold the planning methodology for dependent demand of the solar energy product, let us look at the following aspects of the planning framework.

(a) Multiple level of dependencies of solar energy product
(b) The solar product structure and bill of material
(c) Time phasing assessment of solar energy product
(d) Incorporating the lead time information of solar energy product
(e) Determining the lot size of solar energy product
(f) Establishing the planning premises for solar energy product

Clearly, a parent child relationship in dependent demand items causes dependency. In the reality multiple levels of relationships exist in a solar energy product. For the example, proper functioning of the solar power plant, it is necessary to unit or optimum sizing of solar panel, battery, inverter and different types of auxiliary equipment's.

The solar product structure graphically depicts the dependency relationship between various items that make up the final product or create final structure of solar energy system.

4.5.1 Planning a framework: solar panel building blocks

Semiconductor material such as silicon are used in photovoltaic solar cell and construction of solar panel is completed in the number of stages.

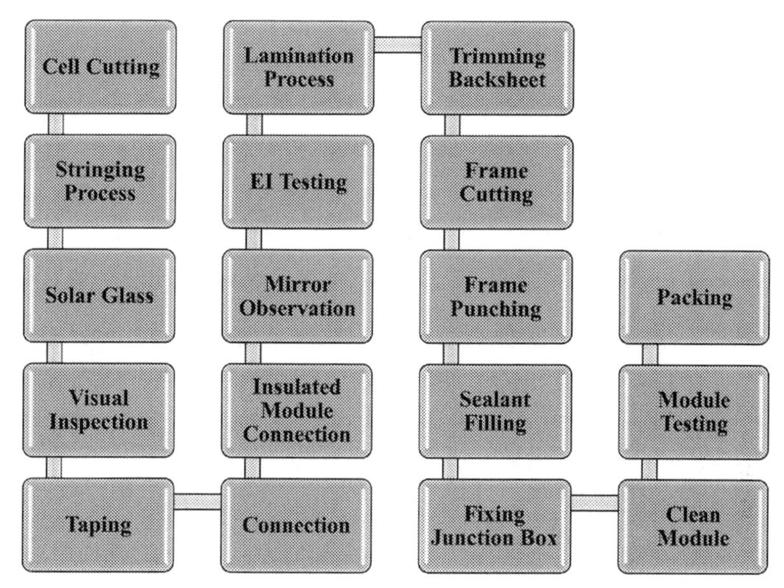

Figure 4.4 Solar panel building blocks.

Fig. 4.4 shows the solar panel building blocks. Following are the different stages of the construction of solar panel:

- **Cell cutting:** In this process cells are cut out according to the wattage size of the solar panel and size of the cell is increased with the increasing of the wattage rating of solar panel.
- **Stringing process:** It's a completely automated procedure and in loom solar, use any cell that is larger than 39 mm in diameter and after that the negative component is the upper Sun facing side (Blue/Black side), whereas the positive part is the bottom white side.
- **Solar glass:** The machine then transfers the cells to tempered glass, which already has an ethylene vinyl acetate (EVA) encapsulating coating on it.
- **Visual inspection:** A technician examines the cells for any faults or errors in the strings.
- **Taping:** A technician tapes the cells into a matrix alignment during taping.
- **Connection:** After that, the connections are soldered together, and any surplus material is removed.
- **Insulate module connection:** The next step is to use a back sheet and EVA encapsulation to insulate the connections. This procedure safeguards the module against dust and moisture.

- **Mirror observation:** The module is visually inspected once more for dust particles, colour mismatches, and other issues.
- **EI Testing:** EI Testing or Electroluminescence test is the real testing of the module made so far. It is a testing process, where the module is kind of scanned in an EI machine. We can easily spot any dead or low power cell, short circuit cells, cracks, etc. If any such error is spotted, the module is sent back for fixing the error.
- **Lamination process:** At 140 degrees Celsius, the module is laminated and such type of procedure takes about 20 minutes. In the next step after the lamination, the modules are allowed to cool for 10-15 minutes until they reach to the room temperature.
- **Trimming back sheet:** This procedure entails trimming the back sheet's superfluous material to create perfectly shaped modules.
- **Frame cutting:** In this step, frames of different sizes are cut out for bordering the panels.
- **Frame punching:** The frames are then punched with holes for mounting and grounding the panels.
- **Sealant filling/framing:** The panels are protected from air, dust, and moisture by a sealant, which also aids in the module's secure attachment to the frame. After the frame has been mounted to the module, it is sent back to the framing machine to be punched to ensure that it is permanently attached to the frame.
- **Fixing junction box:** The sealant is used to firmly bind a junction box to the frame of the module. The connections are then soldered and let to cure for 10-12 hours, ensuring that the structures are completely dry and correctly bonded.
- **Clean module:** The module is then wiped outside to remove any traces of dust, foreign particles or extra sealant.
- **Module testing:** The module is linked in order to check the output current, voltage, and power, among other things. For each module's output data, a report is generated. For the benefit of the users, a back label (with full details) is put behind the module. Finally, the module is delivered to the QC lab for insulation resistance testing. For a minute, a 3000 V DC is sent through it. If the panel can withstand the current, it passes; otherwise, it fails. After that, it is subjected to Mechanical Load Testing.
- **Packing:** After Final Quality Assurance (FQA), this is the last step in the module manufacturing process, where the modules are safely packed into large boxes for transportation and storage.

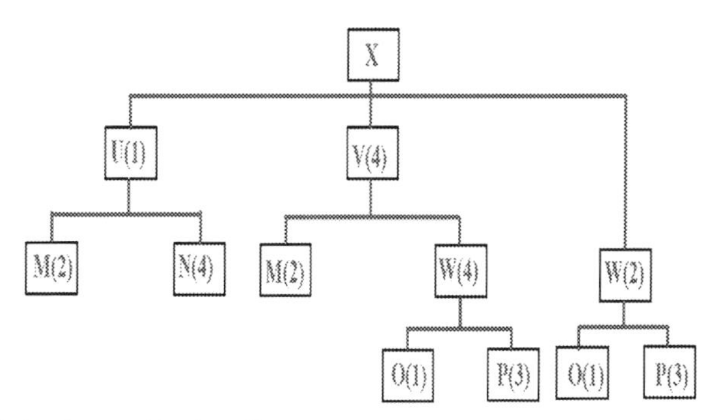

Figure 4.5 Solar System Product Structure.

Example 4.3. Consider a solar panel unit with name 'X' manufactured by one of the solar company, solar panel 'X' made of three sub-assemblies U, V and W. one unit of U, four unit of V and two unit of W are required for assembling one unit of solar product 'X'. Subassembly U is made of two assembly of M and four units of N. subassembly V is made of two units of M and four units of W, whereas subassembly W is made of one unit of O and three units of P.

(a) Develop a product structure for solar panel "X."

(b) How many units of "W" are required to manufacture 10 units of "X."

(c) Will the answer change if there is already an inventory of 10 units each of "V" and "W?"

Solution:

(a) From a description of the product and its sub-assemblies, we note that item W occurs at two different levels. Therefore we employ low-level coding for item 'W'. The product has three levels shown in Fig. 4.5.

(b) If we assume that no inventory of subassemblies or the end product exists in the system, then the number of W's required as follows:

Number of V's required for assembling 10 units of X: 40

Number of W's required for assembling 40 units of V: 160

Number of W's required for assembling 10 units of X: 20

Therefore, the total number of W's required for manufacturing X = 180

(c) Since an inventory of the intermediate sub-assemblies exists, it is important to compute the number of W's required using a level by level approach.

Number of V's required for assembling 10 units of X: 40

Number of V's already available: 10

Net requirement of V's: 30
Number of W's required for assembling 30 units of V: 120
Number of W's required for assembling 10 units of X: 20
Total requirement of W's: 140
Number of W's already available: 10
Net requirement of W's for assembling 10 units of X: 130

The next building block of planning framework is the Bill of Material (BOM) is a list of all the elements, or materials required to assemble or put together one unit of a product. In the solar panel manufacturing plant list of all the ingredients which is required to design the solar panel. There are different types of Bill of materials such as Single way BOM, Indented BOM and Modular BOM. In the single way BOM, list of all the elements, which is required directly for solar panel or solar system related product. In an Intended way of BOM, it is required to develop a list of all the elements, which effect directly or indirectly for the final solar product. Modular BOMs are very useful to represent solar product structures with several alternatives.

4.6 Manufacturing resource planning of solar energy component

In the solar product manufacturing industry MRP is phased order release system that under ideal circumstance which is suitable for solar product such as demand of solar product increase in summer season, schedule the order release for needed demand inventory solar products so that the solar products arrives just as they are required. MRP analyze the date upon which a solar product may be ordered either from solar product manufacturing company or from solar product supplier, by reference to the date on which the solar product needed and solar product arrive. MRP requires backward scheduling of each solar product from its requirements date to launch production and purchase orders accordingly so that they will be completed on time.

➢ **The data use for MRP system:-**

- **Bill of solar product:** The bill of solar product is a list of all solar energy system component and raw material required to manufacture a solar product. It can be broke down in sub-assemblies which turn into require other component of solar energy product.
- **Inventory status file:** The inventory status file keeps a record of amount of stock and usually stock transactions of different types of solar energy

product. Clearly, inventory status file will be incorrect then an incorrect amount will be manufactured. This can create excess stock or shortage.

- **Solar component master file:** This file holds the information related to ordering each solar energy system component. The file normally stores the quantity of each solar product, name of company, price of product and expiry date.
- **Process file:** For each manufactured solar component, this gives sequence of operations and work centers used. If infinite capacity planning was in solar energy operation then predicated or standard times for each operation would be stored here.
- **Service centre file:** This file stores details about each service centre of a particular solar industry, such as costs associated with it, standard set-up available on it.
- **Tool file:** It stores details about tools, which is required for the assembly of solar energy system. This allows the MRP to order tools required for a specific operation of solar power plant.
- **Dealing with orders:** The addition of these functions, where they are needed, makes for a set of repetitive clerical task which lends itself to being performed by computer. It is usually for MRP system also print out the actual paperwork which is sent to solar product suppliers or to solar industry, transmits the information by electronics means, such as e-mail.

Meaning of manufacturing resource planning: Manufacturing resource planning (MRP II) is method for the effective planning of all resource of a solar industry. It is made up of variety of interlinked functions, as shown in such as:

- Strategic and business planning of solar product manufacturing industry and solar energy generation system.
- Load demand management of the electricity consumer, who are taken the electricity from solar power plant.
- Planning of sales of solar energy product and planning of operation of solar power plant.
- Raw material requirements planning for solar energy product.
- Capacity requirements planning of each solar product and solar energy product vendor requirements planning.
- The execution support system for capacity and material.

Sustainable energy planning: More worldwide integration of solar energy supply systems, as well as local and global environmental restrictions, expands the scope of planning, both in terms of subject and time. Sustainable Solar Energy Planning should take into account the environmental consequences of energy use and production, especially in view of the threat of global climate change, which is mostly produced by greenhouse gas emissions from the world's energy systems and is a long-term process. Many OECD countries, as well as some states in the United States, are now taking steps to tighten regulation of their energy systems. Many governments and states, for example, have set targets for CO_2 and other greenhouse gas emissions. As a result of these advances, integrated solar energy planning with a broad scope may become increasingly relevant. Sustainable Energy Planning provides a more comprehensive approach to the problem of anticipating future energy requirements. It is based on a structured decision making process based on six key steps, namely:

- Exploration of the current and future situation's context.
- Identifying specific issues and possibilities that must be addressed as part of the Sustainable Energy Planning process. This could encompass topics like "Peak Oil," "Economic Recession/Depression," and the evolution of energy demand technology.
- Make a variety of models to forecast the impact of various scenarios. This used to just be mathematical modeling, but it's now expanded to include "Soft System Methodologies" like focus groups, peer ethnographic study, and "what if" logical scenarios, among other things.
- The results were analyzed and organized in an easily read style based on the output from a variety of modeling exercises and literature reviews, as well as open forum debate.
- The findings are then analyzed to establish the scope, scale, and potential implementation approaches that will be needed to achieve a successful implementation.
- This stage is a quality assurance procedure that actively interrogates each stage of the Sustainable Energy Planning process to ensure that it was carried out rigorously, without bias, and that it advances rather than hinders the goals of sustainable development.
- Taking action is the final step in the process. This could include the creation, publication, and execution of a variety of policies, rules, processes, or tasks that, when combined, will help to fulfil the Sustainable Energy Plan's objectives.

Sustainable energy planning is particularly appropriate for communities who want to develop their own energy security, while employing best available practice in their planning processes.

Purchase

- Purchase order of solar energy based product

Account Payable and Receivable

- Invoice and account statement of solar energy product

Treasury and Sales

- Cash management forecast and current position of solar energy based product
- Sales order of solar energy product

Figure 4.6 Different level of ERP.

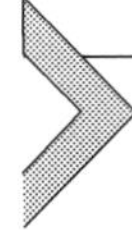

4.7 Enterprise resource planning of solar energy system

In the solar product manufacturing company, Enterprise resource planning is an integrated process that uses common software and an integrated database for planning and control purpose which is shown in Fig. 4.6. The following are the typical module of solar energy system under the enterprise resource planning condition:

- Sales and distribution of solar system based product
- Production planning of solar system based product
- Account payable/receivable of solar system based product
- Operational control of solar product manufacturing company
- Finance and cost control of solar product manufacturing company
- Human resources of solar product manufacturing company

4.7.1 Various types of solar ERP software

ACTIS: For the worldwide solar industry which include solar product manufacturing company or solar power plant, ACTIS provides asset care innovation for proper functioning of a particular system. This software empowers to the solar industry with full control of their portfolios to realize accurate, high-performing solar assets with its integrated suite of operations of solar energy system, maintenance of solar product, and asset management services, and retrofit solutions of all the technical and managerial parameters. Alectris developed ACTIS, the award-winning Solar Enterprise Resource Planning Platform, based on considerable European market solar development experience (Solar ERP). ACTIS is more than just daily power output monitoring based on the given load demand; it's also the monitoring of a set of criteria that ensure accurate and authenticate performance of solar power plant. ACTIS is a real-time monitoring system that aggregates over 70 parameters on each individual solar asset in a solar portfolio, regardless of where the assets are situated. ACTIS is used in hundreds of solar power plants around the world.

Main features of ACTIS monitoring system:

- Full alerting and alert handling features, including individual string configuration and seasonal alerts and it is very important feature of ACTIS because electricity generation capacity of solar power plant varies season to season and location to location.
- This software provide monitoring of all the solar power plant devices such as sensors & actuators, solar inverters, UPS, energy meters, switchgear & relays, and pyrheliometer which is used for measurement of solar radiation.
- This software also worked as data analytics software in a particular situation, because it collect and summarize the data of all the sensitive parameters of solar power plant.
- This software also monitor the communication network quality of IOT based solar power plant, because this software also works with different types of RFID devices.
- ACTIS also used for analyzing the different technical and financial parameters of solar power plant based on the given data sets with the help of graph and charts.
- One of the significant feature of ACTIS software is, it evaluates the different parameters of solar power plant through data comparison and data cleaning.
- Dashboards of ACTIS software with true real-time information on the operation of the solar power plant and their portfolio.

ACTIS monitoring user benefits:

- The plant's operating data is monitored 24 hours a day, seven days a week, and all electromechanical equipment of solar power plant and security systems which directly or indirectly affect the overall performance of solar power plant.

- Remote diagnosis of any defects or malfunctions in any solar component, as well as constant alerting, allows for immediate or predictive corrective action of a particular equipment.
- Incident impact analysis of technical parameters of solar power plant and cross comparison and evaluation of technical and financial data of solar power plant.

ACTIS service management main features:

- Incident identification, handling and statistical evaluation of solar energy data by the skilled worker, who is proficient in this software.
- Management of as-built information of all the parameters of solar power plant.
- Solar component inventory management.
- Spare part management of solar power plant.
- Technician/Subcontractor management and assessment, who is the employee of solar power plant.
- Provide training for the Technician/Subcontractor who is the part of solar power plant.
- This software also create proper service scheduling of whole solar power plant.
- ACTIS software also provide Subcontractor Terms and Conditions in the proper manner.
- This software also provide complete alerting system in the different circumstances.
- Service cost management is also an important feature of this software.

ACTIS service management user benefits:

- Alerting on overall plant and specific device performance in real time.
- Real-time scheduling, execution, and monitoring of maintenance tasks (corrective and preventative), ensuring uninterruptible plant operation on a technical and safety level, while simultaneously maximizing energy production and achieving a high return on investment Local technician and subcontractor control, training and evaluation.
- Complete asset and spare parts control with extensive inventory capabilities [12,13].
- Continuous compliance monitoring of service execution, including quality, cost, and timing, as well as automatic calculation and comparison with relevant contracts.

ACTIS solar asset management main features:

- **Account & contact management:** This software provides account and contact management, which gather all the information of electricity consumer, who take the electricity from the particular solar power plant. This section of ACTIS software also include supplier information, who provide equipment for the solar power plant [14,15].

- **Activity management**: This section of ACTIS software provide details of all the technical and other activity, which is done in the solar power plant. It also gather the information of energy scheduling, electricity trading and in what way plant fulfill the load demand of the consumer.
- **Contract management:** This part of the software provide the information, which type of contract in between solar power plant owner and supplier of the equipment.
- **Warranty management:** This section of ACTIS software elaborate warranty of a different component which is connected in solar power plant.
- **Cash flow management:** This section of this software contain information related to the capital cost, replacement cost and operation and maintenance cost of all the component of solar power plant.
- **Loan management:** Loan management of ACTIS software facilitates the automation of the full loan lifecycle of solar power plant and also supply reliable declarations of all the terms and condition which is done in between loan provider company and solar power plant organization. This software also monitor interest rates and provide collection automation solutions of solar power plant.
- **Business plan management:** The 'management part' of a solar power plant business plan explains your management team, staff, resources, and how your company's ownership is structured. This part should not only list who is on your management team, but also explain how each person's skill set will help your bottom line.

  ACTIS solar asset management user benefits:
- Every PV asset related contract is digitalized, customizable contract types are available and respective parameters are monitored for contract compliance and scheduling of the relevant specific tasks.
- Full and updated view on each equipment warranty and management and alerting according to the relevant expirations dates.
- Management of claims towards insurance companies, history of activities, resolutions and settlement.
- Power Purchase Agreement administration.
- Asset managers can periodically control and review operating costs, financial and cash performances of each PV asset, and the relevant spread

(actual vs budget) with respect of bank/financial institution cash flow expectations [16,17].

- Asset managers are automatically alerted if plant's cash flows are going below expectations in order to re-act promptly to protect investors/banks expositions.

VERTS: VERTS is a fast-growing commercial software firm that specialises in ERP Software Solutions for a variety of industries. VERTS was founded with the goal of making technology simple and affordable for small and medium businesses so that they could optimise their business processes and grow faster. We don't sell technology at VERTS; instead, we work with businesses to provide solutions, efficiency, and growth. The solar sector is quickly growing, and this renewable energy component is reaching out directly to consumers and companies. This industry has its unique set of requirements, and VERTS is dedicated to providing creative solutions for solar energy-related businesses. If you offer solar solutions on a pay-as-you-go basis, we have the ideal solution for you. Our solution can handle the entire customer management lifecycle, from lead management through recurring billing, service activation and deactivation. If you're a manufacturer or offer an integrated solution that includes assembly, our manufacturing solution is ideal for you. You may not only handle complex solar panel manufacture or panel assembly with this, but you can also manage complex production of solar panels.

Octabees: Octabees is the fastest-growing enterprise software with cutting-edge cloud and mobile applications to manage end-to-end processes of all solar industry. It mainly focus on efficiently solving the problems for small-to-large scale business and offer the best solution to the ever-changing market dynamics of all the solar energy products. Following is the features of Octabees:

- **Manage any size and types of projects:** Regulate projects of all sizes, from small to large (KW to MW) scale, with a variety of different types of solar and engineering projects.
- **Robust, flexible, and scalable:** A single, durable, flexible, and scalable platform is designed to meet the needs of all organizational levels.
- **User-friendly:** For simplicity of company operations, an exceptionally user-friendly web and mobile application was designed for any level of staff.
- **On the cloud and on-premises:** A quick, secure, and simple option for delivering software in the cloud or on-premises, depending on your preferences.

- **Customizable workflow:** A workflow that can be fully customized and configured to meet any business operating requirement.

➢ **5 ways ERP reduces solar soft costs:**

Here are five ways that a unified, cloud ERP designed specifically for the solar business may help installers cut soft expenses and increase profitability on every project.

1. **Replaces widely-used, off-the-shelf software:** Incompatible "point" apps that don't scale well or profitably in combination, such as widely-used systems like QuickBooks and Salesforce, must eventually be replaced as firms develop, at increasing financial and disruptive costs. Integrated management solutions for solar energy companies, that support all core company operations, such as accounting, finance, project management, inventory/procurement, sales/marketing, and human resources, are replacing them.
2. **Provides reliable, real-time data:** Scaling productively for solar installers is nearly impossible without good data that provides visibility across all company processes. "Working with incompatible, incomplete, or out-of-date data may soon turn into a nightmare," stated Jan Rippingale, Chief Executive Officer of Blu Banyan. "Solar installers must have real-time visibility over their complete end-to-end operation to optimise efficiency and profitability."
3. **Breaks down information silos:** Even when they scale up to meet the expanding demand for solar electricity, many installers continue to use systems like QuickBooks to run their businesses. Because QuickBooks and Salesforce.com don't "speak" to one other, and Salesforce doesn't interact with job-costing software, an installer's procurement, accounting, and sales teams are all using different playbooks. In today's flat, collaborative world, when both internal and external partners require common technology platforms and data compatibility, a compartmentalized approach just won't cut it.
4. **Integrates all aspects of the installation business:** Solar Success integrates sales pipeline management, CRM, accounting, purchasing, installation project management (including per-project costing and profitability), inventory management, customer invoicing, universal financier connectivity, and company information into one application. The technology gives solar installers complete access into cash flow, profitability, acquisition costs, project tracking and notifications, subcontractor monitoring, and other critical activities. "When you're aggregating a group of point applications and regardless of how good each of those

programs is individually you really need them to be able to talk to one another," said Rippingale. "The only way to make that happen, and to effectively reduce the soft costs identified by the DOE, is with an integrated application suite that provides reliable data and a common interface across all functions."

5. **Supports scale and growth:** Installers may efficiently fulfil the expectations of their growing client bases while also reducing soft expenses to a minimum, ensuring profitable growth for the installer, by using an end-to-end, fully integrated software package that was specifically built for the solar sector. "In fact, without transitioning to a system that allows those to use the same data across all of their departments, many of them will be unable to maintain their existing profitability levels and competitive position."

4.8 Conclusion

In the above discussion, it is find out aggregate operation planning, master operation scheduling and enterprise resource planning is the benchmark for the assessment of solar industry and solar power plant. The manufacturing resource planning will help to solar industry and their supply chain partners to jointly plan and reduce investment in the supply chain. On the other hand enterprise resource planning embeds all the processes into the software and creates a work flow mechanism of solar industry and solar power plant. At the end it is necessary to the effective utilization of productive and manufacturing economics in the field of solar energy system.

4.9 Exercise/question

1. Explain the basic concept of aggregate operation planning w.r.t. the solar industry.
2. What is the need of aggregate operation planning in the solar industry and in solar power plant?
3. What is the meaning of capacity fluctuation and load demand fluctuation?
4. Explain the concept of aggregate operation planning framework of solar energy system.
5. What are the different alternatives to managing demand in solar energy system?

Table 4.9 Data set for solar panel demand.

Month	Solar panel demand	No. of working days
January	900	14
February	930	13
March	1,000	17
April	1,100	17
May	1,050	16
June	1,000	15
July	1,050	16
August	900	12
September	870	13
October	910	12
November	930	13
December	800	10

6. Write the difference between level, chase and mixed strategy.
7. A manufacturer of solar panel is in the process of preparing the aggregate production plan for the next year. Table 4.9 shows the details pertaining to the force demand for the equivalent model of solar panel and the number of working days available during the planning horizon.

   The following relevant details are also available:
   - The solar panel manufacturer currently works on a single-shift basis for 8 hours and employ 200 workers.
   - One unit of solar panel requires 20 hours of production time.
   - It is expected that at the beginning of the planning horizon, there will be a finished goods inventory of 2000, each 400W solar panel.
   - Inventory carrying are Rs. 200 per solar panel per month and unit shortage/backlogging costs are 150 percent of unit carrying cost.
8. A solar panel of one of the Manufacturer Company, produces four varieties of solar panel 'A', 'B', 'C', 'D' and 'D' requires a large amount of customization than the other three. The production planning department has estimated the number of hours required for manufacturing one unit of solar panel of each variety. Type 'A' requires 35 hours; Type 'B' requires 45 hours, Type 'C' requires 50 hours, Type 'D' requires 60 hours. The forecast for the four varieties are given in Table 4.10.

   The overtime premium is estimated to be Rs. 10 per hour, while under time cost is 40% of the overtime premium. The cost of carrying inventory is Rs. 5 per hour per month and the backordering/shortage cost are 150% of inventory carrying cost. The solar panel manufacturing

Table 4.10 Forecast for the four varieties.

Month	Forecast of solar panel demand (units)			
	Type 'A'	Type 'B'	Type 'C'	Type 'D'
January	1000	400	600	100
February	900	300	300	50
March	800	500	500	100
April	1300	400	700	200
May	700	300	300	150
June	900	500	500	50
July	1000	400	500	100
August	1100	300	400	150
September	600	600	800	50
October	1100	200	300	100
November	1200	400	400	50
December	1400	500	700	100

company can subcontract capacity from a reputed vendor and the additional costs related to subcontracting is Rs. 15 per hour of capacity.

Evaluate the following alternative for AOP:

(b) Prepare a level strategy. Use inventory backlogging to absorb the demand-supply mismatch.

9. Explain the master operations scheduling of solar energy system.

10. Write short note on the following w.r.t. the solar energy system:

a. Manufacturing resource planning

b. Enterprise resource planning

References

[1] J. Dahmus, T. Gutowski, Efficiency and production: Historical trends for seven industrial sectors, in: Tacoma: Proceedings of the 3rd Biennial Conference of the US Society for Ecological Economics, 2005.

[2] Y. Seow, S. Rahimifard, E. Woolley, Simulation of energy consumption in the manufacture of a product, Int. J. Comput. Integr. Manuf. 26 (2013) 663–680.

[3] M. Sachidananda, S. Rahimifard, Reduction of water consumption within manufacturing applications, in: 19th CIRP International Conference on Life Cycle Engineering, Berkeley, 2012.

[4] G. Thompson, J. Swain, M. Kay, C. Forster, The treatment of pulp and paper mill effluent: a review, Bioresour. Technol. 77 (2001) 275–286.

[5] J. Allwood, J. Cullen, Sustainable Materials - With Both Eyes Open, UIT Cambridge, Cambridge, 2011.

[6] R. Milford, S. Pauliuk, J. Allwood, D. Müller, The roles of energy and material efficiency in meeting steel industry CO2 targets, Environ. Sci. Technol. 47 (2013) 3455–3462.

[7] U. Wegst, M. Ashby, Materials selection and design of products with low environmental impact, Adv. Eng. Mater. 4 (2002) 378–383.

[8] A. Munoz, P. Sheng, An analytical approach for determining the environmental impact of machining processes, J. Mater. Process. Technol. 53 (1995) 736–758.
[9] P. Sheng, M. Srinivasan, S. Kobayashi, Multi-objective process planning in environmentally conscious manufacturing: A feature-based approach, CIRP Ann. - Manuf. Technol. 44 (1995) 433–437.
[10] G. Mouzon, M. Yildirim, J. Twomey, Operational methods for minimization of energy consumption of manufacturing equipment, Int. J. Prod. Res. 45 (2007) 4247–4271.
[11] K. Fang, N. Uhan, F. Zhao, J. Sutherland, A new approach to scheduling in manufacturing for power consumption and carbon footprint reduction, J. Manuf. Syst. 30 (2011) 234–240.
[12] X. Xu, L. Wang, S. Newman, Computer-aided process planning – a critical review of recent developments and future trends, Int. J. Comput. Integr. Manuf. 24 (2011) 1–31.
[13] T. Peng, X. Xu, Energy-efficient machining systems: a critical review, Int. J. Adv. Manuf. Technol. 72 (2014) 1389–1406.
[14] O. Gould, J. Colwill, A framework for material flow assessment in manufacturing systems, J. Ind. Prod. Eng. 32 (2015) 1–12.
[15] J. Holland, Adaptation in Natural and Artificial Systems: An Introductory Analysis With Applications to Biology, Control, and Artificial Intelligence, U Michigan Press, Oxford, 1975.
[16] T. Murata, H. Ishibuchi, H. Tanaka, Genetic algorithms for flowshop scheduling problems, Comput. Ind. Eng. 30 (1996) 1061–1071.
[17] J-Y. Potvin, Genetic algorithms for the traveling salesman problem, Ann. Oper. Res. 63 (1996) 337–370.

CHAPTER FIVE

Assessment of solar energy system by probability and sampling distribution

Learning objectives

- Understand about discrete v/s continuous distribution of solar energy parameter.
- Know about binomial, poisson, and hypergeometric of solar energy parameter.
- Know about sampling technique of solar energy parameter.
- Understand about the Weibull distribution of solar energy system.

5.1 Introduction

Solar energy system is one of the prominent renewable energy sources and it is also one of the most intermittent renewable energy sources because output of the solar energy system is dependent on the unpredictable nature of day-wise solar radiation. In a solar power plant most of the decision making involves uncertainty, for example the site engineer of a solar power plant does not know definitively whether a solar power plant fulfills the load demand regularly or not. What are the chances that the solar power plant produces the same amount of electricity in the next week, because all the conditions depend on the different atmospheric conditions? Because most of the solar power plant questions do not have accurate assessment and the final outcome is depending on the uncertainty. In many of such circumstances, a probability can be assigned to the likelihood of an outcome. Statistical analysis of solar energy systems is inferential and in this analysis taking a sample of solar energy data from a population, computing a statistic on the sample and inferring from the statistics the value of the corresponding solar energy system parameter of the population. Suppose a design and quality engineer of a solar product manufacturing company randomly selects a sample of 60 solar panels from a population of monocrystalline panels and the number of hours of electricity of each panel is recorded. The average electricity generation of each panel is 6 hours in a day. Because of the solar panels being analyzed are only a sample of the population, the average number of hours of electricity generation for the 60 solar panels may or may

DOI: https://doi.org/10.1016/B978-0-323-85761-1.00003-2

not accurately estimate the average for all solar panels in the population. The results are uncertain and by applying the laws presented in this chapter, the analyzer can assign a value of probability to this estimate [1].

The three general methods of assigning probabilities are classical method, relative frequency of occurrence and subjective probability method. When the outcomes of the probabilities are based on the different laws and identity, then the method is referred to as the classical method of probabilities. When we apply the classical method of probability in the solar energy system, it determines the ratio of the number of items in a population to the total number of population N [2].

In a solar manufacturing company out of 500 solar panels, 300 is monocrystalline solar panel and 200 become polycrystalline solar panel, so probability of randomly selected solar panel is:

Classical method of assigning probability of monocrystalline solar panel is $300/500 = 0.6$

Classical method of assigning probability of polycrystalline solar panel is $200/500 = 0.4$

The range of possible probability of solar power plant is

0 ≤ Probability of solar power plant ≤ 1

For example data gathered from company records shows that the supplier sent the company 80 batches of solar panels in the past and solar power plant engineers rejected 10 of them. By the method of relative frequency of occurrence, the probability of the inspector rejecting the next batch is 10/80 or 0.125. If the next batch is rejected, the relative frequency of occurrence probability for the subsequent shipment would be changed to $11/81 = 0.135$ and so on. The subjective method of assigning probability is based on the feelings or insights of the person determining the probability. Suppose an engineer of a solar power plant asked the probability of getting a shipment of solar panels from Germany to India within three weeks. An engineer who has scheduled many such shipments of solar panels for the company, has a knowledge of German politics and has an awareness of current economic conditions may be able to give an accurate probability that the shipment of solar panels can be made on time [3,4].

For the assessment of the solar energy system four types of probability are presented, which is Marginal, Union, Joint and Conditional probabilities. A marginal probability is usually computed by dividing subtotal by whole. In union probability, if Events A and B are mutually exclusive, P (A ∩ B) … The probability that Events A or B occur is the probability of the union of A and B. The probability of the union of Events A and B is denoted by

Table 5.1 Crosstabulation of solar panel data.

Type of installation		Solar panel		
		Monocrystalline	**Polycrystalline**	**Total**
	Solar farm	8	4	12
	Standalone	31	14	45
	Hybrid system	52	19	71
	Solar power vehicle	10	22	32
		101	59	160

P (A ∪ B). If the occurrence of Event A changes the probability of Event B, then Events A and B are dependent. Joint Probability Joint probability is a statistical measure that calculates the likelihood of two events occurring together and at the same point in time. Joint probability is the probability of event Y occurring at the same time that event X occurs. Conditional probability is the probability of one event occurring with some relationship to one or more other events. After the above four types of probabilities the law of addition and the law of multiplication is one of the most significant parts of probability function [5,6].

The general law of addition is used to find the probability of the union of two events P (XUY) and which is given by P(XUY)=P(X)+P(Y)-P(XY)

Where X,Y are events and (X Y) is the intersection of X and Y.

General law of multiplication $P(X\ Y) = P(X).P(Y\ /\ X) = P(Y).P(X/Y)$

Following example shows the application of different types of probability in the solar energy system [7,8].

Example 5.1. A solar manufacturing company data reveals that 160 solar panels are installed at various locations in the way of solar farm, standalone, hybrid system and for solar powered vehicle. The input is given in the cross tabulation form in the Table 5.1. If the solar panel is selected randomly, what is the probability that the solar panel is monocrystalline and a connected in the way of standalone system?

Solution: According to the probability theory, X denote the event of polycrystalline and Y denote the event of a standalone system. The question is P (XUY) =?

By the general law of addition $P\ (XUY) = P(X) + P(Y) - P(X\ Y)$

According to the data of Table 5.1 out of 160 solar panels, 59 are polycrystalline.

Therefore, $P(X) = 59/160 = 0.368$.

Table 5.2 Resulting data of solar panel.

		Solar panel		
Type of installation		**Monocrystalline**	**Polycrystalline**	
	Solar Ffarm	0.05	0.025	0.075
	Standalone	0.193	0.087	0.281
	Hybrid system	0.325	0.118	0.443
	Solar power vehicle	0.062	0.137	0.2
		0.631	0.368	1

The 160 solar panels include 45 panels for a standalone system.

Therefore, P(Y) = 45/160 = 0.281.

Because 14 solar panels are polycrystalline and used in the standalone system.

$$[P(XY) = 14/160 = 0.0875.]$$

The union probability is solved as

$$\mathrm{P(XUY) = P(X) + P(Y) - P(XY)} = 0.368 + 0.281 - 0.0875 = 0.5615$$

Suppose you want to solve Example using a joint probability concept. A joint probability table for this can be constructed from the contingency table by dividing all the data by 160. The resulting data is shown in Table 5.2.

Solving for the probability that the selected solar panel is polycrystalline or for a standalone system using a joint probability concept is done by adding the values which is shown below:

$$0.025 + 0.087 + 0.118 + 0137 + 0.2 = 0.5615$$

5.2 Discrete v/s continuous distribution of solar energy parameters

A random variable is a variable that contains the outcomes of a chance experiment. For example, suppose an experiment is to measure the arrivals of solar energy system products during a 60 second period. The possible outcomes are 0 solar inverter, 1 solar inverter…, n solar inverter. These numbers are the values of the random variable of the solar inverter. The random variables are categorized into the two forms discrete random variable and continuous random variable.

Table 5.3 Examples of discrete random variables of solar energy system.

Experiment	Random variable (x)	Possible value for the random variable
Contact 10 customer of solar panel	Number of customer who place an order	0, 1, 2, 3, 4, 5, 6, 7, 8, 9, 10
Inspect a shipment of 100 solar inverter	Number of defective solar inverter	0, 1, 2, 3, 4, 99, 100
Operation of solar power plant for one day	Amount of electricity generation in KW	0, 1, 2, 3, 4, 5
Sell the solar panel	Material of the solar panel	0 if Polycrystalline, 1 if Monocrystalline

Table 5.4 Examples of continuous random variables of solar energy system.

Experiment	Random variable (x)	Possible value for the random variable
Operation of solar power plant	Seasonal variation of load demand	$x > 0$
Recharge a solar battery	Number of hours required	$0 < x < 12$
Installation of new solar power plant	Percentage of project after 4 month	$0 < x < 100$
Testing of solar panel	Solar radiation when 4 to 9 kwh/m2/day	$4 < x < 9$

A random variable is a "discrete random variable" if the set of all possible values is at most a finite or a countably infinite number of possible values. For example, if eight locations are randomly selected for the solar power plant installation and how many of them are suitable according to the load demand. So in that case location and load demand is discrete in nature. Tables 5.3 and 5.4 represents the examples of discrete random and continuous random variables of solar energy system. Other examples, which show application of random variables in the field of the solar energy system.

- Randomly selected 35 site engineers for solar power plants and determined how many suitable for solar related job description.
- Identifying the number of defective solar panels in a batch of 1000 solar panels.
- Sampling the number of solar inverters, which provide efficiency more than 98%.

A discrete distribution that contains the probability of a particular location is suitable for a solar power plant. Table 5.5 shows the probability of

Table 5.5 Discrete distribution of suitability occurrence of number of location for solar power plant.

Number of location	Suitability probability
0	0.43
1	0.25
2	0.11
3	0.17
4	0.02
5	0.02

Table 5.6 Computing the mean of suitability occurrence of number of location for solar power plant.

Number of location = x	Suitability probability = P(x)	x.P(x)
0	0.25	0
1	0.43	0.43
2	0.11	0.22
3	0.17	0.51
4	0.02	0.08
5	0.02	0.1
		= 1.34

5 locations and their number of occurrences according to suitability for solar power plants. Table 5.6 shows the computation of the mean of suitability occurrence of number of location for the solar power plant.

Now you can identify mean, variance and standard deviation of given data of the solar energy system in the way of discrete distribution.

The mean or expected value of a discrete distribution is given by $\mu = S(x) = \sum [x.P(x)]$

Where

S(x) = Long run average of solar energy system

x = An outcome of solar system

P(x) = Probability of that outcome

The variance and standard deviation of a discrete distribution are solved by using the outcomes (x) and suitability probability for solar energy system location P(x) in a manner similar to that of computing a mean.

Variance of discrete distribution $\sigma^2 = \Sigma[(x-\mu)^2.P(x)]$

Standard deviation of a Discrete Distribution is given by =

$$\sigma = \sqrt{\Sigma\left[(x-\mu)^2.P(x)\right]}$$

Table 5.7 Computing the mean of suitability occurrence of number of location for solar power plant.

Number of location x	Suitability probability P(x)	(x-μ)^2	(x-μ)^2.P(x)
0	0.25	$(0\text{-}1.34)^2 = 1.795$	0.448
1	0.43	$(1\text{-}1.34)^2 = 0.115$	0.049
2	0.11	$(2\text{-}1.34)^2 = 0.435$	0.047
3	0.17	$(3\text{-}1.34)^2 = 2.755$	0.468
4	0.02	$(4\text{-}1.34)^2 = 7.075$	0.141
5	0.02	$(5\text{-}1.34)^2 = 13.395$	0.267
			= 1.42

Table 5.8 Discrete distribution of suitability occurrence of solar inverter for solar power plant.

Number of location	Suitable probability of solar inverter
0	0.40
1	0.21
2	0.16
3	0.12
4	0.07
5	0.02
6	0.02

The variance and standard deviation of the suitability probability of solar energy system location are calculated and shown in Table 5.7.

Variance of discrete distribution $\sigma^2 = \Sigma[1.42] = 1.42$

Standard deviation of a Discrete Distribution is given by $= \sigma = \sqrt{1.42}$

The standard deviation is = 1.19

Example 5.2. A discrete distribution that contains the probability of a solar inverter is suitable for a solar power plant. Table 5.8 shows the probability of 6 locations and their number of occurrences according to suitability for solar power plants. Table 5.9 shows the computing the mean of suitability occurrence (part 1) of number of location for solar power plant. Table 5.10 shows the computing the mean of suitability occurrence (part 2) of number of location for solar power plant.

Variance of discrete distribution $\sigma^2 = \Sigma[2.98] = 2.98$

Standard deviation of a Discrete Distribution is given by $= \sigma = \sqrt{2.98}$

The standard deviation is = 1.72

Table 5.9 Computing the mean of suitability occurrence (part 1) of number of location for solar power plant.

Number of location x	Suitability probability P(x)	x.P(x)
0	0.40	0
1	0.21	0.21
2	0.16	0.32
3	0.12	0.36
4	0.07	0.28
5	0.02	0.1
6	0.02	0.12
		=1.39

Table 5.10 Computing the mean of suitability occurrence (part 2) of number of location for solar power plant.

Number of location x	Suitability probability P(x)	$(x-\mu)^2$	$(x-\mu)^2.P(x)$
0	0.40	$(0-1.39)^2=1.932$	0.772
1	0.21	$(1-1.39)^2=0.152$	0.031
2	0.16	$(2-1.39)^2=0.372$	0.059
3	0.12	$(3-1.39)^2=2.592$	0.311
4	0.07	$(4-1.39)^2=6.812$	0.476
5	0.07	$(5-1.39)^2=13.032$	0.912
6	0.02	$(6-1.39)^2=21.252$	0.425
			=2.98

The continuous random variables take on values at every point over a given interval. Thus continuous random variables have no gaps or unassumed values. In a continuous system things are measured not counted. For example day-wise measurement of solar radiation through the pyrheliometer. Other examples are

- Day-wise efficiency measurement of solar panel system
- Day-wise charging and discharging capability of battery system
- Month-wise average solar radiation at a particular location.
- Hour wise load demand

5.3 Binomial, poisson, and hypergeometric distribution of solar energy data

The binomial distribution has been used for hundreds of years. Besides its historical significance, the binomial distribution is studied by business statistics students because it is the basis for other important statistics [9,10].

Binomial Formula $P(x) = nC_X.p^x.q^{n-x} = \frac{n!}{x!(n-x)!} p^x.q^{n-x}$
Where n = The number of trials

x = The number of successes desired

p = The probability of getting a success in one trial

q = 1-p = The probability of getting a failure in one trial

Mean and Standard Deviation of a Binomial Distribution is given by

$$\mu = n.p \text{ And } \sigma = \sqrt{npq}$$

Following example shows the application of binomial distribution in the field of solar energy system.

Example 5.3. In the solar energy related survey, it is found out 75% of electricity consumers were very satisfied through the standalone solar energy system, which is connected in their house. Suppose that 35 electricity consumers are satisfied. If the solar energy survey result still holds true today, what is the probability that exactly 21 are very satisfied with their solar energy survey?

Solution: The value of p is .75(very satisfied), the value of q = 1-p = 1-.75 = .25 (not very satisfied), n = 35 and x = 21. The binomial formula yields the final answer:

$$P(x) = nC_X \cdot p^x \cdot q^{n-x}$$

$$35C_{21} \cdot (0.75)^{21}(.25)^{35-21} = 0.0205$$

If 75% of all electricity consumers are very satisfied from a standalone solar energy system, about 2.05% of the time in the survey exactly 21 out of 35 are very satisfied from the standalone solar energy system. If 75% of electricity consumers are very satisfied with their standalone solar energy system, one would expect to get about 75% of 35 or (.75) (35) = 26.25 very satisfied electricity consumers. While in any individual sample of 35 the number of financial consumers who are very satisfied cannot be 26.25, it can be understood that x value near 26.25 is the most likely occurrence.

Example 5.4. One of the solar product manufacturing company test different solar panels in the different critical circumstances, where the amount of solar radiation is very less. Solar panel have a 35% chance of failing to complete the test run without a blowout. 15 solar panel of 300W each go through the test, find the following:

a. The values of the Probability Mass function.
b. The probability that fewer than four solar panel have blowouts.
c. The probability that more than two have blowouts.
d. The expected number of solar panel with blowouts.

Solution:
Assume the blowout are random and independent. The probabilities of failure and success are 35% and 65% respectively.

Since there are fifteen solar panels in the test, the probability mass function is given by

$$P(X = x) = (15Cx) \cdot (0.35)^x (0.65)^{15-x}$$

a) The calculated results are listed below
$P(X=0) = (1)(0.35)^0(0.65)^{15} = 0.00156$
$P(X=1) = (15)(0.35)^1(0.65)^{14} = 0.01261$
$P(X=2) = (105)(0.35)^2(0.65)^{13} = 0.04755$
$P(X=3) = (455)(0.35)^3(0.65)^{12} = 0.11096$
$P(X=4) = (1365)(0.35)^4(0.65)^{11} = 0.17924$
$P(X=5) = (3003)(0.35)^5(0.65)^{10} = 0.21233$
$P(X=6) = (5005)(0.35)^6(0.65)^9 = 0.19056$
$P(X=7) = (6435)(0.35)^7(0.65)^8 = 0.13192$
$P(X=8) = (6435)(0.35)^8(0.65)^7 = 0.07103$
$P(X=9) = (5005)(0.35)^9(0.65)^6 = 0.02975$
$P(X=10) = (3003)(0.35)^{10}(0.65)^5 = 0.00961$
$P(X=11) = (1365)(0.35)^{11}(0.65)^4 = 0.00235$
$P(X=12) = (455)(0.35)^{12}(0.65)^3 = 0.00042$
$P(X=13) = (105)(0.35)^{13}(0.65)^2 = 0.000052$
$P(X=14) = (15)(0.35)^{14}(0.65)^1 = 0$
$P(X=15) = (1)(0.35)^{15}(0.65)^0 = 0$

b) $P(X < 4) = 0.11096 + 0.04755 + 0.01261 + 0.00156 = 0.17268$

c) $P(X > 2) = 1 - P(0) - P(1) - P(2) = 1 - 0.00156 - 0.01261 - 0.04755 = 0.938$

d) $E(X) = (0)(0.00156) + (1)(0.01261) + (2)(0.04755) \ldots + (15)(0) = 5.249$

Example 5.5. One of the solar product manufacturing company test different solar inverters in the different critical circumstances. Solar panel have a 30% chance of failing to complete the test run without a blowout. 18 solar inverter of 1000W each go through the test, find the following:

The values of the Probability Mass Function

a. The probability that fewer than four solar panel have blowouts.

b. The probability that more than two have blowouts.

c. The expected number of solar panel with blowouts.

Solution:

Assume the blowout are random and independent. The probabilities of failure and success are 30% and 70% respectively.

Since there are fifteen solar panels in the test, the probability mass function is given by

$$P(X = x) = (18Cx) \cdot (0.30)^x (0.70)^{18-x}$$

a) The calculated results are listed below

$P(X=0) = (1)(0.30)^0(0.70)^{18} = 0.001628$

$P(X=1) = (18)(0.30)^1(0.70)^{17} = 0.012562$

$P(X=2) = (153)(0.30)^2(0.70)^{16} = 0.045762$

$P(X=3) = (816)(0.30)^3(0.70)^{15} = 0.104598$

$P(X=4) = (3060)(0.30)^4(0.70)^{14} = 0.168104$

$P(X=5) = (8568)(0.30)^5(0.70)^{13} = 0.201725$

$P(X=6) = (18564)(0.30)^6(0.70)^{12} = 0.187316$

$P(X=7) = (31824)(0.30)^7(0.70)^{11} = 0.13762$

$P(X=8) = (43758)(0.30)^8(0.70)^{10} = 0.081098$

$P(X=9) = (48620)(0.30)^9(0.70)^9 = 0.038618$

$P(X=10) = (43758)(0.30)^{10}(0.70)^8 = 0.014895$

$P(X=11) = (31824)(0.30)^{11}(0.70)^7 = 0.004643$

$P(X=12) = (18564)(0.30)^{12}(0.70)^6 = 0.001161$

$P(X=13) = (8568)(0.30)^{13}(0.70)^5 = 0.00023$

$P(X=14) = (3060)(0.30)^{14}(0.70)^4 = 0.000035$

$P(X=15) = (816)(0.30)^{15}(0.70)^3 = 0.000004$

$P(X=16) = (153)\ (0.30)^{16}(0.70)^2 = 0$

$P(X=17) = (18)\ (0.30)^{17}(0.70)^1 = 0$

$P(X=18) = (1)\ (0.30)^{18}(0.70)^0 = 0$

Since there are fifteen trucks in the test, the PMF is given by

b) $P(X < 4) = 0.104598 + 0.045762 + 0.01256 + 0.001628 = 0.164548$

c) $P(X > 2) = 1 - P(0) - P(1) - P(2) = 1 - 0.001628 - 0.012562 - 0.045762 = 0.940048$

d) $E(X) = (0)(0.001628) + (1)(0.012562) + (2)(0.045762) \ldots + (15)(0) = 5.4$

Example 5.6. A sample containing 400 solar panel is taken from the output of a production line. Defective solar panels occur randomly and independently at a rate of 1.6% of the population. What is the probability that the sample will have no more than 2.5% defective items?

Solution:

We have $n = 400, p=0.016$, and $q=0.984$., We want the cumulative probability of 0% through 2.5% defective. 2.5% of 400 items is 10 items, so we want $F(X \leq 10)$. Since the sample size is large, we use a spreadsheet to calculate the probabilities, which are listed below.

$P(X=0) = (1)(0.016)^0(0.984)^{400} = 0.00158$
$P(X=1) = (400)(0.016)^1(0.984)^{399} = 0.01026$
$P(X=2) = (79800)(0.016)^2(0.984)^{398} = 0.03329$
$P(X=3) = (10586800)(0.016)^3(0.984)^{397} = 0.07181$
$P(X=4) = (1050739900)(0.016)^4(0.984)^{396} = 0.11589$
$P(X=5) = (83218600080)(0.016)^5(0.984)^{395} = 0.14924$
$P(X=6) = (5478557838600)(0.016)^6(0.984)^{394} = 0.15976$
$P(X=7) = (308364541201200)(0.016)^7(0.984)^{393} = 0.14621$
$P(X=8) = (15148408086509000)(0.016)^8(0.984)^{392} = 0.11679$
$P(X=9) = (659797329990168000)(0.016)^9(0.984)^{391} = 0.08271$
$P(X=10) = (25798075602615500000)(0.016)^{10}(0.984)^{390} = 0.05259$

The cumulative probability is the sum of P (0) through P (10), or 94.01%

Example 5.7. Data was collected on bolt failures in solar panel connection. A large number of solar panel assemblies, each with the same number of bolts, were tested to find out whether each bolt was a success or a failure. The number of passing bolts in an assembly had a mean value of 4.3 and a variance of 1.65. If failing bolts occur randomly and independently, find the probabilities associated with possible numbers of passing bolts in an assembly. An assembly of solar panel is considered adequate if there is one or fewer bolt failures, what is the probability that a randomly chosen assembly will be inadequate?

Solution:

We know μ and σ^2, and we need to find n, p, and q.

$$\mu = np = 4.3 = \text{Mean Value}$$

$$\sigma^2 = \text{npq} = 1.65 = \text{Variance}$$

$$q = \frac{npq}{np} = \frac{1.65}{4.3} = 0.38$$

Table 5.11 Computation of data.

x	n−x	nCx	P(X=x)
0	7	1	$1\ (0.62)^0(0.38)^7 = 0.00114$
1	6	7	$7(0.62)^1(0.38)^6 = 0.01306$
2	5	21	$21(0.62)^2(0.38)^5 = 0.06396$
3	4	35	$35(0.62)^3(0.38)^4 = 0.17393$
4	3	35	$35(0.62)^4(0.38)^3 = 0.28378$
5	2	21	$21(0.62)^5(0.38)^2 = 0.27780$
6	1	7	$7(0.62)^6(0.38)^1 = 0.15108$
7	0	1	$1(0.62)^7(0.38)^0 = 0.03521$

$$p = 1 - q = 1 - 0.38 = 0.62$$

$$n = \frac{np}{p} = \frac{4.3}{0.62} = 6.93 = 7$$

The probability a bolt fails is 38%, and the probability is passes is 62%. Let x be the number of passing solar panel bolts in an assembly. We can now find the probability mass function using the general formula:

An assembly with 7 bolts that has one or fewer failures means having 6 or 7 successes. The probability an assembly is inadequate is $1 - P\ (x = 6$ or $7) = 1 - (0.15108 + 0.03521) = 0.81371$

There is an 81% chance a random assembly will be inadequate. Computation of data shown in Table 5.11.

a) The calculated results are listed below

$P(R=0) = (1)(0.25)^0(0.75)^{15} = 0.01336$

$P(R=1) = (15)(0.25)^1(0.75)^{14} = 0.06682$

$P(R=2) = (105)(0.25)^2(0.75)^{13} = 0.15591$

$P(R=3) = (455)(0.25)^3(0.75)^{12} = 0.2252$

$P(R=4) = (1365)(0.25)^4(0.75)^{11} = 0.2252$

$P(R=5) = (3003)(0.25)^5(0.75)^{10} = 0.16515$

$P(R=6) = (5005)(0.25)^6(0.75)^9 = 0.09175$

$P(R=7) = (6435)(0.25)^7(0.75)^8 = 0.03932$

$P(R=8) = (6435)(0.25)^8(0.75)^7 = 0.01311$

$P(R=9) = (5005)(0.25)^9(0.75)^6 = 0.0034$

$P(R=10) = (3003)(0.25)^{10}(0.75)^5 = 0.00068$

$P(R=11) = (1365)(0.25)^{11}(0.75)^4 = 0.0001$

$P(R=12) = (455)(0.25)^{12}(0.75)^3 = 0.00001$

$P(R=13) = (105)(0.25)^{13}(0.75)^2 = 0$

$P(R=14) = (15)(0.25)^{14}(0.75)^1 = 0$

$P(R=15) = (1)(0.25)^{15}(0.75)^0 = 0$

b) $P(R < 4) = 0.225 + 0.156 + 0.067 + 0.013 = 0.461$
c) $P(R > 2) = 1 - P(0) - P(1) - P(2) = 1 - 0.236 = 0.764$
d) $E(R) = (0)(0.01336) + (1)(0.06682) + (2)(0.15591) \ldots + (15)(0) = 3.75$
e) $E(R) = 3.75$ and $E(R^2) = (0^2)(0.01336) + (1^2)(0.06682) + (2^2)(0.15591) \ldots + (15^2)(0) = 16.875$
f) CV = standard deviation / mean = 1.68 / 3.75 = 44.7%

Poisson distribution:

A second discrete distribution is the Poisson distribution seemingly different from the binomial distribution, but actually derived from the binomial distribution. The Poisson distribution describes the occurrence of rare events. In fact, the poisson formula has been referred to as the law of improbable events.

Examples of Poisson distribution type situations include the following:

1. Number of times efficiency of solar panel reach more than 20%.
2. Number of times the losses of solar inverter is 0%.
3. Number of times the performance factor of standalone solar power plant is 100%.
4. Number of times the solar radiation is more than 12 kwh/m^2/day.

If a poisson distributed phenomenon is studied over a long period of time, a long run average can be determined. The poisson formula is used to compute the probability of occurrences over an interval for a given lambda value.

$$\text{Poisson Formula} = \frac{\lambda^x e^{-\lambda}}{x!}$$

Where x = 0, 1, 2, 3

λ = long run average

e = 2.7182

Here, x is the number of occurrences per interval for which the probability is being computed, is the long run average and e = 2.7182 is the base of natural logarithms.

Hypergeometric distribution

The hypergeometric distribution applies only to experiments in which the trials are done without distribution. The hypergeometric distribution like the binomial distribution, consist of two possible outcomes: success and failure.

Hypergeometric Distribution is given by $P(x) = [({}_AC_X).({}_{N-A}C_{n-x})]/{}_NC_n$

N = Size of the population

n = Sample Size

Table 5.12 Random values of solar radiation ($kwh/m^2/day$).

3.3	4.3	6.7	7.2	4.0	8.1	7.6	7.9
4.7	8.8	11.2	4.1	8.6	8.3	4.6	9.6
6.9	9.3	2.9	6.7	9.1	2.1	7.7	10.1
7.1	9.1	4.8	7.2	9.9	4.5	8.2	10.9
8.3	9.2	5.3	8.4	9.7	4.9	9.4	10.7
2.9	8.6	8.9	2.3	8.7	6.1	12.3	8.6
3.7	8.3	9.4	3.8	8.4	6.4	13.8	8.9

A = Number of successes in the population
x = Number of successes in the sample, sampling is done without replacement

Example 5.8. Suppose 20 solar product manufacturing companies operate in the India and 12 are located in Bombay. If 4 computer companies are selected randomly from the entire list, what is the probability that one or more of the selected companies are located in the Bombay?

Solution:

$$N = 20, n = 4, A = 12, ANDX \geq 1$$

Hence this problem is a candidate for the hypergeometric distribution.

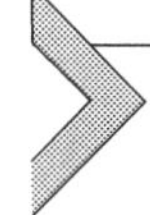

5.4 Assessment of solar energy system by sampling technique

Sampling is widely used as a means of gathering information about different factors of the solar energy system. Data are gathered from samples and conclusions are drawn about the population as a part of the inferential statistics process. The two main types of sampling are random and non-random. In random sampling every unit of the population has the same probability of being selected into the sample. In non-random sampling not every unit of the population has the same probability of being selected by chance. The four basic types of random sampling techniques are simple random sampling, stratified random sampling, systematic random sampling and cluster random sampling.

- **Simple random sampling:** The most elementary random sampling technique is simple random sampling. With simple random sampling each unit of the frame is numbered from 1 to N. Table 5.12 shows the random values of solar radiation ($kwh/m^2/day$) in the particular area.

Table 5.13 Sample of solar radiation (kWh/m2/day) from 1.1 to 11.

1.1	2.1	3.1	4.1	5.1	6.1	7.1	8.1	9.1	10.1
1.2	2.2	3.2	4.2	5.2	6.2	7.2	8.2	9.2	10.2
1.3	2.3	3.3	4.3	5.3	6.3	7.3	8.3	9.3	10.3
1.4	2.4	3.4	4.4	5.4	6.4	7.4	8.4	9.4	10.4
1.5	2.5	3.5	4.5	5.5	6.5	7.5	8.5	9.5	10.5
1.6	2.6	3.6	4.6	5.6	6.6	7.6	8.6	9.6	10.6
1.7	2.7	3.7	4.7	5.7	6.7	7.7	8.7	9.7	10.7
1.8	2.8	3.8	4.8	5.8	6.8	7.8	8.8	9.8	10.8
1.9	2.9	3.9	4.9	5.9	6.9	7.9	8.9	9.9	10.9
2	3	4	5	6	7	8	9	10	11

5.4.1 Simple random sampling

In a simple random sample, every member of the population has an equal chance of being selected. Your sampling frame should include the whole population. To conduct this type of sampling, you can use tools like random number generators or other techniques that are based entirely on chance. For the example to select a simple random sample of 100 employees of solar industry X. You assign a number to every employee in the solar company database from 1 to 1000, and use a random number generator to select 100 numbers.

5.4.2 Systematic sampling

Systematic sampling is similar to simple random sampling, but it is usually slightly easier to conduct. Every member of the population is listed with a number, but instead of randomly generating numbers, individuals are chosen at regular intervals. For the example all employees of the solar industry are listed in alphabetical order. From the first 10 numbers, you randomly select a starting point: number 6. From number 6 onwards, every 10th person on the list is selected (6, 16, 26, 36, and so on), and you end up with a sample of 100 people.

This sampling technique mostly used in the random sampling of solar radiation in the particular location. Table 5.13 shows the data of 100 values of solar radiation in kwh/m^2/day. Here are the steps you need to follow in order to achieve a systematic random sample of 100 solar radiation data:

- Number the units in the population of solar radiation from 1 to N
- Decide on the n (sample size) that you want or need
- $k = N/n =$ the interval size
- Randomly select an integer between 1 to k
- Then take every kth unit

Table 5.14 Output of the sample of solar radiation (kWh/m2/day) from 1.1 to 11.

1.1	2.1	3.1	4.1	5.1	6.1	7.1	8.1	9.1	10.1
1.2	2.2	3.2	4.2	5.2	6.2	7.2	8.2	9.2	10.2
1.3	2.3	3.3	4.3	5.3	6.3	7.3	8.3	9.3	10.3
[1.4]	[2.4]	[3.4]	[4.4]	[5.4]	[6.4]	[7.4]	[8.4]	[9.4]	[10.4]
1.5	2.5	3.5	4.5	5.5	6.5	7.5	8.5	9.5	10.5
1.6	2.6	3.6	4.6	5.6	6.6	7.6	8.6	9.6	10.6
1.7	2.7	3.7	4.7	5.7	6.7	7.7	8.7	9.7	10.7
1.8	2.8	3.8	4.8	5.8	6.8	7.8	8.8	9.8	10.8
[1.9]	[2.9]	[3.9]	[4.9]	[5.9]	[6.9]	[7.9]	[8.9]	[9.9]	[10.9]
2	3	4	5	6	7	8	9	10	11

Let's assume that we have a population that only has N = 100 values of solar radiation in it and that you want to take a sample of n = 20. To use systematic sampling, the population of solar radiation must be listed in a random order. The sampling fraction would be f = 20/100 = 20%. In this case, the interval size, k, is equal to N/n = 100/20 = 5. Now, select a random integer from 1.1 to 1.5. In our example, imagine that you chose 1.4. Now, to select the sample, start with the 4th unit in the list and take every k-th unit (every 5th, because k = 5). You would be sampling units 1.4, 1.9, 2.4, 2.9, and so on to 11 and you would wind up with 20 units in your sample and data is shown in Table 5.14.

Another example of sampling technique w.r.t. the random sampling of data, energy generation from 20KW solar power plant in the particular location. Data of electricity generation of 160 days is arranged in ascending order, which is given in Table 5.15.

Let's assume that we have a population that only has N=160 values of solar energy generation in it and that you want to take a sample of n = 16. To use systematic sampling, the population of solar radiation must be listed in a random order. The sampling fraction would be f = 16/160 = 10%. In this case, the interval size, k, is equal to N/n = 160/16 = 10. Now, select a random integer from 2.1 to 3. In our example, imagine that you chose 2.3. Now, to select the sample, start with the 2.3th unit in the list and take every k-th unit (every 10th, because k = 10). Output of the sample of energy generation is shown in Table 5.16.

5.4.3 Stratified sampling

Stratified sampling involves dividing the population into subpopulations that may differ in important ways. It allows you draw more precise conclusions

Table 5.15 Sample of energy generation (KW).

2.1	4.1	6.1	8.1	10.1	12.1	14.1	16.1
2.2	4.2	6.2	8.2	10.2	12.2	14.2	16.2
2.3	4.3	6.3	8.3	10.3	12.3	14.3	16.3
2.4	4.4	6.4	8.4	10.4	12.4	14.4	16.4
2.5	4.5	6.5	8.5	10.5	12.5	14.5	16.5
2.6	4.6	6.6	8.6	10.6	12.6	14.6	16.6
2.7	4.7	6.7	8.7	10.7	12.7	14.7	16.7
2.8	4.8	6.8	8.8	10.8	12.8	14.8	16.8
2.9	4.9	6.9	8.9	10.9	12.9	14.9	16.9
3	5	7	9	11	13	15	17
3.1	5.1	7.1	9.1	11.1	13.1	15.1	17.1
3.2	5.2	7.2	9.2	11.2	13.2	15.2	17.2
3.3	5.3	7.3	9.3	11.3	13.3	15.3	17.3
3.4	5.4	7.4	9.4	11.4	13.4	15.4	17.4
3.5	5.5	7.5	9.5	11.5	13.5	15.5	17.5
3.6	5.6	7.6	9.6	11.6	13.6	15.6	17.6
3.7	5.7	7.7	9.7	11.7	13.7	15.7	17.7
3.8	5.8	7.8	9.8	11.8	13.8	15.8	17.8
3.9	5.9	7.9	9.9	11.9	13.9	15.9	17.9
4	6	8	10	12	14	16	18

Table 5.16 Output of the sample of energy generation.

2.1	4.1	6.1	8.1	10.1	12.1	14.1	16.1
2.2	4.2	6.2	8.2	10.2	12.2	14.2	16.2
[2.3]	[4.3]	[6.3]	[8.3]	[10.3]	[12.3]	[14.3]	[16.3]
2.4	4.4	6.4	8.4	10.4	12.4	14.4	16.4
2.5	4.5	6.5	8.5	10.5	12.5	14.5	16.5
2.6	4.6	6.6	8.6	10.6	12.6	14.6	16.6
2.7	4.7	6.7	8.7	10.7	12.7	14.7	16.7
2.8	4.8	6.8	8.8	10.8	12.8	14.8	16.8
2.9	4.9	6.9	8.9	10.9	12.9	14.9	16.9
3	5	7	9	11	13	15	17
3.1	5.1	7.1	9.1	11.1	13.1	15.1	17.1
3.2	5.2	7.2	9.2	11.2	13.2	15.2	17.2
[3.3]	[5.3]	[7.3]	[9.3]	[11.3]	[13.3]	[15.3]	[17.3]
3.4	5.4	7.4	9.4	11.4	13.4	15.4	17.4
3.5	5.5	7.5	9.5	11.5	13.5	15.5	17.5
3.6	5.6	7.6	9.6	11.6	13.6	15.6	17.6
3.7	5.7	7.7	9.7	11.7	13.7	15.7	17.7
3.8	5.8	7.8	9.8	11.8	13.8	15.8	17.8
3.9	5.9	7.9	9.9	11.9	13.9	15.9	17.9
4	6	8	10	12	14	16	18

Table 5.17 Data of solar power plant.

Solar power plant	Capacity of plant (MW)	Consumer demand (MW)
A	4	8
B	16	4
C	18	6
D	2	10
E	17	2

by ensuring that every subgroup is properly represented in the sample. To use this sampling method, you divide the population into subgroups (called strata) based on the relevant characteristic (e.g. gender, age range, income bracket,job role). Based on the overall proportions of the population, you calculate how many people should be sampled from each subgroup. Then you use random or systematic sampling to select a sample from each subgroup.

For the example solar panel manufacturing company has produce 800 polycrystalline and 200 monocrystalline solar panel in a day. You want to ensure that the sample reflects the material balance of the company, so you sort the population into two strata based on material. Then you use random sampling on each group, selecting 80 polycrystalline and 20 monocrystalline, which gives you a representative sample of 100 people.

5.4.4 Cluster sampling

Cluster sampling also involves dividing the population into subgroups, but each subgroup should have similar characteristics to the whole sample. Instead of sampling individuals from each subgroup, you randomly select entire subgroups. If it is practically possible, you might include every individual from each sampled cluster. If the clusters themselves are large, you can also sample individuals from within each cluster using one of the techniques above. This method is good for dealing with large and dispersed populations, but there is more risk of error in the sample, as there could be substantial differences between clusters. It's difficult to guarantee that the sampled clusters are really representative of the whole population.

Example 5.9. The generation capacity of solar power plant (X1) and consumer demand (X2) of 5 solar energy site are shown in the Table 5.17.

The numbers are fictitious and not at all realistic, but the example will help us explain the essential features of cluster analysis as simply as possible

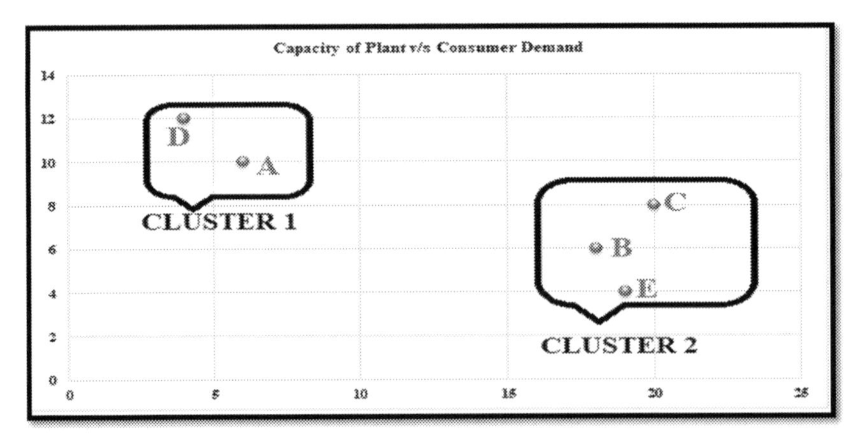

Figure 5.1 Capacity of plant v/s consumer demand.

Table 5.18 Monthwise wave height and wave period.

Month	Wind velocity (m/s)	Solar radiation (kWh/m2/day)
Jan.	2.5	6
Feb.	4.5	7.4
March	5	7.8
April	1.3	6.1
May	5.5	7.5
June	1.2	6.3
July	6	7.6
August	2.2	8.1
Sep.	4	7.3
Oct.	3.3	8.3
Nov.	3.4	8.2
Dec.	3.5	8.4

in the field of ocean power plants. The data of Table 5.17 are plotted in Fig. 5.1.

Inspection of Fig. 5.1 suggests that the five observations from two clusters of solar power plant. The first consists of A and D, and the second of B, C and E. It can be noted that the observations in each cluster are similar to one another with respect to the capacity of the solar power plant and consumer demand in MW, and that the two clusters are quite distinct from each other.

Example 5.10. The Month-wise average wind velocity and their solar radiation of solar power plant data are shown in the Table 5.18.

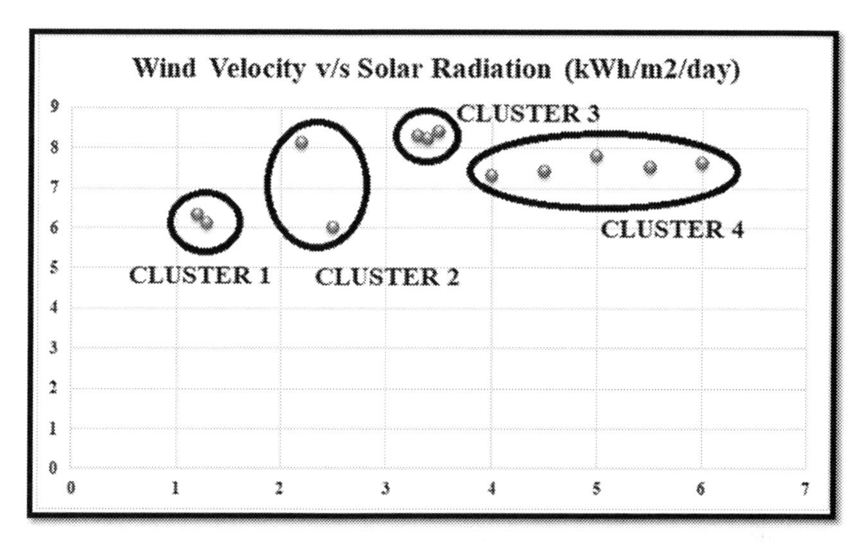

Figure 5.2 Cluster of wind velocity v/s solar radiation.

The numbers are fictitious and not at all realistic, but the example will help us explain the essential features of cluster analysis as simply as possible in the field of solar power plants. The data of Table 5.18 are plotted in Fig. 5.2.

Inspection of figure suggests that the twelve observations from three clusters. The first consists of data of three months, and the second of four month data and third of five month data. It can be noted that the observations in each cluster are similar to one another with respect to the capacity of the plant and consumer demand in MW, and that the three clusters are quite distinct from each other.

K-MEANS cluster sampling:

K-Means clustering intends to partition *n* objects into *k* clusters in which each object belongs to the cluster with the nearest mean. This method produces exactly *k* different clusters of greatest possible distinction. The best number of clusters *k* leading to the greatest separation (distance) is not known as a priori and must be computed from the data. The objective of K-Means clustering is to minimize total intra-cluster variance, or, the squared error function.

Algorithms:

1. Cluster the data into *k* groups where *k* is predefined.
2. Select *k* points at random as cluster centers.
3. Assign objects to their closest cluster center, according to the *Euclidean distance* function.

Table 5.19 Day wise wave periods.

Hours	Energy generation (kW)
1	7.5
2	7.5
3	8
4	9.5
5	9.5
6	10
7	10
8	10.5
9	11
10	14
11	17.5
12	20
13	20.5
14	21
15	21.5
16	22
17	30
18	30.5
19	32.5

Initial cluster (random centroids or average):
K = 2, c1 = 8; c2 = 11
Distance 1 = $|X_1 - C_1|$
Distance 2 = $|X_1 - C_2|$
Iteration 1: c1 = 7.66; c2 = 18.12

4. Calculate the centroid or mean of all objects in each cluster.
5. Repeat steps 2, 3 and 4 until the same points are assigned to each cluster in consecutive rounds.

Hour-wise energy generation in 50KW solar power plant for 19 hours is shown in the Table 5.19. Apply K-means clustering and find centroids. Tables 5.20–5.23 shows data of iteration 1, 2, 3, and 4, respectively.

No change between iterations 3 and 4 has been noted. By using clustering, 2 groups have been identified 15-28 and 35-65. The initial choice of centroids of energy generation can affect the output clusters, so the algorithm is often run multiple times with different starting conditions in order to get a fair view of what the clusters should be and find out the optimum value of energy generation. Similarly, this method is also utilized to find out the optimum value of solar radiation, clearness index, temperature, wind velocity, battery charging & discharging rates and solar inverter efficiency.

Table 5.20 Data of iteration 1.

X1	C1	C2	Distance 1	Distance 2	Nearest Cluster	New Centroid
7.5	8	11	0.5	3.5	1	7.665
7.5	8	11	0.5	3.5	1	
8	8	11	0	3	1	
9.5	8	11	1.5	1.5	2	18.125
9.5	8	11	1.5	1.5	2	
10	8	11	2	1	2	
10	8	11	2	1	2	
10.5	8	11	2.5	0.5	2	
11	8	11	3	0	2	
14	8	11	6	3	2	
17.5	8	11	9.5	6.5	2	
20	8	11	12	9	2	
20.5	8	11	12.5	9.5	2	
21	8	11	13	10	2	
21.5	8	11	13.5	10.5	2	
22	8	11	14	11	2	
30	8	11	22	19	2	
30.5	8	11	22.5	19.5	2	
32.5	8	11	24.5	21.5	2	

Iteration 2: c1 = 9.28; c2 = 22.95

Table 5.21 Data of iteration 2

X1	C1	C2	Distance 1	Distance 2	Nearest Cluster	New Centroid
7.5	7.66	18.12	.16	10.625	1	9.28
7.5	7.66	18.12	.16	10.625	1	
8	7.66	18.12	0.34	10.125	1	
9.5	7.66	18.12	1.84	8.625	1	
9.5	7.66	18.12	1.84	8.625	1	
10	7.66	18.12	2.34	7.625	1	
10	7.66	18.12	2.34	7.625	1	
10.5	7.66	18.12	2.84	7.62	1	
11	7.66	18.12	3.34	7.12	1	
14	7.66	18.12	6.34	4.12	2	22.95
17.5	7.66	18.12	9.84	0.62	2	
20	7.66	18.12	12.34	1.88	2	
20.5	7.66	18.12	12.84	2.38	2	
21	7.66	18.12	13.34	2.88	2	
21.5	7.66	18.12	13.84	3.38	2	
22	7.66	18.12	14.34	3.88	2	
30	7.66	18.12	22.34	11.88	2	
30.5	7.66	18.12	22.84	12.38	2	
32.5	7.66	18.12	24.84	14.38	2	

Iteration 3: c1 = 9.75; c2 = 23.945

Table 5.22 Data of iteration 3.

X1	C1	C2	Distance 1	Distance 2	Nearest Cluster	New Centroid
7.5	9.28	22.95	1.78	15.45	1	9.75
7.5	9.28	22.95	1.78	15.45	1	
8	9.28	22.95	1.28	14.95	1	
9.5	9.28	22.95	.23	13.95	1	
9.5	9.28	22.95	.23	13.95	1	
10	9.28	22.95	0.72	12.95	1	
10	9.28	22.95	0.72	12.95	1	
10.5	9.28	22.95	1.22	12.45	1	
11	9.28	22.95	1.72	11.75	1	
14	9.28	22.95	4.72	8.95	1	
17.5	9.28	22.95	8.22	5.45	2	23.945
20	9.28	22.95	10.72	2.95	2	
20.5	9.28	22.95	11.22	2.45	2	
21	9.28	22.95	11.72	1.95	2	
21.5	9.28	22.95	12.22	1.45	2	
22	9.28	22.95	12.72	0.95	2	
30	9.28	22.95	20.72	7.05	2	
30.5	9.28	22.95	21.22	7.55	2	
32.5	9.28	22.95	23.22	9.55	2	

Iteration 4: c1 = 9.75; c2 = 23.945

Table 5.23 Data of iteration 4.

X1	C1	C2	Distance 1	Distance 2	Nearest Cluster	New Centroid
7.5	9.75	23.945	2.25	16.445	1	9.75
7.5	9.75	23.945	2.25	16.445	1	
8	9.75	23.945	1.75	15.945	1	
9.5	9.75	23.945	0.25	14.445	1	
9.5	9.75	23.945	0.25	14.445	1	
10	9.75	23.945	0.25	13.945	1	
10	9.75	23.945	0.25	13.945	1	
10.5	9.75	23.945	0.75	13.445	1	
11	9.75	23.945	1.25	12.945	1	
14	9.75	23.945	4.25	9.945	1	
17.5	9.75	23.945	7.75	6.445	2	23.945
20	9.75	23.945	10.25	3.945	2	
20.5	9.75	23.945	10.75	3.445	2	
21	9.75	23.945	11.25	2.945	2	
21.5	9.75	23.945	11.75	2.445	2	
22	9.75	23.945	12.25	1.945	2	
30	9.75	23.945	20.25	6.055	2	
30.5	9.75	23.945	20.75	6.555	2	
32.5	9.75	23.945	22.75	8.555	2	

Table 5.24 Data of solar energy system.

Object	Wind velocity (m/s)	Solar radiation (kwh/m^2/day)
A	2	2
B	4	2
C	8	6
D	10	8

K - means clustering in solar energy system: Suppose we have 4 groups of wind velocity (m/s) and solar radiation (kwh/m^2/day) of a particular site are of solar power plants, which is shown in Table 5.24.

Initial value of centroids: Suppose we use wind velocity and solar radiation of the solar power plant as the first centroids. Let c1 and c2 denote the coordinate of the centroids, then c1 = (2, 2) and c2 = (4, 2).

Object centroids distance: We calculate the distance between cluster centroids to each object. Let us use Euclidean distance, then we have a distance matrix at iteration zero is

$$D^0 = \begin{bmatrix} 0 & 2 & 7.2 & 10 \\ 2 & 0 & 5.6 & 8.4 \end{bmatrix} c1 = [2, 2];\ c2 = [4, 2]$$

Each column in the distance matrix symbolizes the object. The first row of the distance matrix corresponds to the distance of each object to the first centroid and the second row is the distance of each object to the second centroid. For example distance from solar plant data c = (8, 6) to the first centroid c1 = (2, 2) is $\sqrt{(8-2)^2 + (6-2)^2} = 7.2$ and its distance to the second centroid c2 = (4, 2) is $\sqrt{(8-4)^2 + (6-2)^2} = 5.66$, etc.

Object clustering: We assign each object based on the minimum distance. Thus wind velocity is assigned to group 1, solar radiation is to group 2. The elements of the group matrix is

$$D^0 = \begin{bmatrix} 2 & 0 & 0 & 0 \\ 0 & 2 & 2 & 2 \\ A & B & C & D \end{bmatrix} \begin{matrix} \text{Group 1} \\ \text{Group 2} \end{matrix}$$

Iteration 1 - Determine Centroids: c1 = (2, 2), the centroid is the average coordinate among the three members:

$$c2 = \left(\frac{4+8+10}{3}, \frac{2+6+8}{3}\right) = \left(\frac{22}{3}, \frac{16}{3}\right)$$

Table 5.25 Final grouping of solar energy data.

Object	Wind velocity (m/s)	Solar radiation (kwh/m²/day)
A	2	2
B	2	2
C	6	4
D	8	4

Iteration 1 objects centroids distances: The next step is to compute the distance of all objects to the new centroids. The distance matrix at iteration 1 is

$$D^1 = \begin{bmatrix} 0 & 2 & 7.23 & 10 \\ 6.28 & 4.72 & 0.94 & 3.8 \end{bmatrix} \begin{matrix} c1 = (2, 2) \text{ Group 1} \\ c2 = \left(\frac{22}{3}, \frac{16}{3}\right) \text{ Group 2} \end{matrix}$$

$$\begin{bmatrix} 2 & 4 & 8 & 10 \\ 2 & 2 & 6 & 8 \end{bmatrix} \begin{matrix} X \\ Y \end{matrix}$$

Iteration 1, object clustering: similar to step 3 we assign each object based on the minimum distance. Based on the new distance matrix, we move the group matrix is shown below:

$$G^1 = \begin{bmatrix} 2 & 2 & 0 & 0 \\ 0 & 0 & 2 & 2 \end{bmatrix} \begin{matrix} \text{Group 1} \\ \text{Group 2} \end{matrix}$$

Iteration 2, determine centroids:

$$c1 = \left(\frac{2+4}{2}, \frac{2+2}{2}\right) = (3, 2)$$

$$c2 = \left(\frac{8+10}{2}, \frac{6+8}{2}\right) = (9, 7)$$

Iteration 2 objects centroids distances: Repeat step 2 again, we have new distance matrix at iteration 2 as

$$D^2 = \begin{bmatrix} 1 & 1 & 6.4 & 9.2 \\ 8.6 & 7 & 1.42 & 1.42 \end{bmatrix} \begin{matrix} c1 = (3, 2) \text{ Group 1} \\ c2 = (9, 7) \text{ Group 2} \end{matrix}$$

We get the final grouping are:

Above Table 5.25 shows that optimum value of wind velocity and solar radiation.

Example 5.11. As a simple illustration of a k-means algorithm, consider the following data set of solar radiation of two different locations which

Table 5.26 Data of solar radiation.

Subject	(A)	(B)
1	1.0	1.0
2	1.5	2.0
3	3.0	4.0
4	5.0	7.0
5	3.5	5.0
6	4.5	5.0
7	3.5	4.5

Table 5.27 Group of solar radiation.

	Individual	Mean vector (centroid)
Group 1	1	(1.0, 1.0)
Group 2	4	(5.0, 7.0)

Table 5.28 Cluster 1 and 2 of tidal current.

	Cluster 1		Cluster 2	
Step	Individual	Mean vector (centroid)	Individual	Mean vector (centroid)
1	1	(1.0, 1.0)	4	(5.0, 7.0)
2	1, 2	(1.2, 1.5)	4	(5.0, 7.0)
3	1, 2, 3	(1.8, 2.3)	4	(5.0, 7.0)
4	1, 2, 3	(1.8, 2.3)	4, 5	(4.2, 6.0)
5	1, 2, 3	(1.8, 2.3)	4, 5, 6	(4.3, 5.7)
6	1, 2, 3	(1.8, 2.3)	4, 5, 6, 7	(4.1, 5.4)

consisting of the scores of two variables on each of seven individuals. Table 5.26 shows the data of solar radiation.

This data set is to be grouped into two clusters. As a first step in finding a sensible initial partition, let the A & B values of the two individuals furthest apart (using the Euclidean distance measure), define the initial cluster means, giving. Table 5.27 shows group of solar radiation.

The remaining individuals are now examined in sequence and allocated to the cluster to which they are closest, in terms of Euclidean distance to the cluster mean. Table 5.28 shows cluster 1 and 2 of tidal current. The mean vector is recalculated each time a new member is added. This leads to the following series of steps:

Table 5.29 Two cluster stage.

	Individual	Mean vector (centroid)
Cluster 1	1, 2, 3	(1.8, 2.3)
Cluster 2	4, 5, 6, 7	(4.1, 5.4)

Table 5.30 Distance to mean of solar radiation data.

Individual	Distance to mean (centroid) of cluster 1	Distance to mean (centroid) of cluster 2
1	1.5	5.4
2	0.4	4.3
3	2.1	1.8
4	5.7	1.8
5	3.2	0.7
6	3.8	0.6
7	2.8	1.1

Table 5.31 Mean vector of solar radiation data.

	Individual	Mean vector (centroid)
Cluster 1	1, 2	(1.3, 1.5)
Cluster 2	3, 4, 5, 6, 7	(3.9, 5.1)

Now the initial partition has changed, and the two clusters at this stage having the following characteristics. Table 5.29 shows the two cluster stage.

But we cannot yet be sure that each individual has been assigned to the right cluster. Table 5.30 shows the distance to mean of solar radiation data. So, we compare each individual's distance to its own cluster mean and to that of the opposite cluster. And we find:

Only individual 3 is nearer to the mean of the opposite cluster (Cluster 2) than its own (Cluster 1). In other words, each individual's distance to its own cluster mean should be smaller than the distance to the other cluster's mean (which is not the case with individual 3). Table 5.31 shows the mean vector of solar radiation data. Thus, individual 3 is relocated to Cluster 2 resulting in the new partition:

The iterative relocation would now continue from this new partition until no more relocations occur. However, in this example each individual is now nearer its own cluster mean than that of the other cluster and the iteration stops, choosing the latest partitioning as the final cluster solution.

Also, it is possible that the k-means algorithm won't find a final solution. In this case it would be a good idea to consider stopping the algorithm after a pre-chosen maximum of iterations.

➢ **Hierarchical cluster sampling of solar energy data:**

In the hierarchical method of clustering, we identify step by step process of data gathering. In this method data collection and decomposition is done in two ways, first one is agglomerative and another one is divisive approach. In the agglomerative process merging the object and group that are close to one another and in the case of solar energy system data are collected in the following manner:

1. Collection of pre-feasibility data of solar energy system
2. Collection of data for modeling of solar energy system
3. Collection of data for controlling of solar energy system
4. Collection of data for reliability assessment of solar energy system

In divisive approaches, we start with all of the objects in the same cluster and in the continuous iteration, a cluster is split into smaller clusters. According to the divisive approach agglomerative data is distributed and divide into following manner.

1. Collection of prefeasibility data of solar energy system
- **(i)** Location of the site
- **(ii)** Geographical condition of the site
- **(iii)** Data of solar radiation
- **(iv)** Data of clearness index
- **(v)** Data of temperature
- **(vi)** Data on wind velocity
- **(vii)** Data of temperature
- **(viii)** Data of humidity
- **(ix)** Data on consumer demand

2. Collection of data for modeling of solar energy system
- **(i)** Data of electricity required
- **(ii)** Data of specification of solar panel
- **(iii)** Data of specification of solar inverter
- **(iv)** Data of specification of battery
- **(v)** Data of specification of solar controller

3. Collection of data for controlling of solar energy system
   (i) Data of different types of errors
   (ii) Data of different control strategies of solar power plant
   (iii) Data for stability analysis of solar power plant
   (iv) Data of different control strategies of solar energy conversion system
4. Collection of data on reliability assessment of solar energy system
   (i) Data of failure distribution model of solar energy system
   (ii) Data of time dependent failure model of a solar energy system
   (iii) Data of failure distribution model of solar energy system
   (iv) Data of time dependent failure model of a solar energy system
   (v) Data of constant failure rate model of solar energy system

5.5 Weibull distribution of solar energy parameters

The Weibull Distribution is a continuous probability distribution used to analyze life data, model failure times and access product reliability. It can also fit a huge range of data from many other fields like economics, hydrology, biology, engineering sciences. It is an extreme value of probability distribution which is frequently used to model the reliability, survival, wind speeds and other data. The only reason to use Weibull distribution is because of its flexibility. Because it can simulate various distributions like normal and exponential distributions. Weibull's distribution reliability is measured with the help of parameters. The two versions of Weibull probability density function (pdf) are

- Two parameter pdf
- Three parameter pdf

$$f(x) = \frac{\gamma}{\alpha}\left(\frac{(x-\mu)}{\alpha}\right)^{\gamma-1} exp\left(-\left(\frac{(x-\mu)}{\alpha}\right)^{\gamma}\right) \quad x \geq \mu;\ \gamma, \alpha > 0$$

Where

- γ is the **shape parameter**, also called as the Weibull slope or the threshold parameter.
- α is the **scale parameter**, also called the characteristic life parameter.
- μ is the **location parameter**, also called the waiting time parameter or sometimes the shift parameter.

Table 5.32 Data of solar panel.

Area required for solar panel in square feet (x_1)	Life length (year) (x_2)
1602	34
2387	44
1453	21
2303	33
1984	24
1678	13
3015	9
1682	12
2380	29
1780	32
2043	3
2012	7
3450	15
2200	8
2817	5
2363	7
3078	4
3367	7
4875	9
4450	8

The standard Weibull distribution is derived, when μ=0 and α =1, the formula is reduced and it becomes

$$f(x) = \gamma x^{\gamma-1} exp^{(-x)^{\gamma}}, \; x \geq 0, \; \gamma > 0$$

5.5.1 Two-parameter Weibull distribution

The formula is practically similar to the three parameters Weibull, except that μ isn't included:

$$f(x) = \frac{\gamma}{\alpha}\left(\frac{(x-\mu)}{\alpha}\right)^{\gamma-1} \exp\left(-\left(\frac{(x)}{\alpha}\right)^{\gamma}\right) \; x \geq 0$$

The failure rate is determined by the value of the shape parameter γ

- If γ < 1, then the failure rate decreases with time
- If γ = 1, then the failure rate is constant

If γ > 1, the failure rate increases with time

Example 5.12. Find the statistics of following data set of solar panel (Table 5.32) through the NCSS tool.

Table 5.33 Weibull statistics of area required for solar panel.

Parameter	Probability plot estimate	Maximum likelihood estimate	Standard error	95% lower conf. limit	95% upper conf. limit
B (shape)	3.741408	2.921116	0.4711373	2.12942	4.007157
C (scale)	2784.432	2855.262	232.436	2434.181	3349.184
Log likelihood		-164.2555			
Mean	2514.363	2546.775			
Median	2524.6	2518.575			
Mode	2562.345	2473.675			
Sigma	749.2819	947.988			

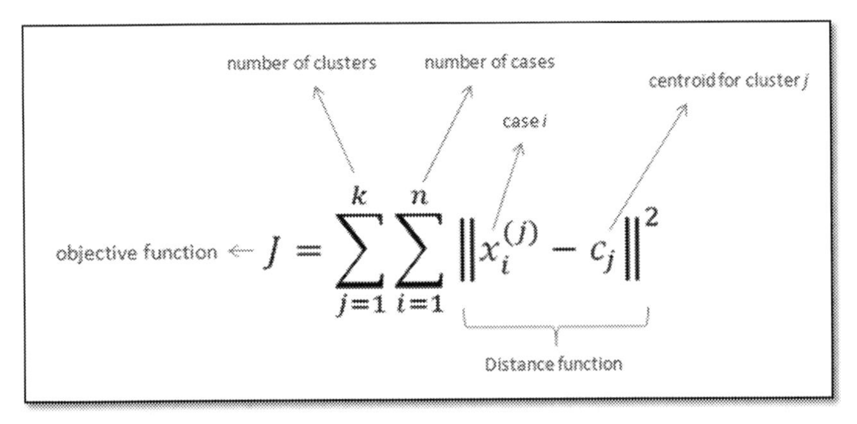

Figure 5.3 Formulization of K-means.

Solution: There are two data set of solar panel, in which x1 is area required for solar panel in square feet and x2 is life length of solar panel. The Weibull statistics of area required for solar panel is shown in Table 5.33. Fig. 5.3 and Fig. 5.4 shows Weibull survival and Weibull probability plot of area required for solar panel. Survival analysis is a branch of statistics designed for analyzing the expected duration until an event of interest occurs. In general, our "event of interest" is the failure of a solar panel. We use the Weibull distribution to model the distribution of failure times for a fleet of solar panel. The Weibull statistics of life length for solar panel is shown in Table 5.34. Figs. 5.4 and 5.5 shows Weibull survival and Weibull probability plot of life length for solar panel. Fig. 5.6 shows the Weibull probability plot of life length for solar panel.

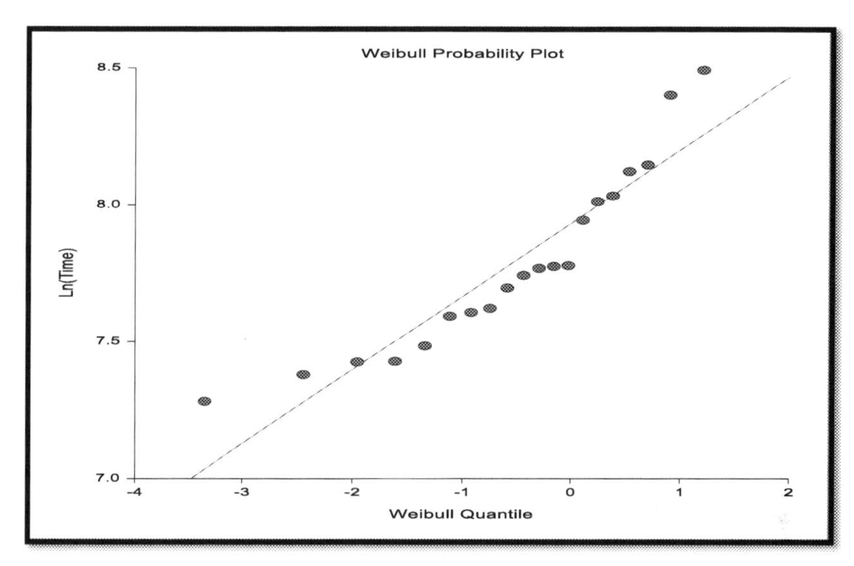

Figure 5.4 Weibull probability plot of area required for solar panel.

Table 5.34 Weibull statistics of life length of solar panel.

Parameter	Probability plot estimate	Maximum likelihood estimate	Standard error	95% lower conf. limit	95% upper conf. limit
B (Shape)	1.558611	1.429855	0.2471972	1.018906	2.00655
C (Scale)	17.36781	17.9498	2.974551	12.97188	24.83799
Log-Likelihood		-73.88662			
Mean	15.61145	16.30775			
Median	13.72838	13.89113			
Mode	8.991487	7.744775			
Sigma	10.23261	11.57369			

5.6 Conclusion

According to the above discussion it is find out, a novel framework for incorporating resource risk into utility decision-making processes, based on statistical ideas related to "likelihood of exceedance." In the solar energy business, the probability of exceedance is already frequently used to describe the uncertainty surrounding central or estimations of yearly energy production at solar installations. It is also find out, Binomial, Poisson and Hypergeometric distribution play vital role to identify the different parameters of solar energy system.

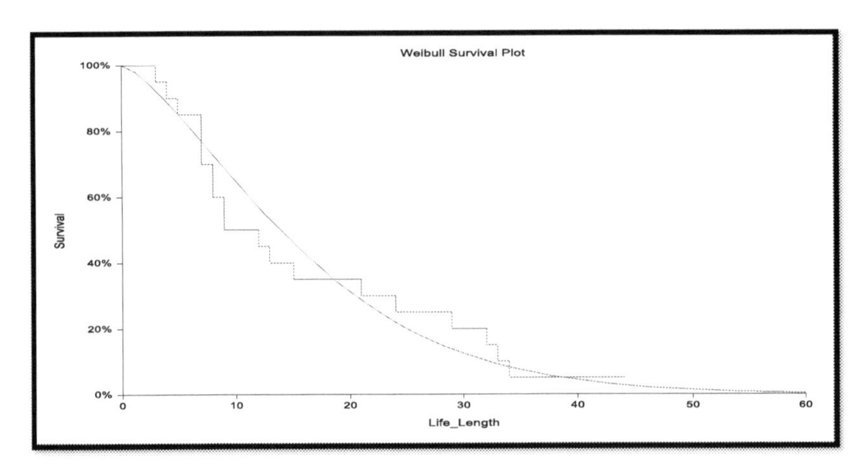

Figure 5.5 Weibull survival plot of life length for solar panel.

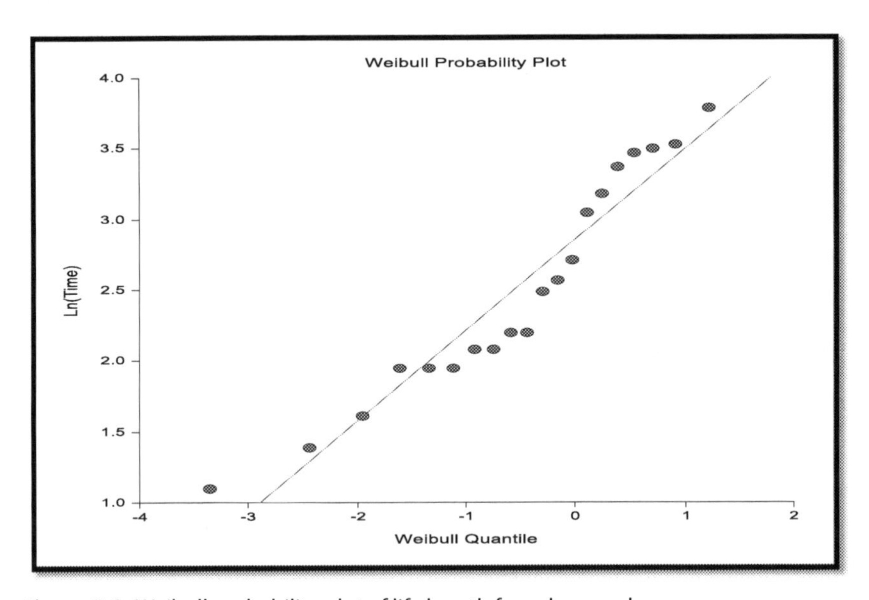

Figure 5.6 Weibull probability plot of life length for solar panel.

5.7 Exercise/question

1. What is the use of probability distribution in the field of solar energy system?
2. What is the use of sampling distribution in the field of solar energy system?

3. Write the difference between continuous and discrete distribution parameters of solar energy system.
4. What is the significance of Binomial distribution in the field of solar energy system?
5. What is the significance of Poisson distribution in the field of solar energy system?
6. What is the significance of hypergeometric distribution in the field of solar energy system?
7. In the solar energy related survey, it is found out 55% of electricity consumers were very satisfied through the standalone solar energy system, which is connected in their house. Suppose that 45 electricity consumers are satisfied. If the solar energy survey result still holds true today, what is the probability that exactly 21 are very satisfied with their solar energy survey?
8. One of the solar product manufacturing company test different solar panels in the different critical circumstances, where the amount of solar radiation is very less. Solar panel have a 45% chance of failing to complete the test run without a blowout. 25 solar panel of 300W each go through the test, find the following:
   a. The values of the Probability Mass function.
   b. The probability that fewer than four solar panel have blowouts.
   c. The probability that more than two have blowouts.
   d. The expected number of solar panel with blowouts.
9. One of the solar product manufacturing company test different solar inverters in the different critical circumstances. Solar panel have a 40% chance of failing to complete the test run without a blowout. 19 solar inverter of 2000W each go through the test, find the following:
   a. The values of the Probability Mass Function
   b. The probability that fewer than four solar panel have blowouts.
   c. The probability that more than two have blowouts.
   d. The expected number of solar panel with blowouts.
10. A sample containing 500 solar panel is taken from the output of a production line. Defective solar panels occur randomly and independently at a rate of 1.8% of the population. What is the probability that the sample will have no more than 2.7% defective items?
11. Data was collected on bolt failures in solar panel connection. A large number of solar panel assemblies, each with the same number of bolts, were tested to find out whether each bolt was a success or a failure. The number of passing bolts in an assembly had a mean value of 5.3 and a

variance of 2.65. If failing bolts occur randomly and independently, find the probabilities associated with possible numbers of passing bolts in an assembly. An assembly of solar panel is considered adequate if there is one or fewer bolt failures, what is the probability that a randomly chosen assembly will be inadequate?

12. Suppose 30 solar product manufacturing companies operate in the India and 14 are located in Bombay. If 5 computer companies are selected randomly from the entire list, what is the probability that one or more of the selected companies are located in the Bombay?

13. Explain the concept of K-means clustering in the field of solar energy system.

14. What is the use of Weibull distribution in the field of solar energy system?

References

[1] A. Ashtari, E. Bibeau, S. Shahidinejad, T. Molinski, PEV charging profile prediction and analysis based on vehicle usage data, IEEE Trans. Smart Grid 3 (2011) 341–350.

[2] S. Ali, S.-M. Lee, C.-M. Jang, Statistical analysis of wind characteristics using Weibull and Rayleigh distributions in Deokjeok-do Island–Incheon, South Korea. Renew. Energy 123 (2018) 652–663.

[3] M. Shoaib, I. Siddiqui, Y.M. Amir, S.U. Rehman, Evaluation of wind power potential in Baburband (Pakistan) using Weibull distribution function, Renew. Sustain. Energy Rev. 70 (2017) 1343–1351.

[4] W. Labeeuw, G. Deconinck, Customer sampling in a smart grid pilot, in: Proceedings of the 2012 IEEE Power and Energy Society General Meeting, San Diego, CA, USA, 2012, pp. 1–7. 22–26 July.

[5] H.K. Elminir, Y.A. Azzam, F.I. Younes, Prediction of hourly and daily diffuse fraction using neural network, as compared to linear regression models, Energy 32 (2007) 1513–1523.

[6] A. Rahimik hoob, Estimating global solar radiation using artificial neural network and air temperature data in a semi-arid environment, Renew. Energy 35 (2010) 2131–2135.

[7] I. Korachagaon, Estimating Global Solar Radiation Potential for Brazil by Iranna-Bapat's Regression Models, Int. J. Emerg. Technol. Adv. Eng. 2 (2012) 178–186 34.

[8] D.L. Woodruff, J. Deride, A. Staid, J.-P. Watson, G. Slevogt, C. Silva-Monroy, Constructing probabilistic scenarios for wide-area solar power generation, Sol. Energy 160 (2018) 153–167.

[9] A. Hammer, D. Heinemann, E. Lorenz, B. Lückehe, Short-term forecasting of solar radiation: a statistical approach using satellite data, Sol. Energy 67 (1999) 139–150.

[10] E. Lorenz, J. Remund, S.C. Müller, W. Traunmüller, G. Steinmaurer, D. Pozo, J.A. Ruiz-Arias, V.L. Fanego, L. Ramirez, M.G. Romeo, Benchmarking of different approaches to forecast solar irradiance, in: Proceedings of the 24th European Photovoltaic Solar Energy Conference, Hamburg, Germany, 2009, pp. 21–25.

CHAPTER SIX

Application of regression analysis and forecasting techniques in solar energy system

Learning objectives

- Understand about correlation and simple regression of solar energy parameter.
- Know about the multiple regression model of solar energy parameter.
- Know about time series component and forecasting of solar energy parameter.
- Understand about the Autocorrelation and auto regression between different parameter.

6.1 Introduction

Regression analysis is one of the statistical tools which gives the relation between a dependent variable and one or more predictor/independent variables of solar energy system and it may be variables of solar product manufacturing company or may be variable of solar power plant. Usually, the researcher who is efficient in the solar energy system field, is going to determine the causal effects of one variable of solar energy system on another. The aim of the study of regression is to express the variety of response as a function of the variables of the predictor of different factors of solar energy system. The duality of suit and precision the results depend on the details used. Thus, non-representational or wrongly recorded bad fits and assumptions are the product of evidence [1,2]. Therefore, for successful use of regression analysis, one must have

- Investigate the method of processing of data of solar product manufacturing companies and solar power plant
- Discover some shortcomings in the collected data of solar energy system
- Restrict inferences accordingly.

After a relationship in regression analysis is obtained, it can be used to forecast values of the output variable, the factors that influence the outcome most, or the hypothesized verification causal simulations of a reaction. For the example it is necessary to forecast the solar radiation for

Decision Science and Operations Management of Solar Energy Systems.
DOI: https://doi.org/10.1016/B978-0-323-85761-1.00006-8

the perfect functioning of solar power plant. It is possible to evaluate the value of each predictor variable through statistical checks on the predictor variables' approximate coefficients (multipliers). For example, a company in the distribution of solar panel may determine that there is a relationship between the price of panel and company's transportation costs. To determine this relationship, a financial expert studying the various factors likes the location to transport the panel, behaviour of market scenario, and sales of a company etc [3,4].

Equations in regression analysis are structured to allow estimates only. If the model is not sufficiently described and the parameter precision is not guaranteed, successful forecasts of solar energy system would not be possible. Accurately forecast and model parameters, however, enable the data to account for all applicable variables and the forecast function to be specified in the prediction equation. The calculation of parameters is the most difficult to do because not only is the model needed to be accurately defined, the calculation must also be precise and effective calculation should be possible with the details. A solar product enterprise, for instance, causes a dilemma which demands that any estimators not be used. Therefore, data constraints and failure to calculate anything in sample, relevant predictor variables limit the use of prediction equations. In this chapter, also assessment of different features of solar energy system through the NCSS tool. To analyze and visualize your data, NCSS software delivers a comprehensive and user-friendly collection of hundreds of statistical and graphical tools.

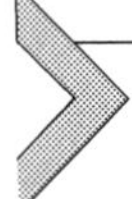

6.2 Correlation and simple regression of solar energy parameter

6.2.1 Correlation

Correlation is a mathematical approach that measures the degree of association between two distinct variables. It is often referred to as a "bivariate," with bimeaning two and variate implying variable, or variance. For a person or object, the two variables are typically a pair of points. For example, whether the stocks of two solar panels of different material may rise and fall in any manner. Correlation analysis can yield a numerical value for a set of data pairs that reflects the degree of interaction of the two stock prices over time. In the solar energy system, is a correlation evident between the size or price of the panel and the availability of energy source at a particular location? If yes, then how strong is the relation or correlation?

This suggests that knowing the score of a person or entity on one variable helps to predict their score on the second variable when a relationship is good. If the correlation between two variables is week then a score of first variable does help to predict the score the second variable. On the basis of this correlation coefficient is range from -1 to +1. The value outside this range is not valid [5,6].

Following are the possible correlation of different parameters of solar energy system.

- Total current and reverse saturation current of solar cell.
- Fill factor and maximum value of voltage of solar panel.
- Fill factor and open circuit voltage of solar panel.
- Fill factor and maximum value of current of solar panel.
- Fill factor and short circuit current of solar panel.
- Hour angle and solar time.
- Solar radiation and clearness index of a particular location.
- Solar radiation and wind velocity of a particular location.
- Solar radiation and temperature of a particular location.
- Solar radiation and humidity of a particular location.
- Solar plant capacity and load demand.

Correlation coefficient: A correlation coefficient, denoted by r, measures the degree of interaction. It is also called the correlation coefficient of ***Pearson*** (r) after its originator and is a linear interaction measure. If there is a need for a curved line to convey the relationship, other and more complex correlation methods may be used. On a scale that ranges from +1 via 0 to -1, the correlation coefficient is calculated. Either +1 or -1 reflects the full association between the two variables for example number of solar panel and load generated by the solar system. The association is positive if one variable increases as the other increases; if one decreases as the other increases, it is negative. The total lack of correlation is expressed by 0.0. The Pearson correlation coefficient (r) of any two variable says a and b can be written as

$$r = \frac{\left(\sum (a - \bar{a})(b - \bar{b})\right)}{\left(\sqrt{\sum (a - \bar{a})^2 (b - \bar{b})^2}\right)}$$

Or

$$r = \frac{\sum ab - \frac{(\sum a \sum b)}{n}}{\sqrt{\left(\sum a^2 - \frac{(\sum a)^2}{n}\right)\left(\sum b^2 - \frac{(\sum b)^2}{n}\right)}}$$

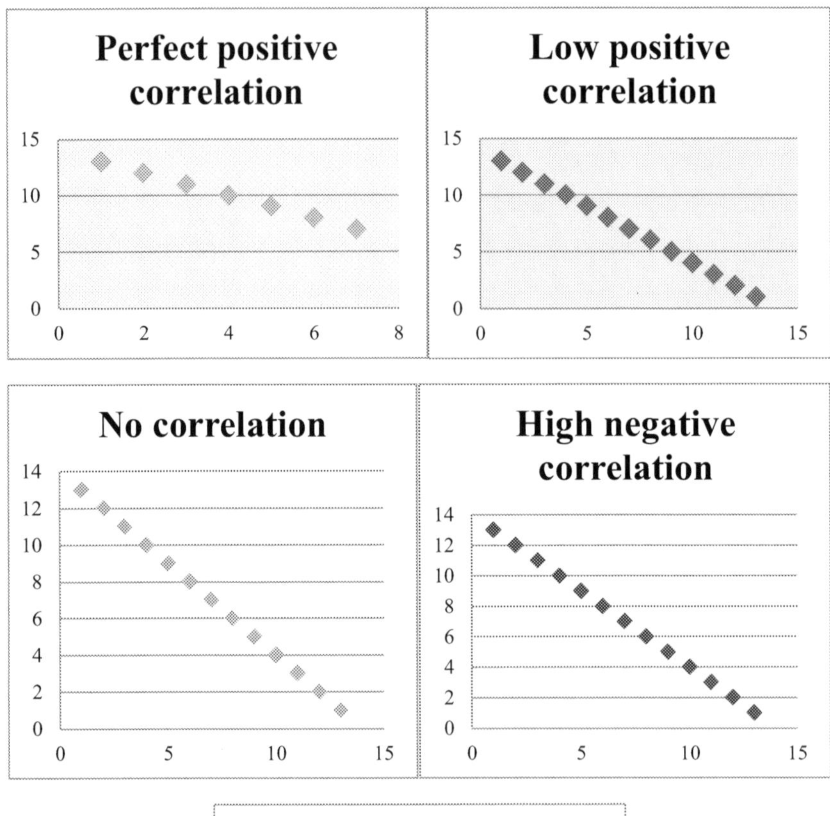

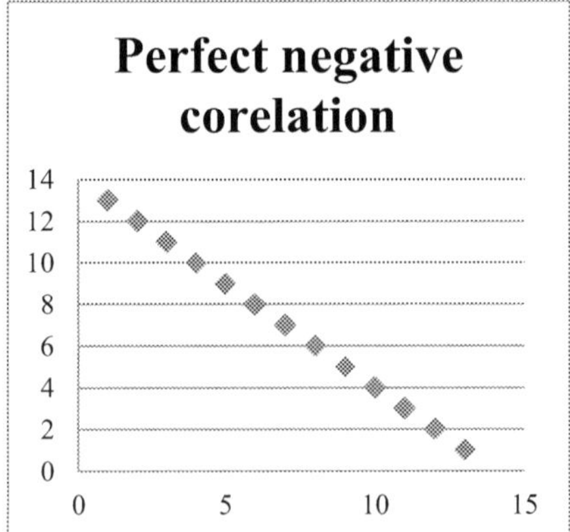

Figure 6.1 Graphical representation of correlation.

A graphical correlation representation is given in Fig. 6.1.

If two sets of findings have been gathered by an investigator and he or she needs to see if there is a relationship between them, he or she may first create a scatter diagram. One range of scales reflects the vertical scale and the other the horizontal scale. It is normal to position the experimental findings on the vertical axis where one collection of measurements consists of experimental results and other consists of a time scale or observed description of some kind. This represents what the 'dependent variable' (number of cells required) is called. The 'independent variable' (load requirement) is measured along the horizontal axis, or mean, such as time or height or some other known form. The terms "independent" and "dependent" might puzzle the beginner, since what depends on what is often not obvious. This misunderstanding is a victory of common sense over confusing terms, as each variable is always dependent on a third variable that may or may not be referred to in the analysis.

6.2.2 Simple regression analysis

Regression analysis is the process of constructing a mathematical model or function that can be used to predict or design a solar system by another variety of solar energy system. The most elementary regression model is called simple regression or bivariate regression involving two variables in which one variable is predicted by another variable. For the example, it is possible to identify the energy generation through the solar panel based on the solar radiation adopted on the solar panel. In simple regression, the variable to be predicted is called the dependent variable. The predictor is called the independent variable, or explanatory variable. On the solar energy system, energy generation through the solar panel is depends on the capacity of solar panels, so in this case capacity of the solar panel is independent variable and quantity of energy generation is dependent variable. In simple regression analysis, only a straight-line relationship between two variables is examined. Nonlinear relationship and regression models with more than one independent variable can be explored by using multiple regression models [7,8].

Can the cost of the floating solar panel be predicted using regression analysis? If so, what variables are related to such cost? A few of the many variables that can be potentially contribute are type of panel material, weather condition, direction of water flowing, location etc. Can the number of panel predict the cost of installation? It seems logical that more panel results in more weight and more required area, which could, in turn, result in an increased requirement of controlling the generation and power demand.

Actually, the first step in simple regression analysis is to construct a scatter plot i.e. graphing the data in this way yields preliminary information about the shape and spread the data as shown in Fig. 6.2, which is based on the

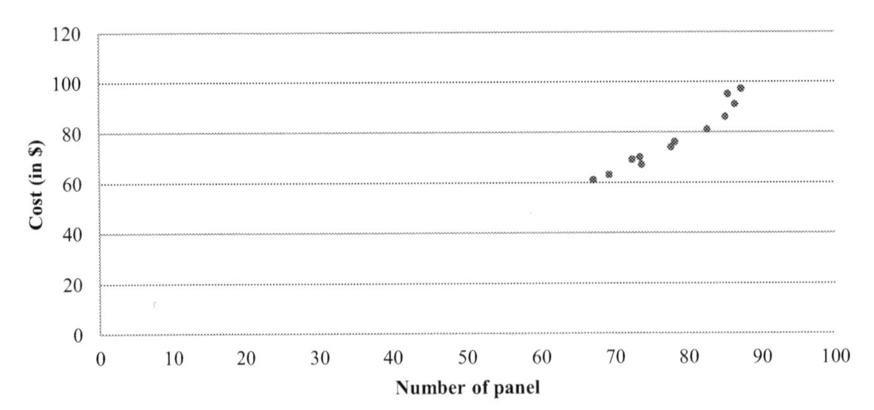

Figure 6.2 Scatter plot of solar data.

Table 6.1 Floating solar panel data.

Number of panel, each of 1 KW	Cost (in $) of per KW Solar Panel
61	67.1
63	69.3
67	73.7
69	72.45
70	73.5
74	77.7
76	78.28
81	82.62
86	85.14
91	86.45
95	85.5
97	87.3

data of Table 6.1.

Try to imagine a line passing through the points. Is linear fit possible? Would a curve fit the data better? The scatter plot gives some idea of how well a regression line fits the data of a number of solar panel and cost of the solar panel.

6.2.3 Determining the equation of regression

The first step in determining the equation of regression line that passes through the sample data is to establish in the equation's form. A line is a two point form, the point slope form, and the slope intercept form.

$$y = \beta_0 + \beta_1 x \tag{a}$$

Where β_0 is the population 'y' intercept and β_1 is the population slope. For a specific dependent variable say y_s,

$$y_s^i = \beta_0 + \beta_1 x_i + \epsilon_i \tag{b}$$

Where,

$x_i = i^{\text{th}}$ value of the independent variable

$y_s{}^i = i^{\text{th}}$ value of the dependent variable

$\epsilon_i =$ error in the prediction for the ith value.

The regression line would skip at least some of the points unless the points being fitted by the regression equation are perfectly aligned. The error of the regression line in fitting these points is represented by ϵ_i in the above equation. If there is a point on the regression line then, $\epsilon_i = 0$.

The following steps are including the regression analysis:

Step 1: Problem statement

The first step in performing any regression analysis in solar energy system is to define the issue and the goals that the regression analysis will discuss. Erroneous statistical inferences will result from the incorrect formulation or interpretation of the problem. The variables chosen are determined by the study's objectives and understanding of the problem. For example, the amount of solar radiation and temperature of the particular city for solar power plant installation is correlated [9,10].

There may be two ways to deal with this problem.

Case 1: Calculating solar radiation for a given temperature, or

Case 2: Calculating temperature for a given solar radiation.

In case 1, the response variable is solar radiation, while in case 2, the response variable is temperature. In cases 1 and 2, the roles of explanatory variables are also reversed.

Step 2: Selection of appropriate variables

The next step is to select the appropriate variables after the problem has been carefully formulated and the goals have been determined. It's important to remember that the variables of solar energy system choose will decide how accurate your statistical inferences are.

Step 3: Collection of data of variables

The next question is how to collect data on such related variables once the study's goal has been clearly stated and the variables have been selected. Data is basically a calculation of these variables. For example, the data of solar radiation for different years to be recorded through the pyrheliometer instrument and this can give the exact amount of intensity of a specific time. On the other hand it is also important to choose that the data of variable to be collected as quantitative or qualitative. If the amount of solar radiation ($kwh/m^2/day$) is $10, 4, 9, 2, 6$ then these are quantitative data. If the data are collected as value $= 0$ for intensity is less than or equal to 20 and values $=$ 1 in intensity greater than 20. So that data recorded is converted as $0, 1, 0, 1,$ 1. But in this way there is possible to lost information of quantitative data.

Qualitative and quantitative data have different methods and approaches. Logistic regression is used when the research variable is binary. The study of variance methodology is used if all explanatory variables are qualitative. The study of covariance technique is used when certain explanatory variables are qualitative and others are quantitative [11–13].

Step 4: Model specification

The model's shape is normally determined with the assistance of the experimenter or the individual working on the subject. The tentative model's shape can only be determined, and it will be dependent on certain unknown parameters. There are two types of mathematical models: deterministic and probabilistic. For a given input, deterministic models are mathematical models that generate exact output. For example output of solar cell (b) will depend on solar radiation, which is represented by regression line as:

$$b = 1.5 + 2.3(\text{solar radiation})$$

For value of solar radiation =5, the exact value of y will be:

$$b = 1.5 + 2.3(5) = 13$$

We realize that the values of b will rarely match the values generated by the equation exactly. Since the variable of solar radiation is unlikely to explain all of the variability of the variable b, random error may occur in the estimation of values of b for solar radiation values. For this reason, a general model is represented in regression analysis as given in eq. a and b.

Usually, all regressions of business data use sample data, not a population data. Due to this β_0 and β_1 are unfeasible and must be assessed by using sample statistics b_0 and b_1. Hence the equation of regression line will become:

$$b = b_0 + b_1 x$$

Where $b_0 =$ the sample of y intercept
And $b_1 =$ the slope of the sample.

A simple regression model is one in which there is only one explanatory variable. A multiple regression model is used for more than one independent variable. When there are just a few options, then univariate regression refers to a regression with only one research component. When there are several studies the regression is referred to as multivariate regression because there are several variables.

The slope of the line (b_1) can be evaluated as:

$$b_1 = \frac{\sum (a - \bar{a})(b - \bar{b})}{\sum (a - \bar{a})^2}$$

Now y intercept of the regression line is:

$$b_0 = \bar{b} - b_1\bar{a} = \frac{\sum b}{n} - b_1\frac{\sum a}{n}$$

Step 5: Select the method to fit the data

The next step is to approximate the model's parameters based on the collected data after the model has been identified and the data have been collected. Parameter estimation or model fitting are other words for this method. For the example collect the data of hours' wise electrical energy generation and data on load demand after the installation of solar power plant and in that case the regression model is constructed using least squares analysis, which produces the smallest number of squared error values. A specific series of equations has been created to generate components of the regression model based on these premises and calculus.

Step 6: Model fitting

The values of uncertain parameters can be found by using the required approach to approximate them. We will use the formula by substituting these values into the equation. Model fitting is the expression for this. Fig. 6.3 examines that the line actually passes through some of the points. The vertical distance of a point to the line is actually the error of the fitting. The least square regression has a line that result in the minimum sum of the errors squared.

Step 7: Validation of model

The reliability of the predictive approach for regression analysis is contingent on a number of assumptions. These hypotheses become the model's and data's assumptions for solar system, respectively. Whether or not these assumptions are followed, which has a huge effect on the accuracy of statistical inferences. Care must be taken from the start of the experiment

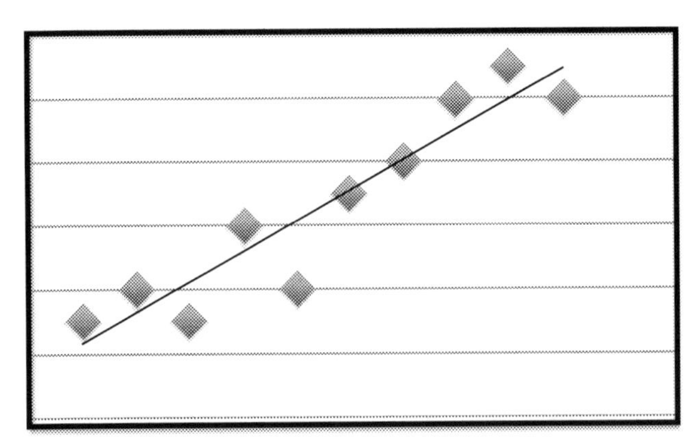

Figure 6.3 Model fitting of line.

to ensure that these hypotheses are true and fulfilled. One must be cautious when selecting the necessary assumptions, as well as determine whether the assumptions are correct for the given experimental conditions. Before drawing any statistical conclusions, the hypotheses must be validated. Any deviation from the assumptions' validity would be expressed in statistical inferences. Regression analysis is, in reality, an iterative method in which the outputs are used to diagnose, validate, critique, and adjust the inputs.

6.2.4 Standard error

The residuals are measurement errors for particular points. When dealing with massive data sets, residual computations become time-consuming. Also with machines, it can be difficult for a researcher to wade through pages of residuals in order to grasp the regression model's mistake. The standard error in the calculation, which gives a single estimation of the regression error, is another method for analysing model error. Attempting to calculate the overall amount of error by summing the residuals is futile since the sum of the residuals is zero. By squaring the residuals and then summing them, the zero sum properties of residuals can be avoided. Table 6.2 contains the solar panel cost data along with the residuals and the residuals squared. The total of the residuals squared column is called the **sum of squares of error (SSE).**

$$\text{SSE} = \sum \left(b - \hat{b}\right)^2$$

Where $\hat{b} = b_0 + b_1 x$

Table 6.2 Floating solar panel data.

Number of panels (a)	Cost (in $) (b)	Residuals $b - \hat{b}$	
61	67.1	-2.00709	4.028394
63	69.3	-0.91572	0.838539
67	73.7	1.267018	1.605335
69	72.45	-1.09161	1.191621
70	73.5	-0.59593	0.355133
74	77.7	1.386806	1.923231
76	78.28	0.858174	0.736463
81	82.62	2.426594	5.888358
86	85.14	2.175014	4.730686
91	86.45	0.713434	0.508988
95	85.5	-2.45383	6.021282
97	87.3	-1.76246	3.106272
Total=930	939.04	0.0004	30.9343

An easier way to compute the SSE rather than the residuals is:

$$\text{SSE} = \sum b^2 - b_0 \sum b - b_1 \sum ab$$

For the above example:

$$\sum b^2 = 74032.92$$

$$b_1 = 0.554316$$

$$b_0 = 35.29381$$

$$\sum b = 939.04$$

$$\sum ab = 73711.84$$

$$\text{SSE} = 74032.92 - 35.29381 \times 939.04 - 0.554316 \times 73711.84 = 30.9683$$

The small difference between this value and the computed value (30.9343) is due to the rounding error. In this way researchers can find out the sum of squares of error of any type of data of solar energy system.

6.2.5 Coefficient of determination

The coefficient of determination is a common measure to design solar systems to fit for regression models. The coefficient of the decision is the percentage of the dependent variable's uncertainty described by the independent variable. The coefficient of determination will vary between 0 and 1. A coefficient of determination of zero means that the predictor accounts for none of the variability of the dependent variable and there is no regression prediction. A value equal to 1 means perfect prediction and that 100% of the variability.

Low coefficient values are more likely to be suitable in exploratory science where the factors are not understood than in areas of research where the parameters are more developed and understood. For example, an ISRO researcher who uses a solar system to forecast mission costs looks for regression models with coefficients of 0.9 or higher. In a regression model, the dependent variables have an action that is determined by the number of squares of the dependent variables as:

$$\text{sum of square of dependent variable (SSDV)} = \sum \left(b - \bar{b}\right)^2 = \sum b^2 - \frac{\left(\sum b\right)^2}{n}$$

The sum of the dependent values' squared deviations from the mean value of the dependent value. This variance is divided into two additive components: explained variation (measured by the sum of squares of regression) and unexplained variation (measured by the sum of squares of error).

$$\text{SSDV} = \text{sum of square of regression} + \text{sum of square of error}$$

Or

$$1 = \frac{\text{sum of square of regression}}{\text{SSDV}} + \frac{\text{sum of square of error}}{\text{SSDV}}$$

Since correlation coefficient is the proportion of the dependent variable which is expressed as:

$$r^2 = \frac{\text{SSR}}{\text{SSDV}} \text{ Or } 1 = r^2 + \frac{\text{sum of square of error}}{\text{SSDV}}$$

Following examples shows the application of the determination of coefficient in the field of solar energy system.

Example 6.1. Evaluate the coefficient of correlation for the tariff of distribution of solar power to different sectors to fulfil the load requirements as given in Table 6.3.

Table 6.3 Tariff of distribution of solar power to different sectors.

Price of solar power in ($)/kW	5	8	12	15	18	22	25
Load (KW)	15	13	18	21	19	24	20

Table 6.4 Calculation of coefficient of correlation.

S. No.	Price of power in ($) (a)	Load (KW) (b)	a²	b²	ab
1	5	15	25	225	75
2	8	13	64	169	104
3	12	18	144	324	216
4	15	21	225	441	315
5	18	19	324	361	342
6	22	24	484	576	528
7	25	20	625	400	500
Sum	105	130	1891	2496	2080

Also find out different statistical assessment through the NCSS data analysis tool.

Solution: According to the data of Table 6.3, calculate a^2, b^2 and ab based on the price of power in ($) and load in KW which is shown in Table 6.4.

From Table $\Sigma a = 105$, $\Sigma b = 130$, $\Sigma a^2 = 1891$, $\Sigma b^2 = 2496$, $\Sigma ab = 2080$, $n = 7$.

$$\text{coefficient of correlation } (r) = \frac{\sum ab - \frac{(\sum a \sum b)}{n}}{\sqrt{\left(\sum a^2 - \frac{(\sum a)^2}{n}\right)\left(\sum b^2 - \frac{(\sum b)^2}{n}\right)}}$$

$$\text{coefficient of correlation } (r) = \frac{2080 - \frac{(105)(130)}{7}}{\sqrt{\left(1891 - \frac{105^2}{7}\right)\left(2496 - \frac{130^2}{7}\right)}}$$

$$r = \frac{130}{17.7764 \times 9.0395} = 0.809$$

According to the analysis the value of 'r' is 0.809, which shows positive correlation between the above two parameters.

Assessment through NCSS tool: Data assessment of Example 6.1 through NCSS tool is represented in the Table 6.5. The equation of the straight line relating Price of power ($) and Load (kW) is estimated as: Price of power ($)= (-14.5455) + (1.5909)

Table 6.5 Linear regression report.

Parameter	Value	Parameter	Value
Dependent variable	Price of power	Rows processed	7
Independent variable	Load kW_	Rows used in estimation	7
Frequency variable	None	Rows with X missing	0
Weight variable	None	Rows with freq missing	0
Intercept	-14.5455	Rows prediction only	0
Slope	1.5909	Sum of frequencies	7
R-squared	0.6545	Sum of weights	7.0000
Correlation	0.8090	Coefficient of variation	0.3115
Mean square error	21.83636	Square root of MSE	4.672939

Load (kW) using the 7 observations in this dataset. The y-intercept, the estimated value of Price of power ($) when Load (kW) is zero, is -14.5455 with a standard error of 9.7615. The slope, the estimated change in Price of power ($) per unit change in Load (kW), is 1.5909 with a standard error of 0.5169. The value of R-Squared, the proportion of the variation in Price of power ($) that can be accounted for by variation in Load (kW), is 0.6545. The correlation between Price of power ($) and Load (kW) is 0.8090. A significance test that the slope is zero, resulted in a t-value of 3.0775. The significance level of this t-test is 0.0275. Since 0.0275 < 0.0500, the hypothesis that the slope is zero is rejected. The estimated slope is 1.5909. The lower limit of the 95% confidence interval for the slope is 0.2621 and the upper limit is 2.9197. The estimated intercept is -14.5455. The lower limit of the 95% confidence interval for the intercept is -39.6381 and the upper limit is 10.5472. Fig. 6.4 shows the regression plot between price of power and load in KW.

Example 6.2. The following data are the total cost (in $) for generation of power using solar farm benefits for nine states/country, along with the cost to export the power (in $) that the generation company had in assets in those states given in Table 6.6.

Use the data to compute a correlation coefficient to determine the correlation between above two parameters.

Solution: According to the data of Table 6.7, calculate a^2, b^2 and ab based on cost for solar power generation and cost for export the solar power.

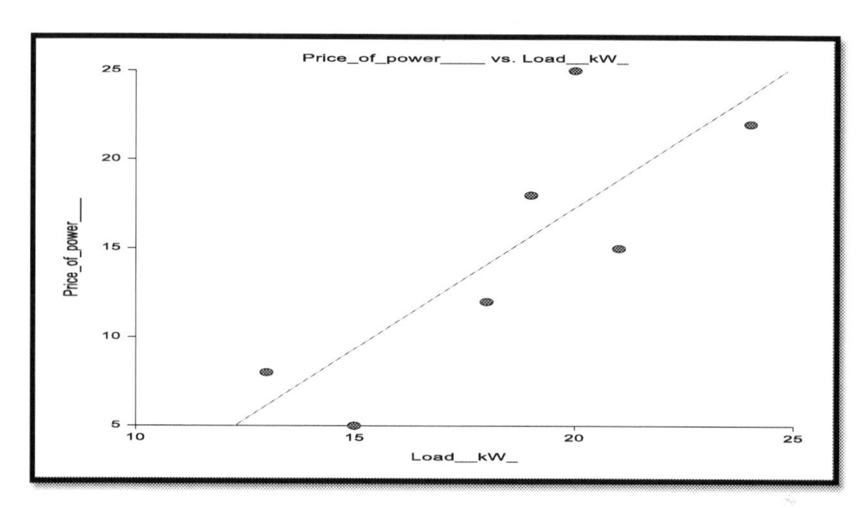

Figure 6.4 Regression plot between price of power and load.

Table 6.6 Data for total cost and export cost for generation of power.

State	Cost of solar power generation	Cost for export the solar power
Colorado	10422	267
Ireland	9082	100
Florida	5048	130
Spain	8734	234
Texas	4578	90
Germany	7806	160
Denmark	6523	89
India	5896	80
China	9543	245

Table 6.7 Calculation of correlation coefficient.

S.N.	Cost of solar power generation (a)	Cost for export the solar power (b)	a^2	b^2	ab
1	10422	267	108618084	71289	2782674
2	9082	100	82482724	10000	908200
3	5048	130	25482304	16900	656240
4	8734	234	76282756	54756	2043756
5	4578	90	20958084	8100	412020
6	7806	160	60933636	25600	1248960
7	6523	89	42549529	7921	580547
8	5896	80	34762816	6400	471680
9	9543	245	91068849	60025	2338035
Total	67632	1395	543138782	260991	11442112

Table 6.8 Data for utilization of solar power in recent three years.

Industries	Solar power (in KW) utilization in recent three years		
	Year 1	**Year 2**	**Year 3**
Textile	2	3	3.5
Chemical	1.5	1.5	2
Communication	3	3.7	4.2
Machinery	3.5	4	5
Food	2	2	3
Government sectors	3	4	4
Pharmaceutical	2.3	2.5	3
Hotels	2.5	3	3

From table $\Sigma a = 67632$, $\Sigma b = 1395$, $\Sigma a^2 = 543138782$, $\Sigma b^2 = 260991$,

$$\sum ab = 11442112, \quad n = 9.$$

$$\text{coefficient of correlation } (r) = \frac{\sum ab - \frac{(\sum a \sum b)}{n}}{\sqrt{\left(\sum a^2 - \frac{(\sum a)^2}{n}\right)\left(\sum b^2 - \frac{(\sum b)^2}{n}\right)}}$$

$$\text{coefficient of correlation } (r) = \frac{11442112 - \frac{(67632)(1395)}{9}}{\sqrt{\left(543138782 - \frac{(67632)^2}{9}\right)\left(260991 - \frac{(1395)^2}{9}\right)}}$$

$$r = \frac{959152}{5908.202 \times 211.5798} = 0.76728$$

According to the analysis coefficient of determination is 0.76728 between two financial parameters of solar energy system.

Example 6.3. The Bureau of Energy Efficiency of India released the following data on the utilization of power (in KW) from the solar energy system for several industries in three recent years given in Table 6.8.

Compare coefficient of correlation for each pair of year and evaluate which years are most highly correlated.

Table 6.9 Calculation of correlation coefficient for solar power generation in recent three years.

S.No.	Year 1 (a)	Year 2 (b)	Year 3 (c)	a^2	b^2	c^2	ab	bc	ac
1	2	3	3.5	4	9	12.25	6	10.5	7
2	1.5	1.5	2	2.25	2.25	4	2.25	3	3
3	3	3.7	4.2	9	13.69	17.64	11.1	15.54	12.6
4	3.5	4	5	12.25	16	25	14	20	17.5
5	2	2	3	4	4	9	4	6	6
6	3	4	4	9	16	16	12	16	12
7	2.3	2.5	3	5.29	6.25	9	5.75	7.5	6.9
8	2.5	3	3	6.25	9	9	7.5	9	7.5
Total	19.8	23.7	27.7	52.04	76.19	101.89	62.6	87.54	72.5

Solution: Table 6.9 shows the evaluation of year-wise data solar power utilization in recent three years.

Correlation between year 1 and year 2:

$$r = \frac{\sum ab - \frac{\sum a \sum b}{n}}{\sqrt{\left(\sum a^2 - \frac{(\sum a)^2}{n}\right)\left(\sum b^2 - \frac{(\sum b)^2}{n}\right)}}$$

$$r = \frac{62.6 - \frac{(19.8)(23.7)}{8}}{\sqrt{\left(52.04 - \frac{19.8^2}{8}\right)\left(76.19 - \frac{23.7^2}{8}\right)}} = \frac{3.9425}{(3.035)(5.97875)}$$

$$r = 0.2176$$

Correlation between year 1 and year 3:

$$r = \frac{\sum ac - \frac{\sum a \sum c}{n}}{\sqrt{\left(\sum a^2 - \frac{(\sum a)^2}{n}\right)\left(\sum c^2 - \frac{(\sum c)^2}{n}\right)}}$$

$$r = \frac{72.5 - \frac{(19.8)(27.7)}{8}}{\sqrt{\left(52.04 - \frac{19.8^2}{8}\right)\left(101.89 - \frac{27.7^2}{8}\right)}} = \frac{3.9425}{(3.035)(5.97875)}$$

$$r = 0.2176$$

Table 6.10 Year-wise number of PV cell.

Year	2016	2017	2018	2019	2020
Number of solar panel	32	45	41	54	67

Table 6.11 Calculation of regression line.

x	y		x ŷ		y²
2016	32	-2	-64	4	1024
2017	45	-1	-45	1	2025
2018	41	0	0	0	1681
2019	54	1	54	1	2916
2020	67	2	134	4	4489
Total	**239**	**0**	**79**	**10**	**12135**

Correlation between year 2 and year 3:

$$r = \frac{\sum bc - \frac{\sum b \sum c}{n}}{\sqrt{\left(\sum b^2 - \frac{\left(\sum b\right)^2}{n}\right)\left(\sum c^2 - \frac{\left(\sum c\right)^2}{n}\right)}}$$

$$r = \frac{87.54 - \frac{(27.7)(23.7)}{8}}{\sqrt{\left(76.19 - \frac{23.7^2}{8}\right)\left(101.89 - \frac{27.7^2}{8}\right)}} = \frac{5.47875}{(5.97875)(5.97895)}$$

$$r = 0.1533$$

The result shows that year 1 & 2 and year 1 & 3 is highly correlated.

Example 6.4. Sketch a scatter plot from the following data, and determine the equation of the regression line of the number of solar panels used in the last five years to generate solar energy at a particular location, which provide electricity for the automobile industry as given in Table 6.10.

Solution: Based on the data of Table 6.11, which shows year-wise number of solar panels, evaluate the value b1 and b0.

$$b_1 = \frac{\sum\left(\hat{x} - \bar{\hat{x}}\right)(y - \bar{y})}{\sum\left(\hat{x} - \bar{\hat{x}}\right)^2} = \frac{\sum \hat{x}y - \frac{\sum \hat{x} \sum y}{n}}{\sum \hat{x}^2 - \frac{\left(\sum \hat{x}\right)^2}{n}}$$

$$b_1 = \frac{79}{10} = 7.9$$

$$b_0 = \frac{\sum y}{n} - b_1 \frac{\sum \hat{x}}{n}$$

$$b_0 = \frac{239}{10} = 23.9$$

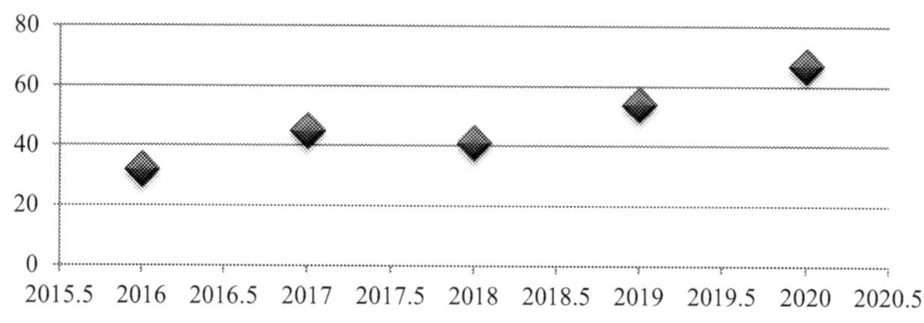

Figure 6.5 Scatter plot of using of PV cell to generate power irrespective to years.

Table 6.12 Data for solar power rating according to number of panel.

Solar panel rating in Watt	Number of panels
200	2000
300	1080
500	900
700	880
800	700
1000	350
1200	298
1500	177

Equation of the regression line is

$$y = b_0 + b_1(x - 2018)$$

$$y = 23.9 + 7.9(x - 2018)$$

Fig. 6.5 shows the scatter plot of using of solar panel to generate power irrespective to years and which also represented through the equation $y = 23.9 + 7.9(x - 2018)$.

Example 6.5. One of the solar manufacturing company design solar panel of different ratings. As an aid in long-term planning they gather the following rating and number of panels required from several sectors. Develop the equation of the simple regression line to predict the number of panels of different rating given in Table 6.12.

Also describe the assessment through the NCSS data analysis tool.

Table 6.13 Evaluation of regression line for solar panel manufacturing company.

S.N.	Solar panel rating in (Watt) X	Number of panels (Y)	XY	X*X	Y*Y
1	200	2000	400000	40000	4000000
2	300	1080	324000	90000	1166400
3	500	900	450000	250000	810000
4	700	880	616000	490000	774400
5	800	700	560000	640000	490000
6	1000	350	350000	1000000	122500
7	1200	298	357600	1440000	88804
8	1500	177	265500	2250000	31329
Total	**6200**	**6385**	**3323100**	**6200000**	**7483433**

Solution: Assessment of data for solar power rating according to a number of panels is shown in Table 6.13.

$$\sum X = 6200 \sum Y = 6385 \sum XY = 3323100$$

$$\sum X^2 = 6200000 \sum Y^2 = 7483433\, n = 8$$

$$b_1 = \frac{\sum (X-\bar{X})(Y-\bar{Y})}{\sum (X-\bar{X})^2} = \frac{\sum XY - \frac{\sum X \sum Y}{n}}{\sum X^2 - \frac{(\sum X)^2}{n}}$$

$$b_1 = \frac{3323100 - \frac{(6200)(6385)}{8}}{6200000 - \frac{(6200)^2}{8}} = -\frac{1625275}{1395000} = -1.2$$

$$b_0 = \frac{\sum Y}{n} - b_1 \frac{\sum X}{n}$$

$$b_0 = \frac{6385}{8} - (-1.2)\frac{6200}{8} = 1728.1$$

$$Y = 1728.1 - 1.2X$$

Number of panels required=1728.1-1.2(rating of the panel)

Assessment through NCSS tool: Data assessment of Example 6.5 through NCSS tool is represented in the Table 6.14. The equation of the straight line relating the number of panels and solar panel rating (W) is estimated as: Number of panel = (1701.0556) + (-1.1651) solar panel rating (W) using the 8 observations in this dataset. The y-intercept, the estimated value of a number of panels when solar panel rating (W) is zero, is 1701.0556 with a standard error of 213.8369. The slope, the estimated change in number

Table 6.14 Parameter assessment through NCSS tool.

Parameter	Value	Parameter	Value
Dependent variable	Number of panels	Rows processed	8
Independent variable	Solar panel rating in Watt	Rows used in estimation	8
Frequency variable	None	Rows with X missing	0
Weight variable	None	Rows with freq missing	0
Intercept	1701.0556	Rows prediction only	0
Slope	-1.1651	Sum of frequencies	8
R-Squared	0.7931	Sum of weights	8.0000
Correlation	-0.8906	Coefficient of variation	0.3595
Mean square error	82307.16	Square root of MSE	286.8922

Table 6.15 Data for market price of control panel for solar plant.

Market price ($100) (y)	Area required for solar panel in square feet (x_1)	Life length (year) (x_2)
62	1602	34
64.2	2387	44
68.3	1453	21
75.4	2303	33
71.1	1984	24
74.8	1678	13
79.2	3015	9
78.3	1682	12
82.5	2380	29
79.7	1780	32
83.9	2043	3
78.2	2012	7
85.3	3450	15
94	2200	8
100.7	2817	5
105.3	2363	7
120.4	3078	4
105.6	3367	7
129.5	4875	9
116.8	4450	8

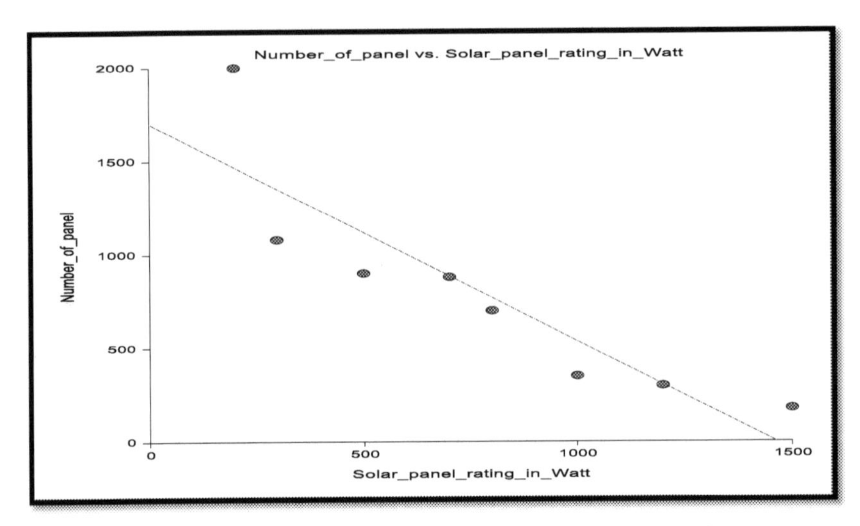

Figure 6.6 Relationship between number of solar panel and solar panel in Watt.

of panels per unit change in solar panel rating (W), is -1.1651 with a standard error of 0.2429. The value of R-Squared, the proportion of the variation in the number of panels that can be accounted for by variation in solar panel rating (W), is 0.7931. The correlation between number of panels and solar panel rating (W) is -0.8906.

A significance test that the slope is zero, result in a t-value of -4.7965. The significance level of this t-test is 0.0030. Since $0.0030 < 0.0500$, the hypothesis that the slope is zero is rejected. The estimated slope is -1.1651. The lower limit of the 95% confidence interval for the slope is -1.7594 and the upper limit is -0.5707. The estimated intercept is 1701.0556. The lower limit of the 95% confidence interval for the intercept is 1177.8156 and the upper limit is 2224.2955. Fig. 6.6 shows the relationship between number of solar panels and solar panel in watt.

6.3 Multiple regressions

A simple regression analysis is a bivariate linear regression in which one dependent variable predicts one independent variable. Multiple regression analysis, on the other hand, refers to regression analysis with two or more independent variables or with at least one nonlinear predictor. For the example generation of electricity through solar power plant is depends on the capacity of the plant as well as the amount of solar radiation at a particular location, so in this case amount of electricity generation is dependent variable and plant capacity and amount of solar radiation is two independent variables. In concept, multiple regression analysis is identical to the basic regression analysis. However, theoretically and computationally, it is more complicated. The probabilistic simple regression model's equation is:

$$y = b_0 + b_1 x + \in$$

Where,

$y =$ dependent variable
$b_0 =$ the intercept
$b_1 =$ the slope
$\in$ = the error of prediction

In general, the equation for multiple regression models can be written as:

$$y = b_0 + b_1 x_1 + b_2 x_2 + - \ - - - - + b_k x_k + \in$$

Where

K = the number of independent variables

The dependent variable is frequently referred to as the response variable in multiple regression analysis. The partial regression coefficient of an independent variable b1 represents the increase that will occur in the value of y from a one unit increase in that independent variable if all other variables are held constant. Because more than one predictor is included in a model, partial regression coefficients appear. The partial regression coefficients are similar to b1, the simple regression model's slope. In a multiple regression model, the partial regression coefficients and the regression constant are population values that are unknown. These values are almost often calculated using sample data in the study.

6.3.1 Regression model with two independent variables

A multivariate regression model with two independent variables i.e. number of panels and its cost of the solar system is the most basic. The constants and coefficients are calculated using sample data, resulting in the model below.

$$\hat{y} = b_0 + b_1 x_1 + b_2 x_2$$

Similarly, determining formulae to solve for multiple regression coefficients follows a similar approach. The formulae were created with the goal of reducing the sum of squares of error for the model. As a result, the regression technique depicted above is known as least squares analysis. The calculus approach is used, resulting in a $i + 1$ equation for $i + 1$ unknowns for multiple regression with i independent variables. For regression model with two independent variables, result in three simultaneous equations can be written as:

$$b_0 n + b_1 \sum x_1 + b_2 \sum x_2 = \sum y$$

Now multiply x_1 on both sides.

$$\sum x_1 y = b_0 \sum x_1 + b_1 \sum {x_1}^2 + b_2 \sum x_1 x_2$$

Further multiply x_2 on both sides.

$$\sum x_2 y = b_0 \sum x_2 + b_1 \sum x_1 x_2 + b_2 \sum {x_2}^2$$

Following examples shows the application of regression analysis in the field of solar energy system.

Example 6.6. A study was conducted in a small Louisiana city to determine what variable, if any are related to the market price of control panel for solar power plant. Several variables were explored, including number of solar panels, square feet of space available and life length of panel etc. Suppose the researcher wants to develop a regression model to predict the market price of a panel by two variable i.e area in square feet and the life length. The data for these three variables are given in the Table.

Also estimation done through the NCSS tool.

Solution: Table 6.16 shows the calculation of multi-regression model based on the data of Example 6.6.

From table:

$$\sum x_1 = 50919, \ \sum y = 1755.2, \ \sum x_2 = 324, \ \sum x_1^2 = 146045929$$

$$\sum x_2^2 = 8088, \sum x_1 x_2 = 744282, \sum x_1 y = 4740283, \sum x_2 y = 25521.3$$

Table 6.16 Calculation of multi regression model.

S.N.	y	x_1	x_2	x_1^2	x_2^2	$x_1 x_2$	$y x_1$	$y x_2$
1	62	1602	34	2566404	1156	54468	99324	2108
2	64.2	2387	44	5697769	1936	105028	153245	2824.8
3	68.3	1453	21	2111209	441	30513	99239.9	1434.3
4	75.4	2303	33	5303809	1089	75999	173646	2488.2
5	71.1	1984	24	3936256	576	47616	141062	1706.4
6	74.8	1678	13	2815684	169	21814	125514	972.4
7	79.2	3015	9	9090225	81	27135	238788	712.8
8	78.3	1682	12	2829124	144	20184	131701	939.6
9	82.5	2380	29	5664400	841	69020	196350	2392.5
10	79.7	1780	32	3168400	1024	56960	141866	2550.4
11	83.9	2043	3	4173849	9	6129	171408	251.7
12	78.2	2012	7	4048144	49	14084	157338	547.4
13	85.3	3450	15	11902500	225	51750	294285	1279.5
14	94	2200	8	4840000	64	17600	206800	752
15	100.7	2817	5	7935489	25	14085	283672	503.5
16	105.3	2363	7	5583769	49	16541	248824	737.1
17	120.4	3078	4	9474084	16	12312	370591	481.6
18	105.6	3367	7	11336689	49	23569	355555	739.2
19	129.5	4875	9	23765625	81	43875	631313	1165.5
20	116.8	4450	8	19802500	64	35600	519760	934.4
Total	1755.2	50919	324	146045929	8088	744282	4740283	25521.3

Put these values in the following equation:

$$b_0 n + b_1 \sum x_1 + b_2 \sum x_2 = \sum y$$

$$20b_0 + 50919b_1 + 324b_2 = 1755.2$$

$$\sum x_1 y = b_0 \sum x_1 + b_1 \sum x_1^2 + b_2 \sum x_1 x_2$$

$$50919b_0 + 146045929b_1 + 744282b_2 = 4740283$$

$$\sum x_2 y = b_0 \sum x_2 + b_1 \sum x_1 x_2 + b_2 \sum x_2^2$$

$$324b_0 + 744282b_1 + 8088b_2 = 25521.3$$

Solve these three equations simultaneously and evaluate coefficient as

$$b_0 = 64.2$$

$$b_1 = 0.0134$$

$$b_2 = -0.6461$$

Table 6.17 Parameter assessment through NCSS tool.

Item	Value
Dependent variable	Market Price 100
Number ind. Variables	16
Weight variable	None
R^2	0.9230
Adj R^2	0.5124
Coefficient of variation	0.1532
Mean square error	180.8333
Square root of MSE	13.44742
Ave abs pct error	3.025
Completion status	Normal Completion

Now the regression equation for estimating output is:

$$\hat{y} = 64.2 + 0.0131x_1 - 0.6x_2$$

According to the this assessment it is finding out

$$\textbf{Market Price } (\$100) = 64.2 + 0.0131$$
$$\textbf{Area Required for solar panel in square feet} - 0.6\,\textbf{Life length}\,(\textbf{year})$$

Assessment through NCSS tool: Data assessment of Example 6.6 through NCSS tool is represented in the Table 6.17. Fig. 6.7 represented histogram of residual of market price. Fig. 6.8 shows a normal probability plot of residuals of market price. Fig. 6.9 shows the residuals of market price v/s area required for solar panel in squares. Fig. 6.10 shows residual of market price v/s life length of solar panel.

6.3.2 Error of the estimate

The fact that the residuals accumulate to zero is one of the characteristics of a regression model. The zero sum property can be avoided by squaring the residuals and then summing the squares in order to obtain a single statistic that can reflect the error in a regression study. The sum of squares of error (SSE) is produced by such an operation.

For multiple regressions, the sum of squares error is computed in the same way as for simple regression.

$$\text{SSE} = \sum (y - \hat{y})^2$$

For the control panel of wind form example, sum of square error can be evaluated as:

Table 6.18 Estimated and actual output from regression model.

S.N.	Actual output y	Estimated output $\hat{y}$		$(y-\hat{y})^2$
1	62	64.7862	-2.7862	7.76291
2	64.2	69.0697	-4.8697	23.71398
3	68.3	70.6343	-2.3343	5.448956
4	75.4	74.5693	0.8307	0.690062
5	71.1	75.7904	-4.6904	21.99985
6	74.8	78.3818	-3.5818	12.82929
7	79.2	98.2965	-19.0965	364.6763
8	78.3	79.0342	-0.7342	0.53905
9	82.5	77.978	4.522	20.44848
10	79.7	68.318	11.382	129.5499
11	83.9	89.1633	-5.2633	27.70233
12	78.2	86.3572	-8.1572	66.53991
13	85.3	100.395	-15.095	227.859
14	94	88.22	5.78	33.4084
15	100.7	98.1027	2.5973	6.745967
16	105.3	90.9553	14.3447	205.7704
17	120.4	102.1218	18.2782	334.0926
18	105.6	104.1077	1.4923	2.226959
19	129.5	122.6625	6.8375	46.75141
20	116.8	117.695	-0.895	0.801025
SUM	1755.2	1756.64	-1.4389	1539.56

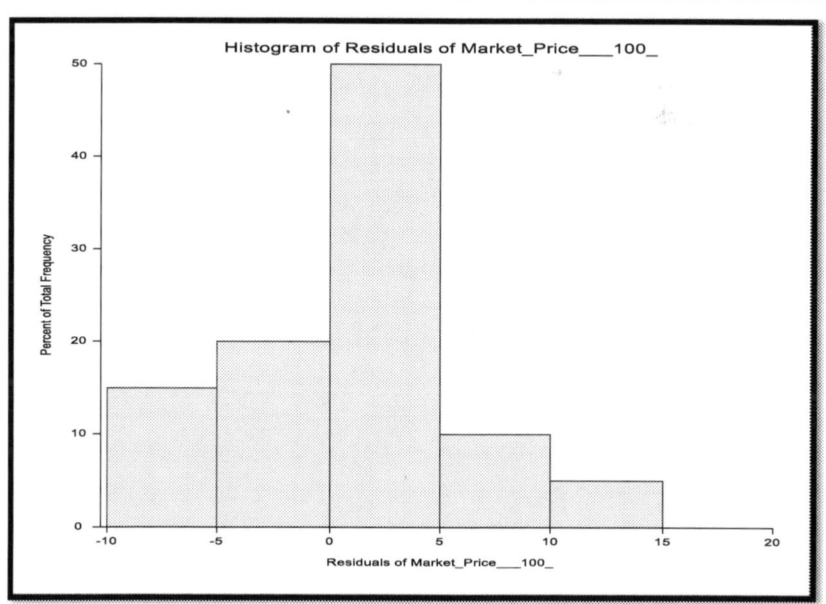

Figure 6.7 Histogram of residual of market price.

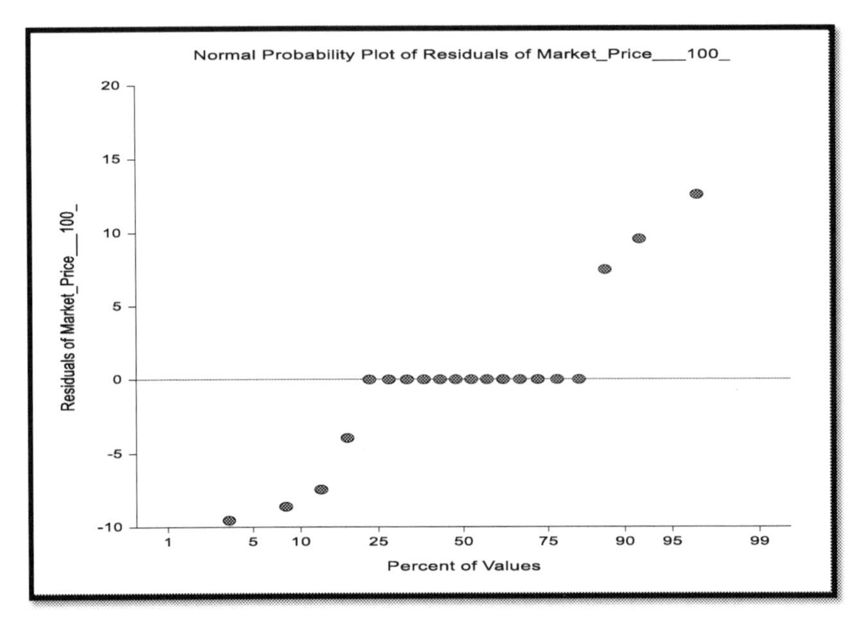

Figure 6.8 Normal probability plot of residuals of market price.

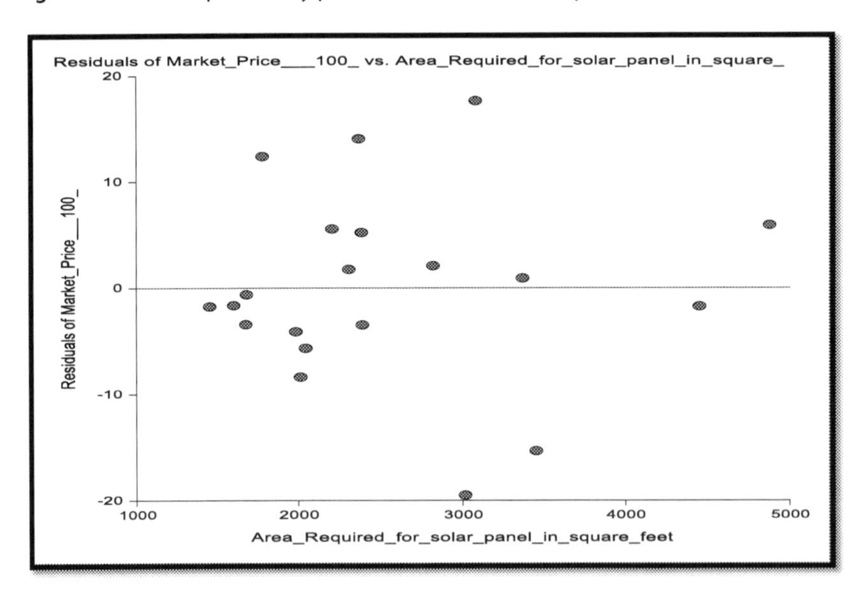

Figure 6.9 Residuals of market price v/s area required for solar panel in squares.

To estimate error at first the residuals are evaluated which is the difference between estimated and actual output. Suppose actual output is representing as y and estimated output those calculated from regression mode as given in

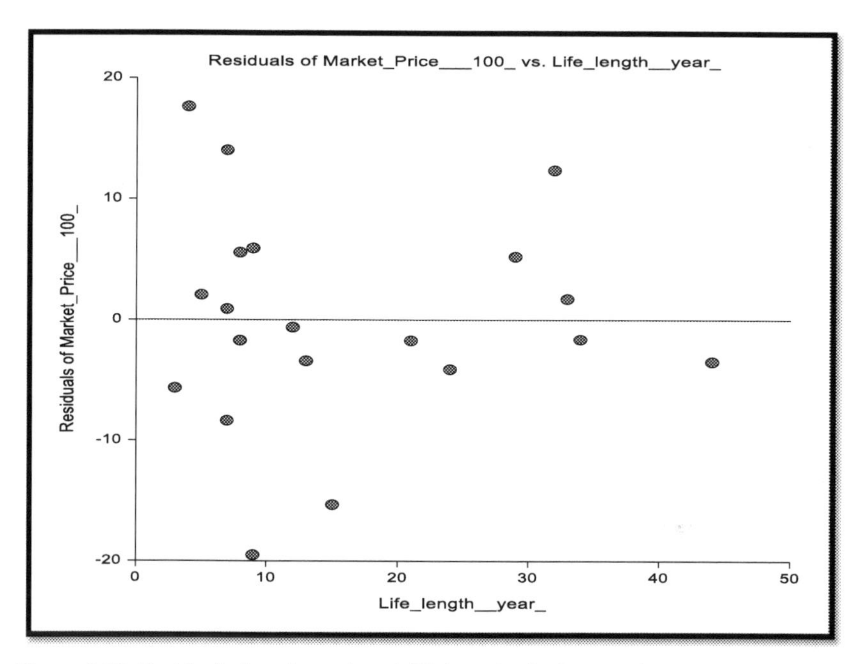

Figure 6.10 Residual of market price v/s life length of solar panel.

table- is represented as$\hat{y}$. The residuals will become:

$$\text{Residuals} = y - \hat{y}$$

From this table it is clear that SSE is

$$SSE = 1539.56$$

As a measure of error, SSE has a limited application. It is, nevertheless, a tool for determining other, more helpful metrics. The standard error ($\text{std}_{\text{error}}$) of the estimate, which is basically the residual standard deviation from the regression model, is one of them. The error terms are assumed to be nearly normally distributed with a mean of zero in regression analysis. With this data and the empirical criterion, about 68 percent of the residuals should be within ± 1 $\text{std}_{\text{error}}$, and 95 percent within ± 2 $\text{std}_{\text{error}}$. Because of this characteristic, the standard error of the estimate may be used to evaluate how well a regression model fits the data.

The $\text{std}_{\text{error}}$ of estimated output of regression models is illustrated by the following equation:

$$\text{std}_{\text{error}} = \sqrt{\frac{\text{SSE}}{n - d - 1}}$$

Table 6.19 Data for pricing of solar-wind hybrid renewable energy system.

Hybrid renewable y ($100)	Solar panel x_1 ($)	Wind turbine x_2 ($100)	Energy devices x_3 ($)
160.1	63.12	3.4	38.8
307	92.3	10.1	60
612	100.2	19.6	70.6
459	83.2	9.5	75
375	71.8	7	76.8
423	75.5	10.4	80
360	65.7	9.1	80
317	68	7.1	80
370	65	4.5	73.3
447	83.5	6	111.1
439	121.4	5.5	88.8
378.6	129.9	4.5	89

Where,
n = number of data/observation
d = number of independent variables
For the control panel of solar form example the std_{error} as follows:

$$std_{error} = \sqrt{\frac{1539.56}{20 - 2 - 1}}$$

$$std_{error} = 9.52$$

Now, according to empirical rule, around 68 percent of the residual should be lying between $\pm 1\ std_{error} = \pm 9.52$ and around 95 percent of the residuals lies between $\pm 2\ std_{error} = \pm 19.03$.

Example 6.7. The bureau of energy efficiency produces data on the price of a hybrid renewable system. Use these data and multiple regression to produce model to predict the average price of a hybrid renewable system from the other variables i.e. price of solar panel, price of energy storing/releasing devices as shown in Table 6.19.

Solution: Table 6.20 shows the assessment of different parameters of a hybrid renewable energy system. To design a regression model of such three independent variables extent model of two variable regressions as:

$$\hat{y} = b_0 + b_1 x_1 + b_2 x_2 + b_3 x_3$$

Table 6.20 Assessment of different parameter.

S.N.	y	x_1	x_2	x_3	x_1y	x_1^2	x_1x_2	x_1x_3	x_2y	x_2^2	x_2x_3	x_3y	x_3^2
1	160.1	63.12	3.4	38.8	10105.51	3984.134	214.608	2449.056	544.34	11.56	131.92	6211.88	1505.44
2	307	92.3	10.1	60	28336.1	8519.29	932.23	5538	3100.7	102.01	606	18420	3600
3	612	100.2	19.6	70.6	61322.4	10040.04	1963.92	7074.12	11995.2	384.16	1383.76	43207.2	4984.36
4	459	83.2	9.5	75	38188.8	6922.24	790.4	6240	4360.5	90.25	712.5	34425	5625
5	375	71.8	7	76.8	26925	5155.24	502.6	5514.24	2625	49	537.6	28800	5898.24
6	423	75.5	10.4	80	31936.5	5700.25	785.2	6040	4399.2	108.16	832	33840	6400
7	360	65.7	9.1	80	23652	4316.49	597.87	5256	3276	82.81	728	28800	6400
8	317	68	7.1	80	21556	4624	482.8	5440	2250.7	50.41	568	25360	6400
9	370	65	4.5	73.3	24050	4225	292.5	4764.5	1665	20.25	329.85	27121	5372.89
10	447	83.5	6	111.1	37324.5	6972.25	501	9276.85	2682	36	666.6	49661.7	12343.21
11	439	121.4	5.5	88.8	53294.6	14737.96	667.7	10780.32	2414.5	30.25	488.4	38983.2	7885.44
12	378.6	129.9	4.5	89	49180.14	16874.01	584.55	11561.1	1703.7	20.25	400.5	33695.4	7921
sum	**4647.7**	**1019.62**	**96.7**	**923.4**	**405871.6**	**92070.9**	**8315.378**	**79934.19**	**41016.84**	**985.11**	**7385.13**	**368525.4**	**74335.58**

Using the following equation to obtain variables b_0, b_1, b_2 and b_3 as:

$$b_0 n + b_1 \sum x_1 + b_2 \sum x_2 + b_3 \sum x_3 = \sum y$$

$$\sum x_1 y = b_0 \sum x_1 + b_1 \sum x_1^2 + b_2 \sum x_1 x_2 + b_3 \sum x_1 x_3$$

$$\sum x_2 y = b_0 \sum x_2 + b_1 \sum x_1 x_2 + b_2 \sum x_2^2 + b_3 \sum x_2 x_3$$

Further multiply x_3 on both the side,

$$\sum x_3 y = b_0 \sum x_3 + b_1 \sum x_1 x_3 + b_2 \sum x_2 x_3 + b_3 \sum x_3^2$$

Using the table, equation for regression can be written as:

$$12b_0 + 1019.62b_1 + 96.7b_2 + 923.2b_3 = 4647.7$$

$$1019.62b_0 + 92070.9b_1 + 8315.378b_2 + 79934.19b_3 = 405871.6$$

$$96.7b_0 + 8315.378b_1 + 985.11b_2 + 7385.13b_3 = 41016.84$$

$$923.4b_0 + 79934.19b_1 + 7385.13b_2 + 74335.58b_3 = 368525.4$$

Solve these equations simultaneously

$$b_0 = -73.919$$

$$b_1 = 0.8029$$

$$b_2 = 17.7832$$

$$b_3 = 3.246$$

So the regression model is:

$$\hat{y} = -73.919 + 0.8029x_1 + 17.7832x_2 + 3.246x_3$$

According to the above assessment the relationship between different parameters is given by

Pricing of hybrid renewable (\$100) – 73.919 + 0.8029 Solar Panel +17.7832 Wind Turbine + 3.246 Energy Devices

Example 6.8. The USA department of agriculture publishes data annually on various selected farms whose use solar pumping system. Shown here is

Table 6.21 Data of using solar pumping system in last 10 year in terms of number of panel and area of farm.

Power saving (KW) y	Number of solar Panel x_1	Area of farm (Square meter) x_2
133.4	23	1500
120.4	34	2400
177.3	30	1600
132.6	49	1884
160.4	31	1789
191.5	75	3100
188.7	67	2560
210.9	78	2188
230.8	110	3079
110.2	90	4756

the power saving for 10 years corresponding to the number of panels and area of farm in square feet. Use these data (Table 6.21) and multiple regression analysis to predict power saving by the number of solar panels according to size of farm.

Solution:

The regression model is evaluated using Table 6.22:

$$\hat{y} = b_0 + b_1x_1 + b_2x_2$$

Using this table written following equations:

$$b_0n + b_1\sum x_1 + b_2\sum x_2 = \sum y$$

$$10b_0 + 587b_1 + 24856b_2 = 1656.2$$

$$\sum x_1y = b_0\sum x_1 + b_1\sum {x_1}^2 + b_2\sum x_1x_2$$

$$587b_0 + 42345b_1 + 1653289b_2 = 102712.2$$

$$\sum x_2y = b_0\sum x_2 + b_1\sum x_1x_2 + b_2\sum {x_2}^2$$

$$24856b_0 + 1653289b_1 + 70370698b_2 = 4082430$$

Solve these simultaneous equations:

$b_0 = 171.0673$

$b_1 = 1.7931$

$b_2 = -0.0445$

Table 6.22 Illustration of regression model.

S.N.	y	x1	x2	x1y	x2y	x1ˆ2	x2ˆ2	x1 × 2
1	133.4	23	1500	3068.2	200100	529	2250000	34500
2	120.4	34	2400	4093.6	288960	1156	5760000	81600
3	177.3	30	1600	5319	283680	900	2560000	48000
4	132.6	49	1884	6497.4	249818.4	2401	3549456	92316
5	160.4	31	1789	4972.4	286955.6	961	3200521	55459
6	191.5	75	3100	14362.5	593650	5625	9610000	232500
7	188.7	67	2560	12642.9	483072	4489	6553600	171520
8	210.9	78	2188	16450.2	461449.2	6084	4787344	170664
9	230.8	110	3079	25388	710633.2	12100	9480241	338690
10	110.2	90	4756	9918	524111.2	8100	22619536	428040
Sum	**1656.2**	**587**	**24856**	**102712.2**	**4082430**	**42345**	**70370698**	**1653289**

Table 6.23 Data for solar radiation according to weather condition.

Power output (MW/year)	Solar radiation in summer (kwh/m²/day)	Solar radiation in raining (kwh/m²/day)
100	10.9	6.6
300	12.3	10.9
150	7.8	4.3
210	11.9	7.8
220	9.8	6.9
310	14.3	9.7
290	10.3	7.7
190	8.2	5.7
124	8.8	6.2
254	13.7	8.9

Table 6.24 Calculation of regression of solar radiation according to weather condition.

S.N.	y	x_1	x_2	yx_1	yx_2	$x_1{}^2$	$x_2{}^2$	x_1x_2
1	100	10.9	6.6	1090	660	118.81	43.56	71.94
2	300	12.3	10.9	3690	3270	151.29	118.81	134.07
3	150	7.8	4.3	1170	645	60.84	18.49	33.54
4	210	11.9	7.8	2499	1638	141.61	60.84	92.82
5	220	9.8	6.9	2156	1518	96.04	47.61	67.62
6	310	14.3	9.7	4433	3007	204.49	94.09	138.71
7	290	10.3	7.7	2987	2233	106.09	59.29	79.31
8	190	8.2	5.7	1558	1083	67.24	32.49	46.74
9	124	8.8	6.2	1091.2	768.8	77.44	38.44	54.56
10	254	13.7	8.9	3479.8	2260.6	187.69	79.21	121.93
Sum	**2148**	**108**	**74.7**	**24154**	**17083.4**	**1211.54**	**592.83**	**841.24**

The regression model is: Power saving (KW) = 187–0.2 (number of solar panels) + 0.000104 (area of farm in square feet)

Example 6.9. A Solar farm organising company of Gujrat in India publishes data about power generation according to solar radiation. Among the variables reported by this organising are the weather conditions i.e. summer and raining. Shown here are the data for these three variables over a ten year of periods. Use the data (given in Table 6.23) to develop a regression model to predict the solar radiation according to weather condition. Comment on regression model and its standard error of estimation.

Solution: Table 6.24 shows the calculation of regression of solar radiation, according to weather condition.

Using this table written following equations:

$$b_0 n + b_1 \sum x_1 + b_2 \sum x_2 = \sum y$$

$$10b_0 + 108b_1 + 74.7b_2 = 2148$$

$$\sum x_1 y = b_0 \sum x_1 + b_1 \sum {x_1}^2 + b_2 \sum x_1 x_2$$

$$108b_0 + 1211.54b_1 + 841.24b_2 = 24154$$

$$\sum x_2 y = b_0 \sum x_2 + b_1 \sum x_1 x_2 + b_2 \sum {x_2}^2$$

$$74.7b_0 + 841.24b_1 + 592.83b_2 = 17083.4$$

$$b_0 = 14.5$$

$$b_1 = -6.6$$

$$b_2 = 36.3$$

The regression model will be

Power Generation = 14.5 − 6.6 (Wind Velocity in Summer)
+ 36.3 (Wind Velocity in Raining)

To determine the standard error evaluate the sum of square error as given in Table 6.25:

Table 6.25 Calculation of error between actual and estimated outputs.

S.N.	y		Residuals	
1	100	182.14	-82.14	6746.98
2	300	328.99	-28.99	840.4201
3	150	119.11	30.89	954.1921
4	210	219.1	-9.1	82.81
5	220	200.29	19.71	388.4841
6	310	272.23	37.77	1426.573
7	290	226.03	63.97	4092.161
8	190	167.29	22.71	515.7441
9	124	181.48	-57.48	3303.95
10	254	247.15	6.85	46.9225
Sum	2148	2143.81	4.19	18398.2

The result of SSE=18398.2.

6.4 Time series forecasting

Many prediction issues include a temporal component, necessitating the extrapolation or forecasting of time series data for solar energy system. One of the most often used data science approaches which is used to predict the value of solar radiation, wind velocity, energy generation through solar power plant and load demand of the consumer. Forecasting entails fitting models to past data on solar energy system and then utilizing those models to forecast future events of solar energy system. On the basis of what has already occurred, the future is forecasted or predicted. The addition of a time order dependency between observations is added by using a time series. This dependency serves as a constraint as well as a structure that offers extra information. Time series forecasting is a method for anticipating occurrences over a period of time to estimate the power generation using the solar system. It forecasts future occurrences by studying previous trends and assuming that future trends will follow in the footsteps of existing trends.

6.4.1 Time series components

The components of a time series are the different reasons or forces that impact the values of an observation in a time series. The four types of time series components are as follows, which is used in the solar energy system.

a) **Trends:** The trend depicts the data's overall propensity to rise or decrease over a lengthy period of time. A trend is an average, smooth, long-term tendency. It is not necessarily required for the growth or drop to be in the same direction throughout a certain time period. In different periods of time, the tendency might be shown to rise, diminish, or remain steady. The general trend, on the other hand, must be upward, downward, or steady. Solar photovoltaic capacity climbed by 95 GW in 2017, with new installations increasing by 34% year over year. By the end of the year, total installed capacity had surpassed 401 GW, enough to meet 2.1 percent of global electricity demand. This increase was enormous, and experts saw it as a critical step toward meeting the world's climate change goals. Machine learning and artificial intelligence are two trends to watch in emerging solar energy technology solutions, such as microgrid controllers and artificial intelligence (AI). This enables the technology to adapt to the needs of the business and evolve in tandem with changing solar energy trends. New software is helping to shape the future of how businesses might use artificial intelligence and machine learning in solar energy technology as new technology emerges to satisfy the rising needs

Table 6.26 Load demand data of a solar power plant.

2.1	5.1	14.1	13.1	8.1	7.1	8.2	15.4	10.4	9.4	10.5	9.5
2.2	5.2	14.2	13.2	8.2	7.2	8.3	15.5	10.5	9.5	10.6	10.7
2.3	5.3	14.3	13.3	8.3	7.3	8.4	15.6	10.6	9.6	10.7	10.8
2.4	5.4	14.4	13.4	8.4	7.4	8.5	15.7	10.7	9.7	10.8	10.9
2.5	5.5	14.5	13.5	8.5	7.5	8.6	15.8	10.8	9.8	10.9	11
2.6	5.6	14.6	13.6	8.6	7.6	8.7	15.9	10.9	9.9	11	11.1
2.7	5.7	14.7	13.7	8.7	7.7	8.8	16	11	10	11.1	11.2
2.8	5.8	14.8	13.8	8.8	7.8	8.9	12.1	11.1	6.1	11.2	11.3
2.9	5.9	14.9	13.9	8.9	7.9	9	12.2	11.2	6.2	11.3	11.4
3	6	15	14	9	8	9.1	12.3	11.3	6.3	11.4	11.5
3.1	16.1	15.1	10.1	9.1	10.2	9.2	12.4	11.4	6.4	11.5	11.6
3.2	16.2	15.2	10.2	9.2	10.3	9.3	12.5	11.5	6.5	11.6	11.7
3.3	16.3	15.3	10.3	9.3	10.4	9.4	12.6	11.6	6.6	11.7	11.8
3.4	16.4	15.4	10.4	9.4	10.5	9.5	12.7	11.7	6.7	11.8	11.9
3.5	16.5	15.5	10.5	9.5	10.6	9.6	12.8	11.8	6.8	11.9	12
3.6	16.6	15.6	10.6	9.6	10.7	9.7	12.9	11.9	6.9	12	8.1
3.7	16.7	15.7	10.7	9.7	10.8	9.8	13	12	7	8.1	8.2
3.8	16.8	15.8	10.8	9.8	10.9	9.9	13.1	8.1	7.1	8.2	8.3
3.9	16.9	15.9	10.9	9.9	11	10	13.2	8.2	7.2	8.3	8.4
4	17	16	11	10	11.1	6.1	13.3	8.3	7.3	8.4	8.5
4.1	17.1	12.1	11.1	6.1	11.2	6.2	13.4	8.4	7.4	8.5	8.6
4.2	17.2	12.2	11.2	6.2	11.3	6.3	13.5	8.5	7.5	8.6	8.7
4.3	17.3	12.3	11.3	6.3	11.4	6.4	13.6	8.6	7.6	8.7	8.8
4.4	17.4	12.4	11.4	6.4	11.5	6.5	13.7	8.7	7.7	8.8	8.9
4.5	17.5	12.5	11.5	6.5	11.6	6.6	13.8	8.8	7.8	8.9	9
4.6	17.6	12.6	11.6	6.6	11.7	6.7	13.9	8.9	7.9	9	9.1
4.7	17.7	12.7	11.7	6.7	11.8	6.8	14	9	8	9.1	9.2
4.8	17.8	12.8	11.8	6.8	11.9	6.9	10.1	9.1	10.2	9.2	9.3
4.9	17.9	12.9	11.9	6.9	12	15.2	10.2	9.2	10.3	9.3	9.4
5	18	13	12	7	8.1	15.3	10.3	9.3	10.4	9.4	9.5
9.1	9.2	9.3	9.4	9.5							

of the solar energy industry and the businesses who use it as an energy-efficient source. Table 6.26 shows the data of load demand of 365 days, which is fulfilled by the particular solar power plant. Fig. 6.11 shows the trend analysis of a given data set of load demand, which is developed by the NCSS tool.

b) Cycles: Cyclic variations are changes in a time series that occur over a longer period of time than a year. This rhythmic movement has a year-long duration of the oscillation. A cycle is made up of one whole era. The 'Business Cycle' is a term used to describe this periodic movement. Prosperity, recession, depression, and recovery are the four phases of

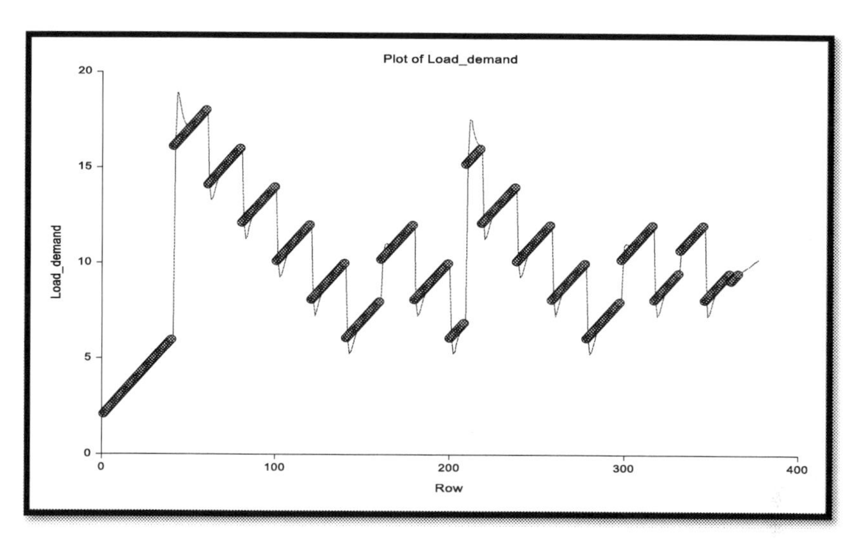

Figure 6.11 Uneven trend of 365 days of load demand (kW).

this four-phase cycle. Although the cyclic fluctuation is regular, it is not periodic. The nature of the economic forces and their interaction determine the ups and downs in business.

c) **Seasonal effects**: These are the rhythmic forces that act in a predictable and predictable pattern during a period of less than a year. During a 12-month period, they have the same or almost the same pattern. If the data is captured hourly, daily, weekly, quarterly, or monthly, this fluctuation will be evident in the time series. Natural forces or man-made norms are both responsible for these variances. Seasonal fluctuations are influenced by the different seasons or climatic conditions. The sale of umbrellas and raincoats increases during the rainy season, whereas the sale of electric fans and air conditioners increases during the summer.

d) **Irregular spectral fluctuation:** Another factor contributes to the variance in the variable under investigation. They are not regular variations, but rather random or irregular variances. Unexpected, uncontrolled, unpredictable, and erratic variations characterize these oscillations. Earthquakes, wars, floods, famines, and other natural calamities are examples of these forces.

6.4.2 The forecasting error measurement

The discrepancy between an actual value and its predicted is referred to as a forecast "error." The term "error" here does not refer to a blunder;

Table 6.27 Data for actual and forecasted solar radiation.

year	Actual radiation (average)	Forecast radiation	Error
2011	14		
2012	14.32	14	0.32
2013	15.32	14.12	1.2
2014	16.11	15.9	0.21
2015	16.5	16.1	0.4
2016	16.9	16.7	0.2
2017	17	17.5	-0.5
2018	18.1	18.5	-0.4
2019	18.3	17.89	0.41
2020	19.1	18.99	0.11

rather, it refers to the unpredictability of an observation. Forecast mistakes are distinguished from residuals in two ways. The training set is used to compute residuals, whereas the test set is used to calculate forecast errors. Second, residuals are based on single-step forecasts, whereas forecast mistakes might entail several steps.

We may assess forecast accuracy by combining forecast mistakes in a variety of ways.

(a) Mean absolute deviation (MAD): The mean absolute deviation, or MAD, is one measure of total predicting inaccuracy. The mean or average of the absolute values of the mistakes is known as the MAD. Table 6.27 represents the average solar radiation data in India over a period of 10 years along with the forecast data of each year and error between the actual and forecast value. The Table 6.27 demonstrates that some of the errors are positive and some are negative. By taking the absolute value of the error measurement and evaluating the size of the prediction mistake without respect to direction, the mean absolute deviation solves this problem.

$$\text{MAD} = \frac{\sum |\text{error}_i|}{\text{number of forecast values}}$$

Now MAD for the forecast error of the above table is as follows:

$$\text{MAD} = \frac{|0.32|+|1.2|+|0.21|+|0.4|+|0.2|+|-0.5|+|-0.4|+|0.41|+|0.11|}{9}$$

$$= 0.417$$

So in this case mean absolute deviation of solar radiation is 0.417.

(b) Mean square error (MSE): Another method to avoid the cancelling effects of positive and negative forecast mistakes to estimate forecast data i.e. Solar radiation is to use the MSE. By squaring each error and averaging the squared errors, the MSE is calculated. The formula below officially expresses it.

$$\text{MSE} = \frac{\sum \text{error}_i^2}{\text{number of forecasts values}}$$

The mean square error for the above forecast error is as:

$$\text{MSE} = \frac{2.3767}{9} = 0.2641$$

6.4.3 Smoothing techniques

There are various approaches to forecasting time series data that are stationary or do not have a strong trend. Seasonal impacts of cyclical ore because they create predictions by smoothing out the uneven fluctuation effects in time series data, these approaches are commonly referred to as smoothing techniques. Table 6.28 shows the data of 365 days of solar radiation (kwh/m^2/day). Table 6.29 shows the exponential smoothing report of forecast summary of solar radiation, which is developed through the NCSS tool. Fig. 6.12 shows the uneven fluctuation of solar radiation in terms of exponential smoothing plot of solar radiation, which is developed through the NCSS tool.

The common categories of smoothing techniques are as follows:

(a) Average methods: Some predictions have a function of irregular data fluctuations, resulting in over-steered forecasts. A forecaster who uses the average technique inserts data from various time periods into the forecast and smoothes it out. The average technique calculates the forecast for the following time period by averaging data from many time periods and using the average as the forecast.

Simple averages & moving averages: The simple average model is the most basic of the averaging models to forecast the power generation using solar energy system. The prediction for the time period 'n' in this model is the average of the values for a specified number of prior time periods:

Assume we're trying to estimate the price of a PV cell in June of year 3 using the averages forecasting approach. Would we still anticipate for May of year 3 using the simple average from May of year 2 to April of the year 3 as we did in May of year 2? Instead of utilizing the same 12-month average that was used to anticipate a May of the year 3, it would appear to make

Table 6.28 Data of 365 days of solar radiation (kwh/m²/day).

3.7	4.9	4.15	4.35	4.55	4.75	4.95	5.15	3.36	8.56	4.32	8.32	2.5	2.7	2.9	3.1	3.3	3.5	3.9
3.71	3.5	4.16	4.36	4.56	4.76	4.96	5.16	3.62	8.82	4.52	8.52	2.51	2.71	2.91	3.11	3.31	3.51	3.91
3.72	3.65	4.17	4.37	4.57	4.77	4.97	5.17	3.88	9.08	4.72	8.72	2.52	2.72	2.92	3.12	3.32	3.52	3.92
3.73	3.8	4.18	4.38	4.58	4.78	4.98	5.18	4.14	9.34	4.92	8.92	2.53	2.73	2.93	3.13	3.33	3.53	3.93
3.74	3.95	4.19	4.39	4.59	4.79	4.99	5.19	4.4	9.6	5.12	9.12	2.54	2.74	2.94	3.14	3.34	3.54	3.94
3.75	4.1	4.2	4.4	4.6	4.8	5	5.2	4.66	9.86	5.32	9.32	2.55	2.75	2.95	3.15	3.35	3.55	
3.76	4.25	4.21	4.41	4.61	4.81	5.01	5.21	4.92	10.12	5.52	9.52	2.56	2.76	2.96	3.16	3.36	3.56	
3.77	4.4	4.22	4.42	4.62	4.82	5.02	5.22	5.18	10.38	5.72	9.72	2.57	2.77	2.97	3.17	3.37	3.57	
3.78	4.55	4.23	4.43	4.63	4.83	5.03	5.23	5.44	10.64	5.92	9.92	2.58	2.78	2.98	3.18	3.38	3.58	
3.79	4.7	4.24	4.44	4.64	4.84	5.04	5.24	5.7	2.12	6.12	10.12	2.59	2.79	2.99	3.19	3.39	3.59	
3.8	4.85	4.25	4.45	4.65	4.85	5.05	5.25	5.96	2.32	6.32	10.32	2.6	2.8	3	3.2	3.4	3.6	
3.81	5	4.26	4.46	4.66	4.86	5.06	5.26	6.22	2.52	6.52	10.52	2.61	2.81	3.01	3.21	3.41	3.61	
3.82	5.15	4.27	4.47	4.67	4.87	5.07	5.27	6.48	2.72	6.72	10.72	2.62	2.82	3.02	3.22	3.42	3.62	
3.83	5.3	4.28	4.48	4.68	4.88	5.08	5.28	6.74	2.92	6.92	10.92	2.63	2.83	3.03	3.23	3.43	3.63	
3.84	5.45	4.29	4.49	4.69	4.89	5.09	5.29	7	3.12	7.12	11.12	2.64	2.84	3.04	3.24	3.44	3.64	
3.85	4.1	4.3	4.5	4.7	4.9	5.1	5.3	7.26	3.32	7.32	2.45	2.65	2.85	3.05	3.25	3.45	3.65	
3.86	4.11	4.31	4.51	4.71	4.91	5.11	5.31	7.52	3.52	7.52	2.46	2.66	2.86	3.06	3.26	3.46	3.66	
3.87	4.12	4.32	4.52	4.72	4.92	5.12	5.32	7.78	3.72	7.72	2.47	2.67	2.87	3.07	3.27	3.47	3.67	
3.88	4.13	4.33	4.53	4.73	4.93	5.13	5.33	8.04	3.92	7.92	2.48	2.68	2.88	3.08	3.28	3.48	3.68	
3.89	4.14	4.34	4.54	4.74	4.94	5.14	3.1	8.3	4.12	8.12	2.49	2.69	2.89	3.09	3.29	3.49	3.69	

Table 6.29 Exponential smoothing report of forecast summary of solar radiation (NCSS tool).

Variable	Solar Radiation kwh m² day
Number of Rows	365
Missing Values	None
Mean	4.498849
Pseudo R-Squared	0.861835
Mean Square Error	0.4398433
Mean |Error|	0.1197082
Mean |Percent Error|	3.573057
Alpha Search	Mean Square Error
Alpha	0.9815376
Forecast	3.939812

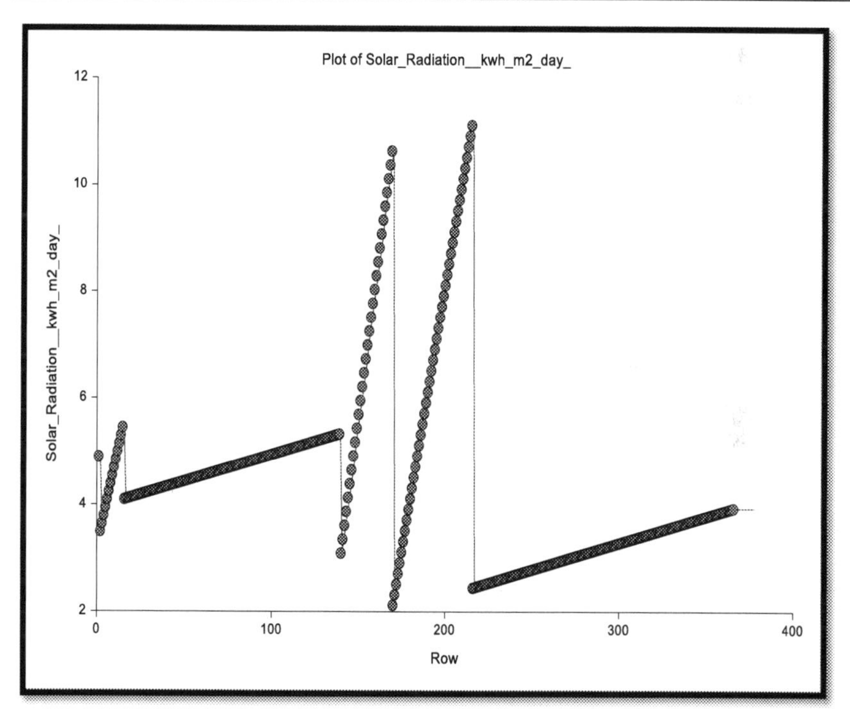

Figure 6.12 Exponential smoothing plot of solar radiation.

more sense to average f for the 12 months preceding June of year 3.

$$M_n = \left(\frac{1}{n}\right)\left[D_{n-(n-1)} + D_{n-(n-2)} + \ldots\ldots\ldots\ldots\ldots + D_{n-2} + D_{n-1} + D_n\right]$$

A moving average is one that is updated or recalculated for each new time period taken into account to estimate the data for solar system. Each

Table 6.30 Demand of number of solar panels for solar power plant for 12 months.

Month	Demand (D_t)
January	70
February	75
March	80
April	98
May	105
June	95
July	85
August	90
September	78
October	75
November	60
December	62

new moving average is based on the most current data. This benefit is counterbalanced by the disadvantages that (1) it is difficult to determine the best amount of time for computing the moving average and (2) moving averages do not generally correct for such time series impacts.

Weighted moving averages: Certain periods of time may be given greater weight in a forecast than others. Weighted moving average is a moving average in which some time periods are weighted differently than others.

$$wt\ MA_t = \frac{\sum_{i=1}^{n} w_i D_i}{\sum_{i=1}^{n} w_i}$$

Where D_i is the actual demand of the period i, n is the number of the periods and W_i is the weight of the data of the time period i.

Example 6.10. The actual demand of Solar panels of each 500W for solar power plant for 12 months is summarized in Table 6.30. Calculate:

(i) Find three month weighted moving averages by assuming the weights $w_1 = 0.2$, $w_2 = 0.3$, $w_3 = 0.4$.

(ii) Compute forecast errors,

(iii) Compute mean square error,

(iv) Mean absolute deviation. Also draw graph to show the actual demand values and the forecast demand value.

Solution: Based on the data of Example 6.10, calculation of moving averages and the error is shown in Table 6.31. Fig. 6.13 shows the graph between actual demand and forecast demand of solar panel.

Table 6.31 Calculation of moving average and error.

Month	Demand (D_t)	Moving average (MA_t)	Forecast (F_t)	Error
January	70			
February	75			
March	80	76.11		
April	98	82.56	76.11	6.45
May	105	97.11	82.56	14.55
June	95	99	97.11	1.89
July	85	92.78	99	-6.22
August	90	89.44	92.78	-3.34
September	78	83.56	89.44	-5.88
October	75	79.33	83.56	-4.23
November	60	69	79.33	-10.33
December	62	64.22	69	-4.78

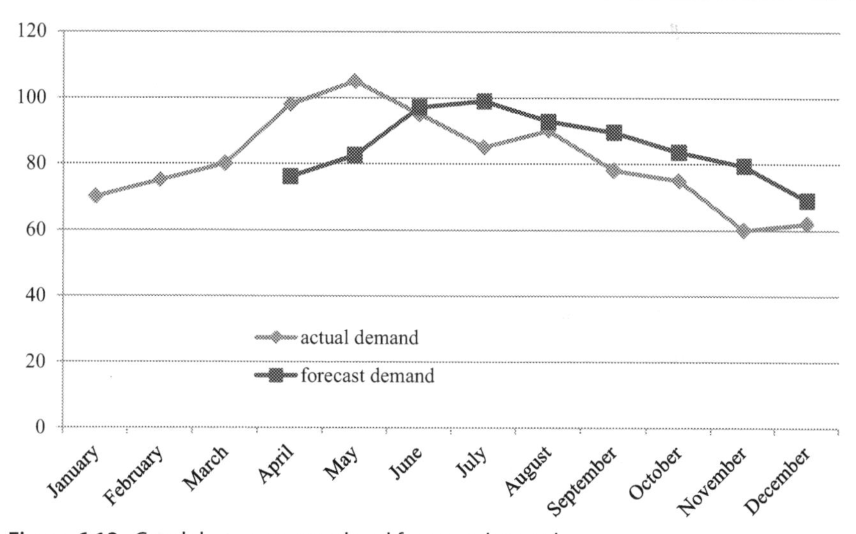

Figure 6.13 Graph between actual and forecast demand.

(i) Three months moving average for solar panel:

$$MA_3 = \frac{w_1 D_1 + w_2 D_2 + W_3 D_3}{W_1 + W_2 + W_3}$$

$$MA_3 = \frac{0.2 \times 70 + 0.3 \times 75 + 0.4 \times 80}{0.2 + 0.3 + 0.4} = \frac{68.5}{0.9} = 76.11 \quad (6.1)$$

$$MA_3 = \frac{0.2 \times 75 + 0.3 \times 80 + 0.4 \times 98}{0.2 + 0.3 + 0.4} = \frac{78.2}{0.9} = 82.56 \quad (6.2)$$

$$MA_3 = \frac{0.2 \times 80 + 0.3 \times 98 + 0.4 \times 105}{0.2 + 0.3 + 0.4} = \frac{87.4}{0.9} = 97.11 \quad (6.3)$$

$$MA_3 = \frac{0.2 \times 98 + 0.3 \times 105 + 0.4 \times 95}{0.2 + 0.3 + 0.4} = \frac{89.1}{0.9} = 99 \quad (6.4)$$

$$MA_3 = \frac{0.2 \times 105 + 0.3 \times 95 + 0.4 \times 85}{0.2 + 0.3 + 0.4} = \frac{83.5}{0.9} = 92.78 \quad (6.5)$$

$$MA_3 = \frac{0.2 \times 95 + 0.3 \times 85 + 0.4 \times 90}{0.2 + 0.3 + 0.4} = \frac{80.5}{0.9} = 89.44 \quad (6.6)$$

$$MA_3 = \frac{0.2 \times 85 + 0.3 \times 90 + 0.4 \times 78}{0.2 + 0.3 + 0.4} = \frac{75.2}{0.9} = 83.56 \quad (6.7)$$

$$MA_3 = \frac{0.2 \times 90 + 0.3 \times 78 + 0.4 \times 75}{0.2 + 0.3 + 0.4} = \frac{71.4}{0.9} = 79.33 \quad (6.8)$$

$$MA_3 = \frac{0.2 \times 78 + 0.3 \times 75 + 0.4 \times 60}{0.2 + 0.3 + 0.4} = \frac{62.1}{0.9} = 69 \quad (6.9)$$

$$MA_3 = \frac{0.2 \times 75 + 0.3 \times 60 + 0.4 \times 62}{0.2 + 0.3 + 0.4} = \frac{57.8}{0.9} = 64.22 \quad (6.10)$$

(ii) Mean forecast error for solar panel:

$$\text{MFE} = \frac{\sum_{i=1}^{n} (D_t - F_t)}{n} = \frac{-11.89}{9} = -1.32$$

(iii) Mean square error for solar panel:

$$\text{MSE} = \text{MFE} = \frac{\sum_{i=1}^{n} (D_t - F_t)^2}{n} = \frac{411.37}{9} = 45.71$$

(iv) Mean absolute deviation for solar panel:

$$\text{MAD} = \frac{\sum_{i=1}^{n} |(D_t - F_t)|}{n} = \frac{57.67}{9} = 6.41$$

Exponential smoothing:

Exponential smoothing is a forecasting approach that uses data from prior time periods to weight items in the prediction with exponentially diminishing relevance. Exponential smoothing is achieved by multiplying the current time period's actual value by a value between 0 and 1, referred to as value 'a,' and then adding the result to the product of the current time period's forecast and the (1-a).

$$F_{t+1} = F_t + a(D_t - F_t)$$

Table 6.32 Data for new units of installed solar energy system.

Year	Total units
2011	567
2012	620
2013	630
2014	780
2015	790
2016	810
2017	830
2018	940
2019	1030
2020	1120

Or

$$F_{t+1} = aD_t + (1 - a)F_t$$

Where, F_{t+1} is the forecast for the next period (t+1), F_t is forecast value, D_t actual value and 'a' is constant between 0 and 1.

Example 6.11. The Canada bureau reports on the total units of new installed solar energy system over a 10 year are given in the Table 6.32. Use exponential smoothing to forecasting the values using a=0.5 and 0.7.

Solution: According to the equation

$$F_{t+1} = F_t + a(D_t - F_t)$$

$$F_3 = 0.5(620) + 0.5(567) = 593.5$$

$$F_4 = 0.5(630) + 0.5(593.5) = 611.75$$

Similarly, all forecast value calculated for both the value of 'a' as shown in Table 6.33.

$$\text{MAD for a} = 0.5 = \frac{\sum_{i=1}^{n} |(D_t - F_t)|}{n} = \frac{936.8086}{9} = 104.08$$

$$\text{MAD for a} = 0.7 = \frac{\sum_{i=1}^{n} |(D_t - F_t)|}{n} = \frac{281.0426}{9} = 31.23$$

Conclusion: After the whole discussion in this chapter, it is finding out regression analysis and forecasting method play very important role in the field of solar energy system. In the solar energy system it is necessary to identify prediction of solar radiation, load demand, number of solar panel

Table 6.33 Calculation of error.

Year	Total units	a = 0.5 Forecast	Error	a = 0.7 Forecast	Error
2011	567				
2012	620	567	53	604.1	15.9
2013	630	593.5	36.5	619.05	10.95
2014	780	611.75	168.25	729.525	50.475
2015	790	695.875	94.125	761.7625	28.2375
2016	810	742.9375	67.0625	789.8813	20.11875
2017	830	776.4688	53.53125	813.9406	16.05938
2018	940	803.2344	136.7656	898.9703	41.02969
2019	1030	871.6172	158.3828	982.4852	47.51484
2020	1120	950.8086	169.1914	1069.243	50.75742
Total	8117	6613.191	936.8086	7268.957	281.0426

required for smooth functioning of solar power plant. It is also identify through this chapter, necessary to develop or measure perfect relationship between different parameters of solar energy system.

6.5 Exercise

1. What is the application of regression analysis in the field of solar energy system?
2. What is the significance of time series assessment in the field of solar energy system?
3. What are the possible correlation of different parameters of solar energy system?
4. Evaluate the coefficient of correlation for the tariff of distribution of solar power to different sectors to fulfill the load requirements as given in **Table**.

Table: Tariff of distribution of solar power to different sectors.

Price of solar power in (\$)/kW	5	8	12	15	18	22	25
Load (KW)	25	23	28	31	29	34	20

5. The following data are the total cost (in \$) for generation of power using solar farm benefits for nine states/country, along with the cost to export the power (in \$) that the generation company had in assets in those states given in table.

Table: Data for total cost and export cost for generation of power.

State	Cost for solar power generation	Cost for export the solar power
Colorado	20422	367
Ireland	8082	100
Florida	6048	130
Spain	7734	334
Texas	5578	90
Germany	8806	260
Denmark	6523	99
India	6896	80
China	9543	245

Use the data to compute a correlation coefficient to determine the correlation between above two parameters.

6. The Bureau of Energy Efficiency of India released the following data on the utilization of power (in KW) from solar energy system for several industries in three recent years given in table.

Table: Data for utilization of solar power in recent three years.

	Solar power (in KW) utilization in recent three year		
Industries	Year 1	Year 2	Year 3
Textile	3	4	5.5
Chemical	2.5	1.7	2
Communication	4	3.8	5.2
Machinery	4.5	4	5
Food	2	2	3
Government sectors	3	4	4
Pharmaceutical	3.3	2.9	4
Hotels	4.5	3	4

Compare coefficient of correlation for each pair of year and evaluate which years are most highly correlated.

7. Sketch a scatter plot from the following data, and determine the equation of the regression line of the number of solar panel used in last five years to generate solar energy at a particular location, which provide electricity for automobile industry as given in Table.

Table: Year-wise number of PV cell.

Year	2010	2011	2012	2013	2014
Number of solar panel	42	55	51	55	68

8. One of the solar manufacturing company design solar panel of different ratings. As an aid in long-term planning they gather the following rating and number of panel required from several sectors. Develop the equation of the simple regression line to predict the number of panel of different rating given in table.

Table: Data for solar power rating according to number of panel.

Solar panel rating in Watt	Number of panel
200	1000
300	2080
500	800
700	980
800	600
1000	450
1200	398
1500	277

9. A study was conducted in a small Louisiana city to determine what variable, if any are related to the market price of control panel for solar power plant. Several variables were explored, including number of solar panel, square feet of space available and life length of panel etc. suppose the researcher wants to develop a regression model to predict the market price of a panel by two variable, i.e., area in square feet and the life length. The data for these three variables are given in the table.

Table: Data for market price of control panel for solar plant.

Market Price ($100) (y)	Area Required for solar panel in square feet (x_1)	Life length (year) (x_2)
73	3603	34
74.3	3387	44
88.3	3453	31
85.4	3303	33
81.1	3984	34
74.8	3678	13
79.3	4015	9
78.3	1683	13
83.5	3380	39
79.7	1780	33
83.9	3043	3
78.3	3013	7

(continued on next page)

Market Price ($100) (y)	Area Required for solar panel in square feet (x_1)	Life length (year) (x_2)
85.3	3450	15
94	3300	8
100.7	3817	5
105.3	3363	7
130.4	3078	4
105.6	3367	7
129.5	4875	9
116.8	4450	8

10. The bureau of energy efficiency produces data on the price of hybrid renewable system. Use these data and multiple regression to produce model to predict the average price of hybrid renewable system from the other variables i.e. price of solar panel, price of energy storing/releasing devices as shown in table.

Table: Data for pricing of solar-wind hybrid renewable energy system.

Hybrid renewable y ($100)	Solar panel x_1 ($)	Wind turbine x_2 ($100)	Energy devices x_3 ($)
160.1	63.13	3.4	38.8
307	93.3	10.1	60
613	100.3	19.6	70.6
459	83.3	9.5	75
375	71.8	7	76.8
433	75.5	10.4	80
360	65.7	9.1	80
317	68	7.1	80
370	65	4.5	73.3
447	83.5	6	111.1
439	131.4	5.5	88.8
378.6	139.9	4.5	89

The USA department of agriculture publishes data annually on various selected farm whose use solar pumping system. Shown here is the power saving for 10 years corresponding to number of panels and area of farm in square feet. Use these data (Table) and multiple regression analysis to predict power saving by the number of solar cell according to size of farm.

Table: Data of using solar pumping system in last 10 year in terms of number of panel and area of farm.

Power saving (KW) y	Number of solar panel x_1	Area of farm (square meter) x_2
233.4	23	2500
220.4	34	2400
277.3	30	2600
232.6	49	2884
260.4	32	2789
292.5	75	3200
288.7	67	2560
220.9	78	2288
230.8	220	3079
110.2	90	4756

A solar farm organizing company of Gujarat in India publishes data about power generation according to solar radiation. Among the variables reported by this organizing are the weather conditions, that is, summer and raining. Shown here are the data for these three variables over a ten year of periods. Use the data (given in Table) to develop a regression model to predict the solar radiation according to weather condition. Comment on regression model and its standard error of estimation.

Table: Data for solar radiation according to weather condition.

Power output (MW/year)	Solar radiation in summer ($kwh/m^2/day$)	Solar radiation in raining ($kwh/m^2/day$)
200	20.9	6.6
300	22.3	20.9
250	7.8	4.3
220	22.9	7.8
220	9.8	6.9
320	24.3	9.7
290	20.3	7.7
290	8.2	5.7
224	8.8	6.2
254	23.7	8.9

References

[1] F.J. Ardakani, G. Riahy, M. Abedi, Design of an optimum hybrid renewable energy system considering reliability indices, in: 18th Iranian conference on electrical engineering, Isfahan University of Technology, Iran, 2010, pp. 842–847. 11–13 May.
[2] JMS. Cristobal, A multi criteria data envelopment analysis model to evaluate the efficiency of the renewable energy technologies, Renew. Energ. 36 (2011) 2742–2746.
[3] R. Madlener, C.H. Antunes, LC. Dias, Assessing the performance of biogas plants with multicriteria and data envelopment analysis, Eur. J. Operat. Res. 197 (2006) 1084–1094.
[4] L. Jose, B. Agustin, R. DufoLopez, Simulation and optimization of standalone hybrid renewable energy system, Renew. Sustain. Energ. Rev. 13 (2009) 2011–2018.
[5] S. Khare Nema, P. Baredar, Optimization of hybrid renewable energy system by HOMER, PSO and CPSO for the study area, Int. J. Sustain. Energ. 36 (2017) 326–343.
[6] S. Khare Nema, P. Baredar, Optimization of hydrogen based hybrid renewable energy system using HOMER, BB, BC AND GAMBIT, Int. J. Hydrog. Energ. 41 (2016) 16743–16751.
[7] H. Checkoway, N. Pearce, J.M. Dement, Design and conduct of occupational epidemiology studies: I. Analysis of cohort data, Am. J. Ind. Med. 15 (1989) 375–394.
[8] E.L. Frome, H. Checkoway, Epidemiologic programs for computers and calculators. Use of Poisson regression models in estimating incidence rates and ratios, Am. J. Epidemiol. 121 (1985) 309–323.
[9] R.M. Hirsch, D.R. Helsel, T.A. Cohn, E.J. Gilroy, Statistical analysis of hydrologic data, in: DR Maidment (Ed.), Handbook of Hydrology, McGraw-Hill, New York, 1993 Chap. 17.
[10] S. Ikeda, An Integrated Risk Analysis Framework for Emerging Disaster Risks: Towards a Better Risk Management of Flood Disaster in Urban Communities, Terrapub, Tokyo, 2006.
[11] S.W. Jung, D.H. Lee, Y.J. Moon, K.H. Kim, Potential flood damage (PFD) assessment, in: Proceedings of the Korea water resources association conference (in Korean), 2001, pp. 601–606.
[12] H.S. Kim, Potential risk and damage estimation of urban flood II: MOCT core technology development project report. Ministry of Construction and Transportation, Korea (2006) (in Korean).
[13] N.T. Kottegoda, R. Rosso, Statistics, Probability, and Reliability for Civil and Environmental Engineers, McGraw-Hill, New York, 1997.

CHAPTER SEVEN

Inventory and total quality management of solar energy system

Learning objective

- Understand the basics of inventory and total quality management.
- Know how inventory management increases the performance of the solar industry.
- Learn total quality management of the solar energy system.
- Understand the concept SCADA w.r.t. the solar power plant.
- Discuss about the quality certification and society of solar energy system

7.1 Introduction

Any idle materials that can be put to some future use are referred to as inventory. The inventory has been heavily invested in by manufacturing and a number of service firms. Inventory investment frequently has a direct impact on a company's profitability. The term "inventory" refers to the goods for sale as well as the raw materials used to make those goods. Inventory is one of a company's most valuable assets because inventory turnover is one of the key sources of revenue production and, as a result, earnings for the company's shareholders. Experiences over the last two decades demonstrate that a company's capacity to minimize inventory investment to extremely low levels is critical to its world-class success. Solar panels and their spare parts are often classified as independent demand items in a solar energy system. Because of storage expenses, spoiling costs, and the possibility of obsolescence, having a large amount of inventory for a long time is usually not advantageous for a corporation. Having too little inventory, on the other hand, has significant drawbacks; for example, the company risks losing market share and profit from possible sales.

Total quality management of an organization is intimately tied to inventory management in the process of operational management. In recent years, the tremendous success of a number of industrial and service companies

Decision Science and Operations Management of Solar Energy Systems.
DOI: https://doi.org/10.1016/B978-0-323-85761-1.00001-9

has been inextricably related to good quality management methods. TQM is defined as an organization-wide effort to develop the systems, tools, techniques, skills, and attitudes required to establish a quality assurance system that is responsive to growing market needs. Such an attempt will give a solar system manufacturing company specific benefits and enable it to compete effectively in the solar energy market. A good TQM program of solar energy company consists of the following important elements:

- Role of senior management in the solar energy company
- Employee contribution in continuous improvement of solar energy company
- Organizing training and orientation program for workers of solar energy company
- Tools and techniques for quality assurance and continuous improvement of solar energy system

In the solar energy system, it is necessary to maintain quality management from supply side to demand side, for proper establishment of an individual solar company in the renewable energy market. Fig. 7.1 shows the elements of total quality management of the solar energy system.

There are lots of researchers who work in the field of total quality management. Ciupageanu et al. [1] described for the first time the latest progress in the field of real-time power management algorithms designed for hybrid renewable energy systems. The findings of this research provide a comprehensive review of the state-of-the-art, individuating specific fields of application and focusing on the gaps that should be further investigated. Abbas et al. [2] explained the impact of total quality management on corporate green performance through the mediating role of corporate social responsibility. This study also investigates the role of corporate social responsibility (CSR) in the relationship between TQM and CGP and examines how CSR mediates the relationship between them. Six TQM practices are taken from the American "Malcolm Baldrige National Quality Award" CGP includes green management strategies, green processes and green products performance; and CSR comprises social, employees and customer's dimensions. Todorut et al. [3] described sustainable development of organizations through total quality management. In a dynamic economic environment, the companies have to focus on the achievement of sustained success, which includes the satisfaction of all the interested parties' requirements and expectations. One of researchers also described a new energy grid-connected power quality management system based on the internet of

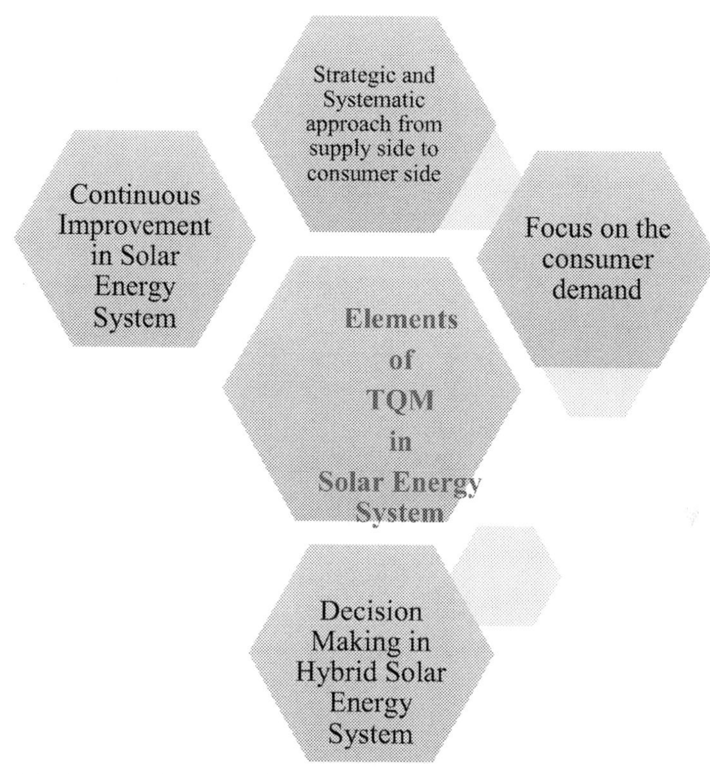

Figure 7.1 Element of TQM in solar energy system.

things. This paper analyzes the improvement of power quality by connecting a group of hybrid reactive power compensation devices which is composed of static reactive power compensator and shunt capacitor. The results show that the voltage deviation is reduced and the harmonic content is reduced. It can be considered that this measure has achieved the purpose of power quality control to a certain extent.

Now this chapter explains the basic concept of Inventory and Total Quality Management in the field of solar energy systems, which is the prominent part of the renewable energy system. This chapter is classified into 7 sections. Section 7.1 describes the introduction of a given objective. Sections 7.2 and 7.3 explained inventory planning and inventory control of the solar energy system respectively. Sections 7.4 and 7.5 explain total quality management and quality certification of solar energy systems. Section 7.6 describes the conclusion of this chapter and chapter end with the exercise and question.

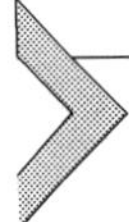

7.2 Inventory planning of independent demand component

Finished solar panels and other solar power plant's spare components are often classified as independent demand items in the solar energy industry. Items that are in constant demand are known as independent demand items. Polycrystalline solar panels will continue to be in high demand and also the demand for polycrystalline solar panels is constant, therefore ensuring continual availability and periodic replacement of a given panel stock is an important part of the planning process.

Appropriate solutions have made inventory management for solar enterprises simple. With barcode scanning capabilities, accurately track goods used in the assembly of solar panels or other equipment. This powerful solution allows you comprehensive control over your stockroom, supplies, and equipment, saving your team time and money when conducting physical inventories. Due to the quick pace of innovation in the solar sector, having too much inventory will be a losing proposition. With each passing year, important developments in solar panel technology continue to be made, resulting in increased efficiency and energy storage. Other required equipment such as batteries and inverters are becoming smarter and cheaper. As a result, outdated equipment will continue to depreciate in your warehouse, and it will be challenging to offload them at significantly discounting with customers wanting the latest technology.

The inventory system of solar energy company is done in a following way:

1. Initial and final solar energy inventory alert
2. Synchronized solar energy tracking
3. Track inventory suppliers and costs
4. Automatic solar energy inventory recorder
5. Energy inventory multi-site tracking
6. Easily import solar energy inventory data
7. Manage solar energy inventory assemblies and kits

Most of the company prepare the following list of items related to the solar energy system:

➢ Solar panel kit
   - Roof top
   - Ground mount
   - Top of pole mount kit

- Solar panel
  - Monocrystalline
  - Polycrystalline
- Inverter
  - Micro
  - String
  - Hybrid
  - Battery based
  - Prewired system
- Inverter monitoring
  - Data loggers
  - Gateways
  - Meters
  - Displays
  - Controllers
  - Communication cards
  - Monitoring cables
  - Monitoring kits
  - Sensors
  - Monitoring accessories
  - Monitoring services
- Inverter accessories
  - Electric boards
  - Optimizers
  - Trunk cable
- Balance of system
  - Adapters
  - Battery BOS
  - Blocks
  - Bus bars
  - Circuit breakers
  - Clips
  - Combiner boxes
  - Connectors
  - Disconnects
  - Fuses
  - Fuse holder
  - Junction box

- Labels
- Load centers
- Panels
- Rapid shutdowns
- Surge devices
- Wiring
- Electrical accessories

➢ Racking and mounting
- Rails
- Flashings
- Splice kits
- Stopper sleeves
- Conduit mounts
- Attachments
- Brace assembly
- Base mount
- Brackets
- Bolts
- Clamps
- Caps
- L-feet
- Washers
- Skirt
- Lugs
- Tilt legs
- Hooks
- Stand-offs
- Ballast bay
- Top of pole mount
- Side of pole mount
- Flush mount kits
- Ground mount kits
- Roof mount kits
- RV racking
- Multi-pole mount
- Adjustable racks

➢ Batteries
- Lithium ion
- Lead acid

- Charge controllers
  - MPPT
  - PWM
  - Converters
  - Accessories

7.2.1 Mobile inventory barcode scanning

To issue and receive equipment such as trailers, rental equipment, and other items, efficient barcode scanning is required. With integrated barcode scanning capabilities, installers can correctly track serial numbers of solar panels or other equipment used in the installation. The cloud-based inventory system gives you complete control over your warehouse, allowing you to save time and money on physical inventory and purchasing decisions. Every house or commercial solar installation includes a warranty for the solar panels, similar to most products with extensive warranties. A solar panel comes with two warranties: one for performance and one for equipment. The performance warranty for a solar panel usually guarantees 90 percent production after 10 years and 80 percent after 25 years. In most cases, an equipment warranty will ensure 10-12 years of trouble-free service. Every solar panel for a house or commercial installation has a barcode engraved or etched on the panel from the manufacturer to track the warranty. As a solar installer, you require records that are easily available, so you must maintain an archive of each solar panel that is related to the installation project and date. You can easily track serial numbers on your products with serial number tracking. We make it simple to track an unlimited number of serial numbers from the time your products are delivered to the time they are installed. For each installation, one could try to write down the serial number with a pen and paper. Manually recording a serial, on the other hand, would be prone to human error. When serial numbers are used for warranty purposes, the amount of errors that a worker can make while recording the serial number is large, and each error has the potential to snowball.

Barcoding helps your firm maintain accuracy in data entry by reducing the chances of forgetting or skipping a number during entry, running into illegible handwriting, or typing in the erroneous number by accident. This saves you time, energy, and money in the long run. Even the most attentive workers will make mistakes when transcribing serial numbers, and barcoding puts you one step ahead of the game by eliminating the possibility of human transcription errors. Finale's inventory management software for solar companies makes it simple to track your products and keep up with these difficult responsibilities. Scanning the serial number on the solar panel is all that is required to record the serial number. The data is subsequently wirelessly or via USB connection sent to the cloud servers. The software produces a full audit trail of stock changes and properly tracks each individual solar panel after the data is stored in the cloud. Multiple reports will be available to business owners. Business owners will be able to quickly locate a certain solar panel or view all of the serial numbers associated with a particular installation.

7.2.2 Types of inventory in solar energy system

Before planning for inventory, it is important to know why solar energy companies carry inventory and what factors influence the level of investment.

Seasonal inventory: Solar energy companies carry inventory to meet fluctuations in consumer demand arising out of seasonality. In the solar energy sector demand for solar energy products is increased in summer and demand is reduced in the rainy season. In the summer season due to the perfect amount of solar radiation and also due to excess electricity bills, consumers want to install solar energy systems to reduce the month wise electricity bill. So when solar energy companies prepare the inventory, this point is kept in mind. Most of the solar company, classify the inventory into the two forms, inventory of summer season and inventory of non-summer season.

Decoupling inventory: Overall solar energy manufacturing to installation stage typically involves a series of production and assembly workstations. Raw material of solar energy products passes through these stages before it is converted into the final product. Solar energy companies are mainly classified into three categories, solar energy product manufacturing company, solar energy system installation company, solar manufacturing to installation company. The inventory of all the mentioned companies is different because in product manufacturing companies inventory is related to the raw material of solar panel, raw material of solar inverter, raw material of battery product, on the other hand solar installation companies prepare the inventory of product which is used during the installation. Solar manufacturing to installation companies prepare the inventory of the cradle to grave process of the solar energy system. When we are not using the decoupling inventory, in that case all the inventory is closely related to each other and affects one to another. In decoupling inventory, one inventory doesn't affect the other

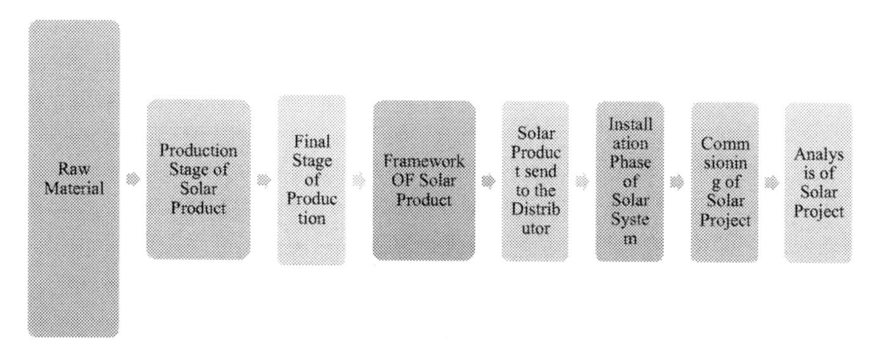

Figure 7.2 Layout of solar system without decoupling.

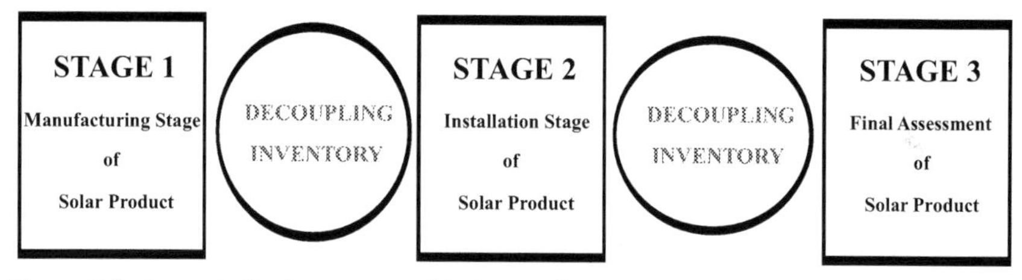

Figure 7.3 Layout of solar system with decoupling.

inventory. Each stage is linked to the other using decoupling inventory. This enables each stage to work with reasonable levels of independence. Further the adverse impact of one stage will not immediately affect the other.

Fig. 7.2 shows the solar system without decoupling, in which inventory of each stage is interlinked to each other and inventory of one stage is directly affected by the next stage. Fig. 7.3 shows the solar system with decoupling, in which the overall process of the solar energy system is classified into three stages: manufacturing, installation and final assessment. Each stage has a different internal process, but those stages decoupled with the individual inventory.

Cyclic inventory: Solar enterprises are accustomed to ordering goods in cycles and consuming it over time. For example, any distributor might place an order for 2000, 300W polycrystalline solar panels. If one order is consumed at a pace of 50 per day, it will take 40 days to finish it. Another order of 2000 will arrive on the forty-first day, and it will be devoured during the next 40 days, and so on. Each cycle of cyclic inventory follows a saw toothed pattern, starting with replenishment and ending with the complete depletion of solar energy system inventory. If 'SQ' is the order quantity of a solar product each cycle, then SQ/2 is the average cycle inventory.

Pipeline inventory: In the solar company pipeline inventory pertains to the level of inventory that organizations carry in the long run due to non-zero lead time for order, transport and receipt of material from the suppliers. Consider the example of the polycrystalline material. Suppose it takes three days to supply 300W polycrystalline solar panels, so at the end of the day there are 37 in order to replenish the stock to 2000 solar panels. In general, if the lead time for supply is L_t and the mean demand per unit time is M_d, then the pipeline inventory for solar system is given by Pipeline Inventory $= L_t \times M_d$

Example 7.1. A manufacturer of polycrystalline solar panels requires polycrystalline material as a key ingredient. The weekly requirement of polycrystalline material is 150 tones. The lead time of supply of polycrystalline material is 3 weeks. If the solar manufacturing company places monthly orders of material, analyze the various types of inventory in the given system.

Solution:

Ordered quantity SQ = 1-month requirement = 600 tones (4 weeks in a month)

Cyclic inventory in the system = SQ/2 = 600/2 = 300 tones

Lead time $L_t = 3$ weeks

Average weekly demand $M_d = 150$ tones

Pipeline inventory $= L_t \times M_d = 150 \times 3 = 450$ tones

Example 7.2. A manufacturer of polycrystalline solar panels requires polycrystalline material as a key ingredient. The weekly requirement of polycrystalline material is 250 tones. The lead time of supply of polycrystalline material is 3 weeks. If the solar manufacturing company places monthly orders of material, analyze the various types of inventory in the given system.

Solution:

Ordered Quantity SQ = 1-month requirement = 1000 tones (4 weeks in a month)

Cyclic inventory in the system = SQ/2 = 1000/2 = 500 tones

Lead time $L_t = 3$ weeks

Average weekly demand $M_d = 250$tones

Pipeline Inventory $= L_t \times M_d = 250 \times 3 = 750$ tones

Pipeline inventory and safety stock are crucial to both the 'How Much' and 'When' questions in inventory planning in the solar energy cyclic inventory. Lead time has a direct impact on the "when' choice and dictates the amount of pipeline inventory in the system. Similarly, cyclic inventory is the result of an inventory planner's "how much" decision. In a roundabout way, safety stocks determine both "how much" and "when."

Solar company inventory costs: There are several costs associated with inventory planning of solar energy companies. These costs could be classified under the three broad categories, the cost of carrying inventory of solar energy products, the cost associated with ordering material and the cost arising out of shortage of material. For an order quantity of solar product SQ, the average inventory carried by a solar energy company is SQ/2. Therefore, cost associated with carrying inventory $=\left(\frac{SQ}{2} \times sc\right)$.

Where SC solar system inventory carrying cost per unit time.

In a larger order quantity of solar product, SQ will require fewer orders to meet a known demand D, the number of orders of solar product to be placed to satisfy a demand of D = D/SQ. If SC_0 denotes the cost of ordering of solar product per order, then total ordering cost of solar inventory is given by

$$\text{Total Ordering Cost} = \frac{D}{SQ} \times SC_0$$

For example if the unit cost of a solar inverter is Rs. 4000, the annual interest charges are 10%, and the other annual costs related to carrying inventory are 2%, then the inventory carrying cost is 12% of the unit cost, that is Rs. 480 per unit per year.

Now we can calculate total cost of overall plan of solar energy system and which is given by

Overall plan of solar energy system = total cost of carrying inventory of solar system + Total cost of ordering of solar system

$$\text{Overall plan of solar energy system } (TS) = \left(\frac{SQ}{2} \times sc\right) + \left(\frac{D}{SQ} \times SC_0\right)$$

When the total cost is minimum, you can obtain the most economic order quantity of the solar system (EOQS).

By taking the first derivative of the above equation with respect to SQ and equating it to zero. We can obtain the EOSQ. Now you obtain $\frac{dTS(SQ)}{dSQ} = \left(\frac{sc}{2}\right) - \left(\frac{SC_0 D}{SQ^2}\right)$

Equating the first derivative to zero and rearranging the terms, we obtain

Denoting EOQS by SQ*, we obtain the expression of SQ* as:

$$SQ^* = \sqrt{\frac{2SC_0D}{SC}}$$

$$\text{Optimal number of order for solar product} = \frac{D}{SQ^*}$$

$$\text{Time between orders} = \frac{SQ^*}{D}$$

SQ* answers the 'how much' question directly. Every time the inventory depletes to zero.

Example 7.3. A solar energy panel manufacturing unit uses large quantities of a component made of polycrystalline material. Although these are production items, the demand is continuous and inventory planning could be done independent of the production plan. The annual demand for the solar panel is 2000. The company procures the item from the supplier at the rate of Rs. 1000 per panel. The solar company estimates the cost of carrying inventory to be 20 percent per unit per annum and the cost of ordering as Rs. 1350 per order. The company works for 200 days in a year. How should the solar company design an inventory control system for this item? What is the overall cost of the plan?

Solution:

Annual demand for the item (D) = 2000 panel
Number of working days = 200
The average daily demand = 2000/200 = 10 panel
Unit cost of the item = Rs. 1350 per panel
Inventory carrying cost = 0.20 × 1000 = 200 per panel per year
Cost of ordering = 1350 per order
The "**how much**" decision:

$$\text{Economic Order Quantity (Q)} = SQ^* = \sqrt{\frac{2SC_0D}{SC}} = \sqrt{\frac{2 \times 1350 \times 2000}{200}}$$

$$= 164.31 = 165$$

Number of orders to be placed = 2000/164.31 = 12.172 = 13
The "**When**" decision:

Time between orders = 200/2000 = 0.1 Years = 0.1 × 200 = 20 days.

Totalcostoftheplan Overall plan of solar energy system (TS)

$$= \left(\frac{SQ}{2} \times sc\right) + \left(\frac{D}{SQ} \times SC_0\right) = \left(\frac{SQ}{2} \times sc\right)$$
$$= \left(\frac{D}{SQ} \times SC_0\right) = \left(\frac{165}{2} \times 200\right) + \left(\frac{2000}{165} \times 1350\right)$$
$$= 16431 + 16432 = \text{Rs. } 32863$$

Hence the solar manufacturer will place an order for 200 panels of the component once in every 20 days and will incur a total cost of Rs. 32863 for the plan.

Example 7.4. Consider the above example, assume that the carrying cost of the solar component remains Rs. 200 per panel per year and the supplier is willing to offer a discount on the unit sprice as per the following structure.

Up to 299 solar panel = No discount
300- 699 solar panel = 2 percent discount
700- 899 solar panel = 3 percent discount
What should the company do in this case?

Solution:

The economic order quantity is 165 boxes and unit price of the panel is Rs. 1000.

There are two other order quantities at which the unit price changes on account of discount.

At Q1 = 400, unit price of the panel = 0.98 × 1000 = Rs. 980

At Q2 = 800, unit price of the panel = 0.97 × 1000 = Rs. 970

Since there is a discount on the price as the order quantity is varied, the total cost comparison between alternatives can be made only after incorporating the purchase price.

Overall plan of solar energy system (TS)

$$= \left(\frac{SQ}{2} \times sc\right) + \left(\frac{D}{SQ} \times SC_0\right) + D \times SC_u$$
$$= \left(\frac{SQ}{2} \times sc\right) + \left(\frac{D}{SQ} \times SC_0\right) + D \times SC_u = \left(\frac{165}{2} \times 200\right)$$
$$+ \left(\frac{2000}{165} \times 1350\right) + 2000 \times 1000 = 16431 + 16432 + 2000000$$
$$= \text{Rs. } 2032863$$

$$TS\,(Q1) = \left(\frac{Q1}{2} \times sc\right) + \left(\frac{D}{Q1} \times SC_0\right) + D \times SC_u\left(\frac{400}{2} \times 200\right) + \left(\frac{2000}{400} \times 1350\right) + 2000 \times 980 = \text{Rs.}\,2006750$$

Total cost of Q2

$$TS\,(Q2) = \left(\frac{Q2}{2} \times sc\right) + \left(\frac{D}{Q2} \times SC_0\right) + D \times SC_u\left(\frac{800}{2} \times 200\right) + \left(\frac{2000}{800} \times 1350\right) + 2000 \times 970 = \text{Rs.}\,2023375$$

Since TS (Q1) is the lowest among the alternatives, the firm can use the discount offered and reset the order quantity of 400 solar panels.

These calculations can be generalized by utilizing a few notations and substituting a theoretical distribution for demand, such as the normal distribution.

Let the demand during lead time follow a normal distribution with:

$\mu(L)$ the mean demand during lead time and

σ (L) the standard deviation of demand during lead time

Let (1-α) denote the desired service level, where α denotes the probability of a stock out.

$Z\alpha$ is the standard normal variate corresponding to an area of (1- α) covered on the left side of the normal curve.

An expression for safety stock (SS) is given by $SS = Z\alpha \times \sigma$ (L)

Features of solar energy inventory tracking system:

- Min-max energy inventory alerts
- Serialized energy stock tracking
- Track inventory suppliers and costs
- Automatic solar energy inventory reorder
- Energy inventory multi-site tracking
- Easily import solar energy inventory data
- Manage solar energy inventory assemblies and kits
- Mobile inventory barcode scanning

7.3 Inventory control system of solar energy system

The inventory control system of the solar energy system, provides organizations with the necessary building blocks to put inventory control systems in place. In a general way solar energy companies apply some methods to manage and control inventory.

Synchronized review (Q) system: In a solar energy company synchronized review system, which is also called a two-basket system. In the operation of the solar energy system available inventory is stored in the two baskets, first is called the mini basket and another one is the major basket. As the material which is used for solar energy products is consumed, the major basket is emptied first. As soon as the larger basket is empty, then further order is placed with a supplier for a predefined quantity and until the raw material for solar energy product in the solar cargo space or solar energy component warehouse, the mini basket is consumed. During the replacement, the smaller basket is filled with first priority and cycle continues. In this review system utilizes the perfect synchronization between mini and major basket.

A synchronized review system: A synchronized review system is an alternative of synchronized review system and it operates in a different way from synchronized review system. In a synchronized system the inventory level of the system is reviewed at the fixed interval of time. The two decisions "when" and "how much" are made in a different fashion compared to the synchronous system.

7.3.1 Selective control of inventory

Selective control of inventory of solar energy companies is categorized into different manner such as basis of unit cost, basis of transfer of inventory and basis of criticality of items. Fig. 7.4 shows the elements of selective control of the solar energy system.

Several times during the selective control of solar energy products, inventory planning for a specific period demand is required. Because the demand ceases to exist after the time during which planning is done, the unfulfilled demand in a single period demand cannot be back-ordered to the next period. The unused inventory may have some salvage value, and the loss incurred as a result of this could be referred to as the cost of overstocking.

C_{os} = Cost of overstocking per unit

C_{us} = Cost of understocking per unit

Q = Optimal number of units to be stocked

d = Single period demand

$P(d \leq Q)$ = The probability of the single period demand being at most 'Q' units.

If $d > Q$, then we incur costs on account of the cost of understocking. On the other hand, if $d < Q$, then we incur the cost of overstocking. At a very low value of Q, we tend to experience costs arising out of understocking and as we increase Q incrementally we will approach optimal Q. At very high

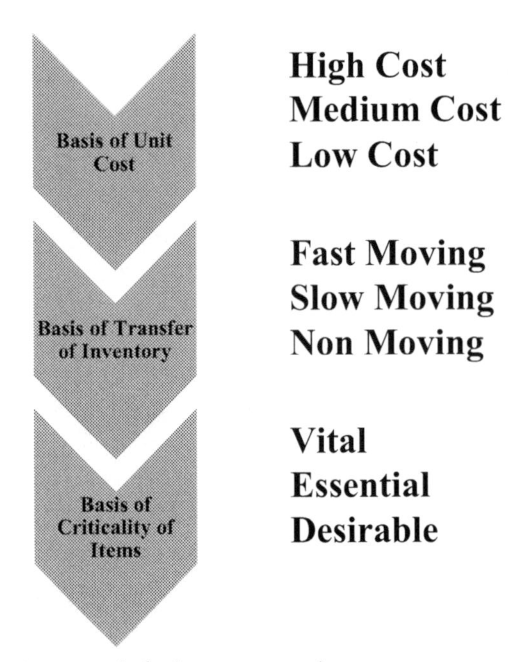

Figure 7.4 Selective control of solar energy product.

value of Q, we will incur overstocking costs. By incremental analysis, we find that while taking a decision to stock Q units, we would like to ensure that:

The expected cost of overstocking ≤ The expected cost of understocking

$$P(d \leq Q) \times C_{os} \leq P(d > Q) \times C_{us}$$

$$P(d \leq Q) \times C_{os} \leq \{1 - P(d > Q)\} \times C_{us}$$

$$P(d \leq Q) \times (C_{os} \times C_{us}) \leq C_{us}$$

$$P(d \leq Q) \leq \frac{C_{us}}{C_{us} + C_{os}}$$

There, we choose the largest value of Q that satisfies the optimal value.

Example 7.5. In India summer season is followed from March to July and in this season demand for solar panel and solar energy system related products has increased in a tremendous way. After the summer season, demand for solar panels drastically decreased. A manufacturer of solar energy products needs to decide on the optimal stock of solar panel that he needs to carry in his inventory to satisfy the demand during the summer season. The item fetches a sales value of ***Rs. 1000*** per panel. The cost of production is ***Rs. 700***

Table 7.1 Distribution of solar panel demand.

Number of solar panel demanded	Probability	Cumulative probability
0	0.05	0.05
200	0.10	0.15
300	0.15	0.30
400	0.25	0.55
500	0.20	0.75
600	0.25	1

per panel. After the summer season salvage at a value of ***Rs. 500*** per panel. Table 7.1 shows the distribution of demand for the solar panel during the summer season. What is the optimal quantity to stock?

Solution:

Since each unfulfilled demand results in a foregone profit, the cost of understock is the profit per panel. Similarly, by salvaging each unsold box after the summer season, the manufacturer losses an amount equal to the difference between the cost of manufacturer and salvage value, which represents the cost of overstocking.

Selling price per panel: Rs. 1000
Cost of Production: Rs. 700
Cost of understocking: Rs. 300
Salvage Value Rs. 500
Cost of overstocking: Rs. 500

$$\text{The optimal quantity to stock is } = P(d \leq Q) \leq \frac{c_{us}}{C_{us} + C_{os}}$$

$$= P(d \leq Q) \leq \frac{300}{500} \leq 0.6$$

On examination of the cumulative probability values in the last column of the demand table, you notice that Q lies between 200 and 300. We round up the value of Q to 300. Therefore, the solar manufacturer company plans for an inventory of 300 solar panels for sale during the summer season.

7.4 Total quality management of solar system

Total quality management can be defined as an organization-wide effort to develop the system, tools, techniques, skills and mindset required to

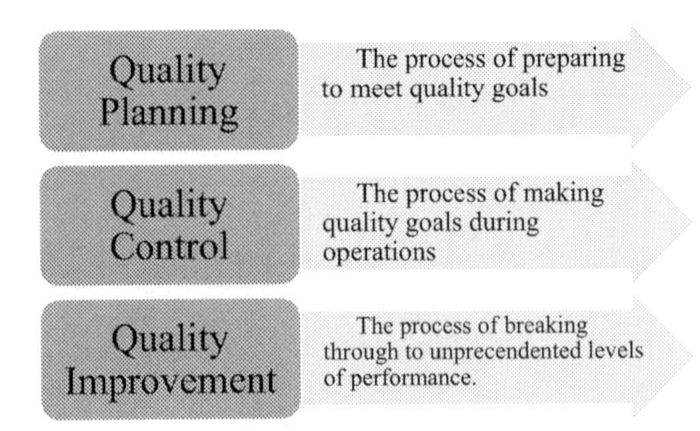

Figure 7.5 Elements of total quality management.

establish a quality assurance system that is responsive to emerging market needs. A good TQM programed consists of the following important elements:

a. Role of the top management
b. Employee involvement for continuous improvement
c. Addressing the training requirement of employee
d. Tools and techniques for quality assurance and continuous improvement

The total quality management of the solar energy system is done through the four step process, plan-do-check-act popularly known as the PDCA cycle. Rather than the PDCA cycle, quality management of the solar energy system is also done through the three step process, which is also known as Juran's Trilogy. The three steps of Juran's trilogy are 'Quality Planning', 'Quality Control' and 'Quality Improvement', which is represented in Fig. 7.5.

The total quality management of the solar energy system is started from the very initial stage at the level of solar cell material. The efficiency and performance of solar panels is largely dependent on the quality of solar cells. So that the construction of solar cells indirectly affects the technical quality management of the overall system. In order to lower the cost of the solar cell, the efficiency should be increased. But there are some losses which cause lower efficiency. Different semiconductor materials are suited only for specific spectral ranges.

Semiconductor materials such as silicon are used in photovoltaic solar cells. In the cell incoming photons separate positive and negative charge carriers. This produces an electrical voltage and electrical current and can drive a load. The solar cells can be connected in series and parallel and incorporated in a module. Several modules may be interconnected to comprise a solar array, but a large land area is required. Solar cell power

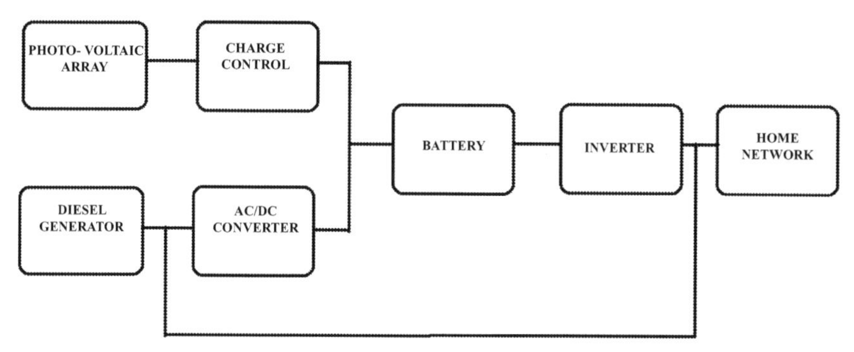

Figure 7.6 Block diagram of an off-grid PV plant.

plants are of two types namely autonomous power plants which are used for local networks and grid connected power plants used for external electrical networks. Photovoltaic array produces dc power and this must be converted into ac power for local use and feeding into the grid. An inverter converts dc voltage to ac and feeds the solar power to the grid or supply to the consumer. Fig. 7.6 shows components of a solar power plant, block diagram of an off-grid power plant. Following are the key parameters of a solar power plant, which is used for the total quality management of a solar power plant.

- ➢ Technical quality parameter of solar panel:
  - Open circuit voltage
  - Short circuit current
  - Maximum voltage
  - Maximum current
  - Solar intensity
  - Required area
- ➢ Technical quality parameter of inverter:
  - Nominal voltage
  - Voltage range
  - Operating range
  - Waveform
  - Harmonics
  - Ripple
  - Efficiency
  - Losses
  - Temperature
  - Humidity
  - Protection
  - Communication interface

- Technical quality parameter of battery
  - State of charge
  - Depth of discharge
  - Terminal voltage
  - Open circuit voltage
  - Internal resistance
  - Cut-off voltage
  - Capacity or nominal capacity
  - Energy or nominal energy
  - Cyclic life
  - Specific energy
  - Specific power
  - Energy density
  - Power density
  - Maximum continuous discharge current
  - Maximum 30-second discharge pulse current
  - Charge voltage
  - Float voltage
  - Charge current
- Technical quality parameter of charge controller:
  - Delay time
  - Peak time
  - Maximum overshoot
- Financial quality parameter of overall solar energy system
  - Capital cost
  - Replacement cost
  - Operational and maintenance cost
  - Interest rate
  - Return of investment
  - Payback period
- Total quality management parameter of overall solar energy system
  - Effective utilization of resources
  - Energy management
  - Transmission and distribution management

7.4.1 Benefits of a quality management in solar energy system

Implementing a quality management system for solar energy systems provides the rigour and discipline required for solar energy product enhancement and validation. It allows a solar energy company to identify system inefficiencies, track and correct deficiencies, and continuously

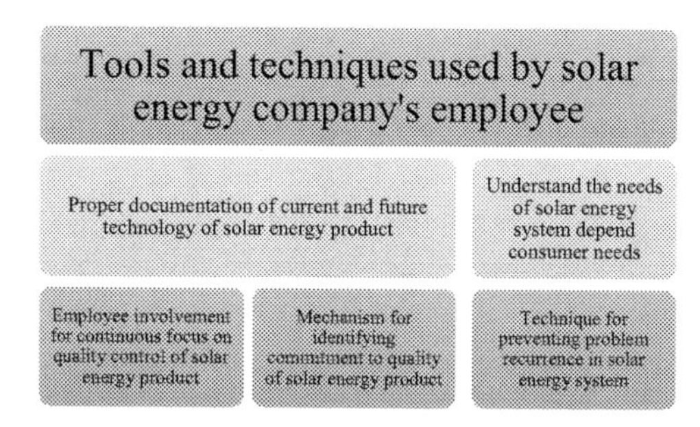

Figure 7.7 Quality assurance from employee's point of view.

improve manufacturing, delivery, training, service, and maintenance processes, all of which has the potential to lower costs and significantly improve profits and product/service quality. Following are some advantages of quality management in solar energy system:

- Demonstrates a solar energy company's commitment to quality;
- Helps to have a better focus on renewable energy business objectives and electricity consumer expectations from the renewable power source;
- Leads to better quality solar energy products and installation and maintenance services;
- Indicates consistency and efficiency of the overall project of the solar energy system.
- Provides a better foundation for continuous improvement of solar energy related technology.
- Improves the performance of the solar energy product and hence leads to a more competitive business environment in the solar energy sector.
- Encourages consumer confidence in the quality of the solar products and solar based electricity services provided by the solar energy company;

In the total quality management system, it is necessary to identify quality assurance from the solar energy company's employee side, which is also shown in Fig. 7.7. It is the responsibility of all the employees to create proper documentation of current and future technology of solar energy products and find out limitations of old technology and benefits of current and future technology of solar energy products. Following is the responsibility of company's employee toward the quality assurance system:

- Proper data assessment of past and current technology of solar energy systems and their products.
- Understand the needs of electricity consumers.
- Identify the mechanism and technique to compensate for the limitation for solar energy systems and their products.

Quality control of the solar energy system is also done through the number of testing, which reduces the chances of error in the solar energy products. The conceptual phase, production phase, transportation and installation phase, commissioning phase, and in-service phase are all stages of the solar module testing life cycle. Different test principles may apply depending on the test phase.

- Conceptual phase: The first stage may include design verification, in which the module's expected output is tested using computer simulation. Temperature, rain, hail, snow, corrosion, dust, lightning, and horizon and near-shadow effects are all examined, as well as the modules' capacity to survive natural environment conditions. At this point, you can test the layout for the module's design and construction, as well as the quality of the components and installation.
- Manufacturing phase: Inspection of component manufacturers is done through the number of visits. Assembly checks, material testing monitoring, and Non Destructive Testing can all be part of the inspection (NDT). The following standards are used for certification: ANSI/UL1703, IEC 17025, IEC 61215, IEC 61646, IEC 61701, and IEC 61730-1/-2.
- Transportation and installation phase: Pre-dispatch inspection, dimensional control, visual control, and damage control are all examples of inspections. It's also a good idea to go over your paperwork and certificates.
- Commissioning phase and in-service phase: Solar module experts will guarantee that the production process is handled correctly and that the start-up is safe. The solar modules are inspected on a regular basis during the in-service phase to ensure they are working properly.

7.4.2 Solar module quality assurance

Solar module quality assurance entails testing and assessing solar cells and panels to ensure that they meet their quality criteria. The service life of solar modules is projected to be between 20 and 40 years. They should be able to transfer and supply the expected power on a consistent and reliable basis. Modules were exposed to a wide range of environment conditions as well as use in a variety of temperatures. Physical testing, laboratory investigations, and numerical analysis can all be used to evaluate solar modules. Furthermore, solar modules must be evaluated at various times throughout their life cycle. Solar module quality assurance is provided by a number of firms, including Southern Research Energy & Environment, SGS Consumer Testing Services, TÜV Rheinland, Clean Energy Associates

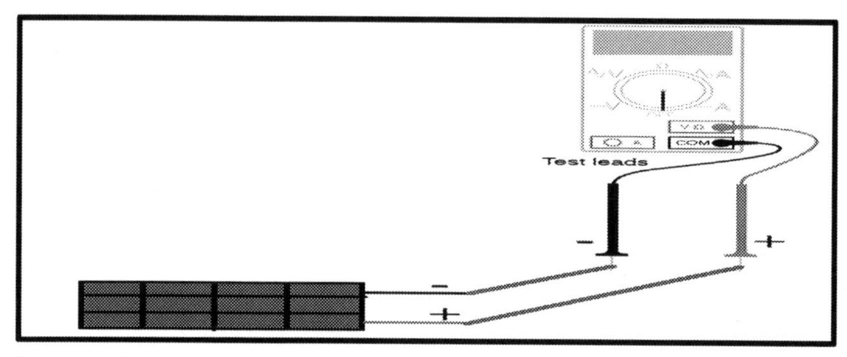

Figure 7.8 Testing solar panel for volts.

(CEA), CSA Solar International, and Enertis. To guarantee and ensure the quality of PV Modules, consistent traceable and reliable production methods must be used.. *Following is the number of testing procedure of solar energy products:*

7.4.2.1 Testing solar panels for volts

To test solar panel voltage output, put your solar panel in direct sunlight, set your multimeter to the "volts" setting shown in Fig. 7.8.

1. Touch the multimeter/voltmeter (red) positive lead to your solar panel's positive wire.
2. Then touch the multimeter/voltmeter (black) negative lead to your solar panel's negative wire.

7.4.2.2 Solar panel testing for amps

To test solar panel amperage output, put your solar panel in direct sunlight, set your multimeter to the "amps" setting, which is also shown in Fig. 7.9.

1. Touch the multimeters (red) positive lead to your solar panel's positive wire.
2. Then touch, the multimeters (black) negative lead to your solar panel's negative wire
3. Testing the charge controller:

   The charge controller must be tested when the solar panels are being tested. Make sure the battery isn't completely charged; otherwise, it won't accept the current. The solar panel is used alone in the first two measurements. Make sure to separate the solar panel from the regulator before connecting the controller, solar panel, and battery. The battery should then be disconnected from the controller/regulator. Connect the controller to the battery first, then the controller to the solar panel during reconnection. You

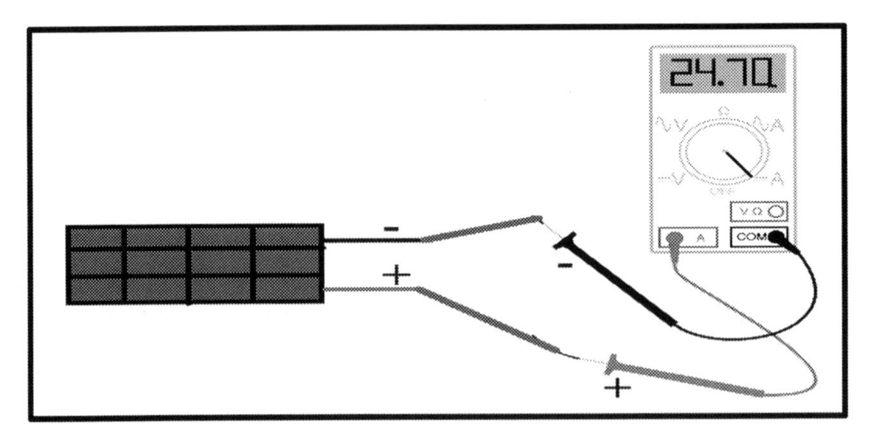

Figure 7.9 Testing solar panel for amps.

might be wondering why not the other way around, however these planned measures will prevent the controller from being damage.

7.4.2.3 You can perform this in a series of steps below

- Set the multimeter's measurements to DC Amps. To locate the DC Amps, make sure the alligator clips are in the correct port.
- Set the multimeter to 10 Amps.
- That being done connect the solar panel to the controller and also the controller to the solar batteries.
- Disconnect the positive cable that is between the battery and controller.
- Connect the positive cable that you recently removed to the lead alligator clips on the multimeter to find the current.
- After that, attach the multimeter's negative lead alligator clips to the positive terminal of the battery as the final step.
- This process will measure the current flowing between the solar panel, controller and solar batteries.

7.4.2.5 UV preconditioning test

The UV preconditioning test is a process for determining in what way ultraviolet radiation affects the solar panels performance and how it is responsible for degradation of the solar panel. UV preconditioning is an irradiance test that determines the extent to which solar panels are sensitive to UV deterioration and, as a result, performance losses. This harsh test condition is normally performed before the thermal cycle and humidity freeze tests, and is largely used in solar panel certification and research. The UV preconditioning test is carried out at a temperature of 60°C 5°C with UV irradiation.

7.4.2.6 Mechanical load test

The PV module's capacity to sustain static snow and wind loads is determined by the Mechanical Load Test. The effects of wind and snow loads on PV modules must be tested as part of the certification process. The mechanical load test is normally carried out at room temperature, with the solar panel installed horizontally and the artificial force applied vertically. Snow loads do not occur at room temperature, and wind does not generally blow vertically on PV modules, so these test conditions are not realistic. The use of a dedicated climate chamber capable of replicating temperatures is one possibility. The test is performed as follows: 3 cycles of 1 hour 2400 pa in pressure, and 3 cycles of 2400 pa in traction. To stimulate large snow loads, 5400 Pa is used during the last pressure cycle.

7.4.2.7 Insulation resistance test

Insulation resistance testing is an electrical safety check that determines whether a solar module is adequately insulated. Manufacturers, installers, and quality testers can use an insulation resistance test to determine whether a solar panel has appropriate insulation between its electricity-conducting components and the module's frame, or the outside world in the event of a frameless panel. Insulation issues may arise from poorly constructed PV modules. Solar cells that are too close to the frame, for example, are a common source of insufficient insulation. Material weakness, such as modules that were poorly laminated with low-quality encapsulates, is another typical reason for insufficient insulating resistance. A dielectric strength tester is used to supply DC voltage to the module of up to 1000 V plus twice the maximum system voltage during the test. The PV module must pass this test without any surface tracking, functional breakdown, or creation of leakage routes. Current leakage can happen on the solar panel's edges or through the backsheet. The resistance of conventional solar panels (modules with an area more than 0.1 m2) should not be less than 40 per square metre.

7.4.2.8 Hail impact test

The hail impact test is a mechanical test that certifies that a solar panel can withstand the impact of hailstones at a temperature of less than 4 degrees Celsius. The hail impact test is an important solar panel quality and safety test since hail can cause severe damage to solar panels in many places of the world. As a result, it's critical to buy solar panels that have been tested for hail impact and have passed the hail impact test for solar panels. In general, the hail impact test is a certification test used for certifying new solar panel types and products, as well as for internal quality testing and research. Hail impact tests are rarely used in quality control during mass manufacture of solar panels. A pneumatic ice ball launcher, which can launch ice balls at different speeds and weights, is commonly used in test labs. Ice balls will be thrown towards the pv module in over ten different locations spot. After the test, the hailstones should have caused no obvious damage to the front of the solar panels. .

7.4.2.9 Damp heat test

The PV Damp Heat Test is an accelerated environmental ageing test that assesses a solar panel's capacity to endure long-term exposure to high temperatures and dampness penetration. The damp heat test is carried out by maintaining an ambient temperature of 85°C and a relative humidity of 85 percent for 1,000 hours, which is well over a month. The to-be-tested solar panel is placed in a moist heat test chamber for this purpose, where it is subjected to these circumstances. This test puts the PV module's lamination process and edge sealing against dampness to the test. The peel strength of the PV modules that are laminated according to the traditional lamination process is decisive.

Problems that can be identified with the PV damp heat test are:

1. Corrosion
2. Delamination
3. Junction box and module connection failure

The IEC 61216 standard for crystalline solar panels and the IEC 61646 standard for thin film solar panels both include damp heat testing. Surprisingly, the test is not included in the UL 1703 standard, which focuses entirely on safety. During the certification procedure, the Damp Heat test is regarded as the most difficult to pass. According to reports, more than 60% of glass solar panels fail this test. .

7.4.2.10 Thermal cycling test

The PV Thermal Cycling Test is an environmental test that uses varying severe temperatures to simulate thermal loads on solar panels. The thermal cycling test is carried out by inserting the solar panel to be evaluated in a heat chamber that is heated and cooled 200 times between C -40 and C +90. The solar panel must be able to withstand large temperature swings on a regular basis. The soldering inside the solar panel may be affected by the thermal expansion coefficients of the encapsulating materials. Major flaws, such as power degradation and electric circuit interruption, may develop as a result of this. Exposing the solar panel towards extreme temperature conditions, problems that can be identified with the thermal cycling test:

1. Broken interconnects
2. Broken cells
3. Solder bond failures
4. Junction box and module connection failures

7.4.2.11 Rubbing test

The Rubbing Test is a quality control method that involves assessing the adhesiveness and conformance of solar product labels. Solar panels and inverters are subjected to extreme external conditions for many years, depending on the geographical location of the installation, resulting in illegible labels and serial numbers connected to these solar devices. In this circumstance, a worn label and serial number may, in the worst-case scenario, affect the warranty terms of that product because it can no longer be traced back to the supplying manufacturer. Furthermore, in a large installation, a non-readable serial number makes it difficult to identify that individual system component and makes performance data comparisons between pre-shipment and post-installation practically impossible.

PV quality auditors use the rubbing test to ensure that the label and serial number on the back of a solar product, especially solar panels and solar inverters, are readable after 25 years. It includes 15 seconds of rubbing with a mixture of water and alcohol on the label and serial number. Quality labels and serial numbers will be printed/produced with a protective coating to ensure that any text, warranty details, or certification logos are not rubbed away by even rigorous rubbing.

7.4.2.12 Electroluminescence crack detection (ELCD)

A PV producer can assess the structural quality of solar cells as well as any other potential flaws resulting from incorrect photovoltaic panel handling. The ELCD test is now included in the manufacturing lines of the majority of leading solar panel manufacturers.

With the help of the ELCD test, a manufacturer can detect defects that are normally not visible. Defects that can be found with an ELCD test are:

1. Broken cells and micro-cracks in the cells.
2. Detection of bus bar contact defects
3. Detection of missing or interrupted screen-printed fingers
4. Detection of non-homogeneity and foreign matter in the crystalline silicon

Microcracks do not always indicate that the cells' performance has been harmed. An ELCD test cannot determine the performance of cells or the impact of micro cracks on cells. A flash test can be used to assess output performance. Professional solar PV quality testing firms would never allow any panels with micro fractures to pass: because the panels' long-term performance may be harmed, they should be replaced. As a result, the ELCD test should be performed prior to the lamination process so that any damaged solar cells can be replaced.

7.4.2.13 Infrared imaging test

The thermographic study of solar cells is referred to as infrared imaging in the solar PV sector. The solar cells' output heat radiation is detected during IR measurements. A thermographic camera detects infrared radiation and creates two-dimensional (2D) images of that radiation. Based on its temperature, any object, including solar panels, emits infrared radiation.

The infrared imaging test is useful to many PV manufacturers since it detects a range of issues. The most logical time to do this test is prior to the lamination of the solar panel. If any problems are discovered, the solar panel can be rectified and tested a second time before being laminated. After lamination, most PV manufacturers, particularly third-party quality testing agencies, do an infrared imaging test. We know from experience that panels are frequently damaged during the laminating process. The cause is often inaccurate connections between the cells, which will crack the cells once pressure is performed through laminating. Of course once the solar panel is laminated, the manufacturer has no chance to repair the panel. The best it can do is sell it as B-grade.

7.4.2.14 Flash test

Solar Flash Tests (also known as Sun Simulator Tests) measure a solar PV module's output performance and are a standard testing process used by manufacturers to assure each PV module's complying operability. A flash test machine, often known as a sun simulator, is a device that is used to assess a solar PV module's output performance conformance. During a flash test, the PV module is subjected to a brief (1–30 ms) and strong (100 mW per sq. cm) burst of light from a xenon-filled arc lamp. This lamp's output spectrum is as close as possible to the spectrum of the sun. A computer collects the data, which is then compared to a precisely calibrated reference solar module. The reference data is based on a power output that is calibrated to a standard solar irradiation level.

At standard test settings, module parameters are measured. Temperature has a significant impact on the performance of PV modules. When a module's temperature rises, two things happen. First, each cell's voltage output diminishes. Second, each cell's current output increases somewhat, lowering the module's maximum power output in Watt (Pm (W)). The criteria utilised may or may not reflect real operating circumstances. A flash test results sheet typically contains all modules that have been tested as well as the individual test results. The yellow indicated panels in the example flash test results sheet below indicate Pm (W) values less than the tested module type's minimum 190 Wp.

7.4.2.15 Solar cell efficiency test

Before the cells are combined into a solar panel or transported to manufacturers for further assembly, they are subjected to a solar cell efficiency test. The efficiency of each solar cell will be tested at the start of the production process. In today's cell manufacturing facilities, the solar cell efficiency test is completely automated and integrated into the production line. The cells are evaluated for efficiency and sorted into groups based on similar power outputs in the last phase of the cell manufacturing process. Many small and medium solar panel manufacturers utilise a simple cell efficiency tester to measure the output of each cell piece by piece. This method of evaluating solar cell efficiency can be time-consuming. It also runs the danger of careless staff causing injury to the delicate cells. This type of solar cell testing is still widely utilised in China, where manual labour costs are still relatively low.

7.4.2.16 Testing of battery

A battery, like a living creature, cannot be measured and must be evaluated with variable degrees of precision depending on accessible symptoms. This simulates a doctor doing a physical examination on a patient by administering a series of tests and employing the law of elimination. Rapid-test methods for batteries have lagged behind other technologies due to their complexity and ambiguous results when evaluating outliers. Cadex recognizes the value of battery diagnostics and has made substantial advancements in quick testing technologies. Diagnostic Battery Management (DBM), a revolutionary approach to battery management and maintenance that enterprising organizations are pursuing, is built on these achievements. DBM is crucial for ensuring the reliability of present battery systems by monitoring capacity and other parameters, rather than developing a new super battery.

Self-discharge is related to mechanical integrity, while internal resistance is related to current delivery. All these properties must be met in order for a battery to be qualified. In addition to these static qualities, a battery contains many state-of-charge (SOC) dynamic factors that affect battery performance and obstruct quick testing. . Even if the charge is low, reliable battery test methods must be able to recognize all battery states and provide reliable results. Because a fully charged good battery behaves similarly to a faded pack that is only partially charged, this is a difficult request. Taking a voltage reading, measuring internal resistance with a pulse or AC impedance technique, coulomb counting, and employing Electrochemical Impedance Spectroscopy to capture a snapshot of the chemical battery are all examples of test procedures (EIS). Estimating capacity using chemical battery decoding is more challenging than using digital monitoring with coulomb counting. The employment of proprietary algorithms and matrices that act as lookup tables, similar to those used in letter or face recognition, is required to investigate chemical batteries. Voltage and internal resistance, especially in Li-ion and lead-acid systems, do not correspond to capacity and are inefficient at predicting the end of battery life. The truth can be found in the chemical battery. A digital measurement alone is subject to failure because the chemical symptoms are not represented. Here are the most common battery test methods, which is shown in Table 7.2:

7.4.3 SCADA based total quality management of solar energy system.

SCADA is the supervisory control and data acquisition system, which is used to mainly supervise, acquire and control the data that is received from a distant source from the SCADA. SCADA systems play a very vital role in the total quality management of the solar power plant. When a solar power plant, connected with the SCADA system, then the SCADA system consists of data collection equipment, data transmission equipment, remote terminal unit, data processing unit and data presentation unit. Data collection equipment of this system collects the year-wise data of solar radiation, wind velocity, temperature, humidity, longitude and latitude of the particular location. It also collects the data of load demand of the consumer, hourly energy generation from the plant, excess energy generation from the solar power plant. In the data transmission process, SCADA covers the data transfer between SCADA control host computers and a number of remote terminal units. The remote terminal units' function as slaves in the master/slave

Table 7.2 Common battery test method.

Voltage	Battery voltage reflects state-of-charge in an open circuit condition when rested. Voltage alone cannot estimate battery state-of-health (SoH).
Ohmic test	When internal resistance is high, it identifies corrosion and mechanical problems. Although these abnormalities signal that the battery is nearing the end of its life, they are not always associated with low capacity. The impedance test is another name for the Ohmic test.
Full cycle	To determine the capacity of a chemical battery, a whole cycle consists of charge/discharge/charge. This gives the most accurate readings and calibrates the smart battery to correct tracking inaccuracies, but it takes time and is stressful.
Rapid-test	Time domain testing involves activating the battery with pulses to examine ion flow in Li-ion batteries, while frequency domain testing involves scanning a battery with numerous frequencies. Complex software containing battery-specific parameters and matrices functioning as lookup tables is required for advanced rapid-test methods.
BMS	Most Battery Management Systems estimate SoC by monitoring voltage, current and temperature. BMS for Li-ion also counts coulombs.
Coulomb counting	A smart battery's full charge capacity (FCC) provides a coulomb count that is related to SoH. Although the FCC reading is instant, the data becomes erroneous with time, and the battery must be calibrated with a full cycle.
Read-and-charge	A charger with RAC technology reads the battery's SoC using a patented filtering algorithm, then counts the coulombs to fill it. For each battery model, RAC requires a one-time calibration; cycling a decent pack supplies this parameter, which is recorded in the battery adapters. Cadex invented the RAC technology.
SOLI	The State-of-Life-Indicator determines battery life by counting the total coulombs a battery may deliver over its lifetime. The amount of delivered coulombs reduces until the allocation is spent and replenishment is required. After calculating the coulomb count of 1 cycle based on the manufacturer's criteria, multiply the number by the stated cycle count to get the entire scale (V, Ah). Cadex's SOLI can be used in wheelchairs, medical devices, traction, and UPS systems, and can be retrofitted or installed when new. Wireless connection allows fleet management to take place.

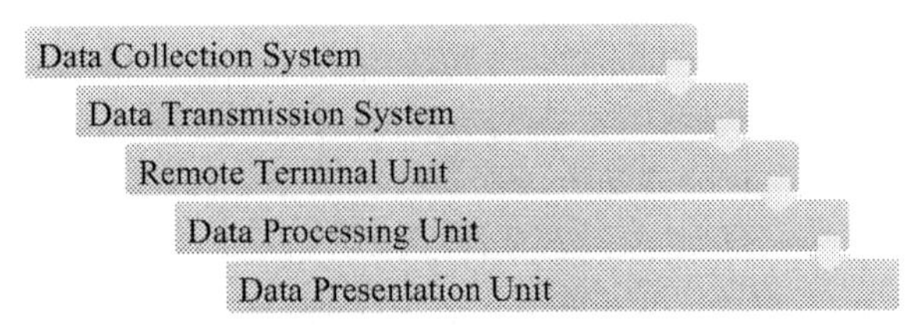

Figure 7.10 General structure of SCADA system in solar power plant.

architecture. The remote terminal unit also connected to the circuit breaker of the solar power plant. Field data interface devices, which gather data of solar radiation and other pre feasibility analysis data are responsible for providing digital inputs to the master station of SCADA. On the other hand, if flow of solar energy in a solar power plant is controlled through SCADA, then power consumption meters, motor control switchboards would be used as field data interface devices. In the data processing and data presentation process, a collection of standard or custom software also referred to as main machine software systems used to present data to the operator and provide for control inputs in a variety of formats, including graphics to support the communication system and monitor and control even that field data device that is located remotely. Fig. 7.10 shows the general structure of the SCADA system in solar power plant.

AGC SCADA is the recent concept of SCADA system, which is the automatic generation control system that involves mainly regional load center and solar power station control systems. The load control center of the solar power plant receives real time data pertaining to how much power is being generated at the solar power plant, what is the system frequency at which generation is taking place and what is the quality of solar power that is flowing through each tie line.

➢ SCADA based solar energy system states:

Based on the SCADA system, the operating states of solar energy system, are mainly classified into the following categories:

- **Normal state:** The solar energy system is said to be in normal state, when there are no contingencies present and the frequency of solar energy supply as well as that of voltage within safe limits. The normal state is also known as a preventive state.
- **Emergency state:** The solar energy system enters into an emergency state when some of the solar components operating limits are violated such that some system variables tend to go outside acceptable ranges and the system frequency starts to decrease. When the solar energy system is in an emergency state as early as possible. If the solar energy system

happens to be in the emergency state for two long then the problem of frequency dropout or grid failure may take place.

- **Restorative state:** The condition when some components or whole of the solar energy system loses power is called restorative state. The objective of this state is to maintain the reliability of the operation of the solar energy system so that by appropriate actions such as load shedding or load scheduling.

Taking this further, an effective SCADA system should incorporate the following functions.

- **Weather forecasts:** By integrating weather forecast service support, along with expected performance, the SCADA system can provide accurate energy forecasts, providing insight into data and sensor calibration.
- **Device-level analysis:** When performance is analysed at the individual device level, such as individual inverter capacity, string counts and installed solar panel specifications, a SCADA system can alert operators and maintenance technicians of device performance issues and the level of impact for quick analysis.
- **Tracker angles:** Adding calculated tracker angles into the SCADA system identifies when trackers are not tracking to the optimal angles, resulting in decreased energy production.
- **Lost energy analysis:** Lost energy, a measure of the amount of energy that would have been produced if all devices were functioning optimally, provides historical details to analyze underperforming devices.
- **Rate schedules:** Including Rate Schedule Awareness enables the SCADA system to know the value of the energy produced by the plant or the value of the energy not produced. This information is used to evaluate a plant's financial performance.

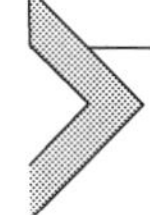

7.5 Quality certification and society of solar energy system

International solar energy society: The International Solar Energy Society (ISES) has been doing product research that has aided the growth of the renewable energy industry for over 60 years. ISES helps its global membership give technical answers to expedite the transition to 100 percent renewable energy and realise the following vision through its knowledge sharing and community building activities. The International Solar Energy

Society (ISES) envisions a society where everyone can use renewable energy wisely and efficiently.

ISES advocates

- Knowledge sharing on renewable energy and energy efficiency technologies
- Universal energy access, including: Decentralized energy systems, Smart grids
- Renewable systems integration
- International cooperation
- Comprehensive assessments/roadmaps

Ministry of new and renewable energy: The Ministry of New and Renewable Energy (MNRE) is the Government of India's main ministry for all new and renewable energy issues. The Ministry's overall goal is to create and deploy new and renewable energy to supplement the country's energy needs.

- Commission for Additional Sources of Energy (CASE) in 1981.
- Department of Non-Conventional Energy Sources (DNES) in 1982.
- Ministry of Non-Conventional Energy Sources (MNES) in 1992.
- The Ministry of Non-Conventional Energy Sources (MNES) was renamed as Ministry of New and Renewable Energy (MNRE) in 2006.

With growing concerns about the country's energy security, the role of new and renewable energy has become increasingly important in recent years. The two oil shocks of the 1970s, energy self-sufficiency was highlighted as a primary driver for new and renewable energy in the country. In March 1981, the Department of Science and Technology established the Commission for Additional Sources of Energy in response to the sharp rise in oil prices, supply uncertainty, and the negative impact on the balance of payments. The Commission was tasked with creating policies and implementing them, as well as programming for the development of new and renewable energy, as well as coordinating and accelerating R&D in the field. In 1982, the Ministry of Energy launched a new agency, the Department of Non-conventional Energy Sources (DNES), which included CASE. DNES changed its name to the Ministry of Non-Conventional Energy Sources in 1992. The Ministry was renamed the Ministry of New and Renewable Energy in October 2006 [10–12].

Intersolar award: Intersolar is dedicated to promoting the solar industry's culture of innovation. Intersolar focuses on the technologies and services that make solar power globally successful, now and in the future, as the world's largest exposition series for the solar sector and its partners. In this sense, the Intersolar award is given to companies that have made a significant contribution to the industry's success. Every year, the solar award recognizes technological breakthroughs and ground-breaking solutions in the field of photovoltaics. Following is the evaluation criteria for Intersolar award:

Degree of technological innovation (30%)
- Technological advancement
- Technological creativity and quality
- Singularity in comparison with other solutions

Safety (15%)
- Compliance with international safety standard

Technical and environmental benefits (20%)
- Technological benefits
- Environmental benefits and socially relevant features

Economic benefits (20%)
- Economic benefits for the user
- Demand, markets and marketing strategy

Proof of innovation (10%)
- Proof of functionality and ingenuity, test results, certifications, patents, references

Presentation (5%)
- Format and quality of description and presentation
- Extent of application documentation

7.6 Conclusion

The culmination of the above discussion is the notion of total quality management. Modern inventory and total quality management are built on the concept of total quality management. In the above discussion it is found out inventory and quality management is very necessary to proper functioning of the solar industry and also for solar power plants.

7.7 Exercise/question

1. What is the role of inventory management in the field of solar energy systems?
2. How total quality management improves the performance of the solar industry?

3. What are the different elements of TQM which directly affect the performance of the solar energy system?
4. What are the different types of inventory used in the solar energy system?
5. Write a short note on the cyclic and pipeline inventory.
6. A manufacturer of polycrystalline solar panels requires polycrystalline material as a key ingredient. The weekly requirement of polycrystalline material is 350 tones. The lead time of supply of polycrystalline material is 2 weeks. If the solar manufacturing company places monthly orders of material, analyze the various types of inventory in the given system.
7. A solar energy panel manufacturing unit uses large quantities of a component made of polycrystalline material. Although these are production items, the demand is continuous and inventory planning could be done independent of the production plan. The annual demand for the solar panel is 2000. The company procures the item from the supplier at the rate of Rs. 2000 per panel. The solar company estimates the cost of carrying inventory to be 20 percent per unit per annum and the cost of ordering as Rs. 1350 per order. The company works for 200 days in a year. How should the solar company design an inventory control system for this item? What is the overall cost of the plan?
8. Explain the concept of inventory control system of solar energy systems.
9. In India summer season is followed from March to July and in this season demand for solar panel and solar energy system related products has increased in a tremendous way. After the summer season, demand for solar panels drastically decreased. A manufacturer of solar energy products needs to decide on the optimal stock of solar panels that he needs to carry in his inventory to satisfy the demand during the summer season. The item fetches a sales value of ***Rs. 900*** per panel. The cost of production is ***Rs. 900*** per panel. After the summer season salvage at a value of ***Rs. 400*** per panel. Table shows the distribution of demand for the solar panel during the summer season. What is the optimal quantity to stock?

Number of solar panel demanded	Probability	Cumulative probability
0	0.05	0.05
200	0.10	0.15
300	0.15	0.30
400	0.25	0.55
500	0.20	0.75
600	0.25	1

10. What are the different benefits of quality management in the solar energy system?
11. What is the requirement of the UV Preconditioning test of a solar module?
12. What is the requirement of the Mechanical load test of solar modules?
13. What is the requirement of the Insulation resistance test of solar modules?
14. What is the requirement of the Hail Impact test of solar modules?
15. What is the requirement of a Damp heat test of a solar module?
16. What is the requirement of Thermal cycling test of a solar module?
17. What is the requirement of Infrared imaging test of solar module?
18. What is the requirement of the Flash test of solar modules?
19. What is the role of the SCADA system in the solar energy system?

References

[1] D.A. Ciupageanu, Real-time stochastic power management strategies in hybrid renewable energy systems: a review of key applications and perspectives, Electr. Power Syst. Res. 187 (2020) 106497.
[2] J. Abbas, Impact of total quality management on corporate green performance through the mediating role of corporate social responsibility, J. Clean. Prod. 242 (2020).
[3] A.V. Todorut, Sustainable development of organizations through total quality management, Procedia Soc. Behav. Sci. 62 (2012) 927–931.
[4] D.I. Prajogo, A.S. Sohal, The integration of TQM and technology/R&D management in determining quality and innovation performance, Omega 34 (2006).
[5] B.T. Qasrawi, S.M. Almahamid, S.T. Qasrawi, The impact of TQM practices and KM processes on organisational performance: an empirical investigation, Int. J. Qual. Reliab. Manag. 34 (2017).
[6] L. Raimi, Understanding theories of corporate social responsibility in the Ibero-American hospitality industry, Dev. Corp. Gov. Responsib. (2017) 11.
[7] L. Rekik, F. Bergeron, Green practice motivators and performance in SMEs: a qualitative comparative analysis, J. Small Bus. Strategy 27 (2017).
[8] M. Río-Rama, C. de la, J. Alvarez-García, J.L. Coca-Perez, 2017. Quality practices, corporate social responsibility and the "society results" criterion of the EFQM model, 19.
[9] W. Rossiter, D.J. Smith, Green innovation and the development of sustainable communities: the case of blueprint regeneration's trent basin development, Int. J. Entrep. Innov. (2018) 19.
[10] J.V. Saraph, P.G. Benson, R.G. Schroeder, An instrument for measuring the critical factors of quality management, Decis. Sci. J. 20 (1989). https://doi.org/10.1111/j.1540-5915.1989.tb01421.x.
[11] H. Sarvaiya, G. Eweje, J. Arrowsmith, The roles of HRM in CSR: strategic partnership or operational support? J. Bus. Ethics 153 (2018).
[12] M. Saunila, J. Ukko, T. Rantala, Sustainability as a driver of green innovation investment and exploitation, J. Clean. Prod. (2018) 179.

CHAPTER EIGHT

Case study: Solar–wind hybrid renewable energy system

Learning objective

- Understand the basic of hybrid renewable energy system
- Know the modeling of solar–wind hybrid renewable energy system by HOMER software
- Learn process of cost analysis of solar–wind hybrid renewable energy system
- Discuss about the life cycle analysis of solar–wind hybrid renewable energy system

Note: The concept of this case study is taken from the following research papers of Author of this book.

1. V Khare, S Nema, P Baredar, "Status of solar wind renewable energy in India", Renewable and Sustainable Energy Reviews, **Elsevier**, 2013, 27, 1-10.
2. V Khare, S Nema, P Baredar, "Optimization of hydrogen based hybrid renewable energy system using HOMER, BB-BC and GAMBIT", International Journal of Hydrogen Energy, **Elsevier**, 2016, 41 (38), 16,743-16,751.
3. V Khare, S Nema, P Barear, "Optimisation of the hybrid renewable energy system by HOMER, PSO and CPSO for the study area", International Journal of Sustainable Energy, **Taylor and Francis**, 2017, 36 (4), 326-343.
4. V Khare, S Nema, P Baredar,"Reliability analysis of hybrid renewable energy system by fault tree analysis", Energy & Environment, **SAGE Publication**, 2019 30 (3), 542-555.

8.1 Introduction

At present, large number of countries produce electricity by conventional or nonrenewable energy systems. These systems produce large amount of atmospheric pollution. This problem is largely overcome by intensive use of alternative or renewable energy system. For many years, several renewable

Decision Science and Operations Management of Solar Energy Systems.
DOI: https://doi.org/10.1016/B978-0-323-85761-1.00009-3

energy sources have been used for electricity generation in all over the world. There are several types of renewable energy sources, such as solar energy, wind energy, geothermal energy, tidal energy, and wave energy. In all of that wind energy and solar energy are ubiquitous, free available, and ecological responsive. But wind energy system may not be technically feasible at all sites because wind velocity is very random in nature and being more irregular than solar energy.

So for better production of electricity, solar and wind integrated systems are preferred. The hybrid system may be one of the alternatives of oil-produced energy. Renewable energy system such as wind and solar system provide a better business opportunity for public and private sector. The common drawback of wind and solar system is their random nature and they rely on the atmospheric change. On the other hand, various sizing, mathematical, and simulation model of these systems have been used in various aspects. The basic structures of component of hybrid system basically depend on the type of application. For proper generation of electricity, it is necessary to observe the energy requirement and availability of different renewable energy system in the study area [1,2].

So based on geological and weather condition of the study area, standalone solar–wind hybrid renewable energy system is proposed for police station Sagar. In which solar system and wind system are main contributor and battery storage, advanced power electronics, and existing diesel generator is working as supporting elements [3,4].

Based on geological and weather condition of the study area, we propose the stand-alone solar–wind hybrid renewable energy system for police station Sagar, India. In this proposed standalone system, solar energy and wind energy worked as the main contributor of electricity along with battery storage, advanced power electronics, and existing diesel generator functioned as supporting elements. The installation of the standalone solar–wind renewable energy system with diesel generator backup system is economical and provides environmental sustainability rather than a conventional diesel generator. The data of hourly wind speed, hourly vertical and horizontal solar radiation, and load during a year are measured in the region with the help solar and wind velocity measuring instruments. HOMER is used to design the appropriate component where net present cost is the main criteria. HOMER microgrid software provides the detailed analysis of chronological simulation & optimization of a model that is relatively simple, easy to use, and adapt for a wide variety of projects. For a village or community-scale power system, HOMER can model the system based on both technical and economic factors which is involved in that project. For larger systems, HOMER can provide an important overview that compares the cost and feasibility of different configurations [1,2,3,4].

Here a case study of Sagar, which is based on the practical data with analysis to be done using HOMER software. HOMER software means hybrid optimization model for electric renewable, and its simulation is carried out with and without grid connection.

8.2 Study area

Sagar district is an extensive, elevated, and in parts tolerably level plain, broken in places by low hills of the Vindhyan sandstone. It is traversed by numerous streams, chief of which are the Sunar, Beas, Dhasan, and Bina rivers, all flowing in a northerly direction toward the valley of the Ganges. In the southern and central parts, the soil is black, formed by decaying trap; to the north and east it is a reddish-brown alluvium. Iron ore of excellent quality is found and worked at Hirapur, a small village in the extreme northeast [1,2].

Sagar region is situated in the north-central part of the state of Madhya Pradesh with distributed over a land area of 1052 square km. The latitude of Sagar city lies between 23.10’ and 24.27’ north and longitude is 78.04’ and 79.21’ east. Sagar district is bound with several other districts. According to the weather condition climate of Sagar region divided into three parts, winter, summer, and monsoon season. March to May comprise summer season and the monsoon period start from the second week of June to end of September. Remaining season constitute with winter atmosphere. Average minimum and maximum temperature of winter season 11.60°C and 24.50°C, respectively, but in the summer season temperature rises up to 42°C. Fig. 8.1 shows the location of the Sagar city on the map of India.

India fulfills the total annual energy requirement by the solar energy system because the country receives solar radiation equivalent to 5000 trillion kWh/yr. The state of MP is blessed with an excellent potential of wind energy as well as an excellent potential of solar energy, in the state approximately 275 sunny days in a year. This brings us to a hybrid regime of PV and winds energy to keep the police station well-lit round the year. The availability of renewable energy resources of police control room sites is an important factor to develop the hybrid system. In many parts of India wind and solar energy is abundantly available. These energy sources are intermittently and naturally available, due to these factor renewable energy sources such as wind and solar are the first choice to power the remotely located Sagar city sites. The availability of renewable energy resources in the study area is an important factor to develop the integrated system. The

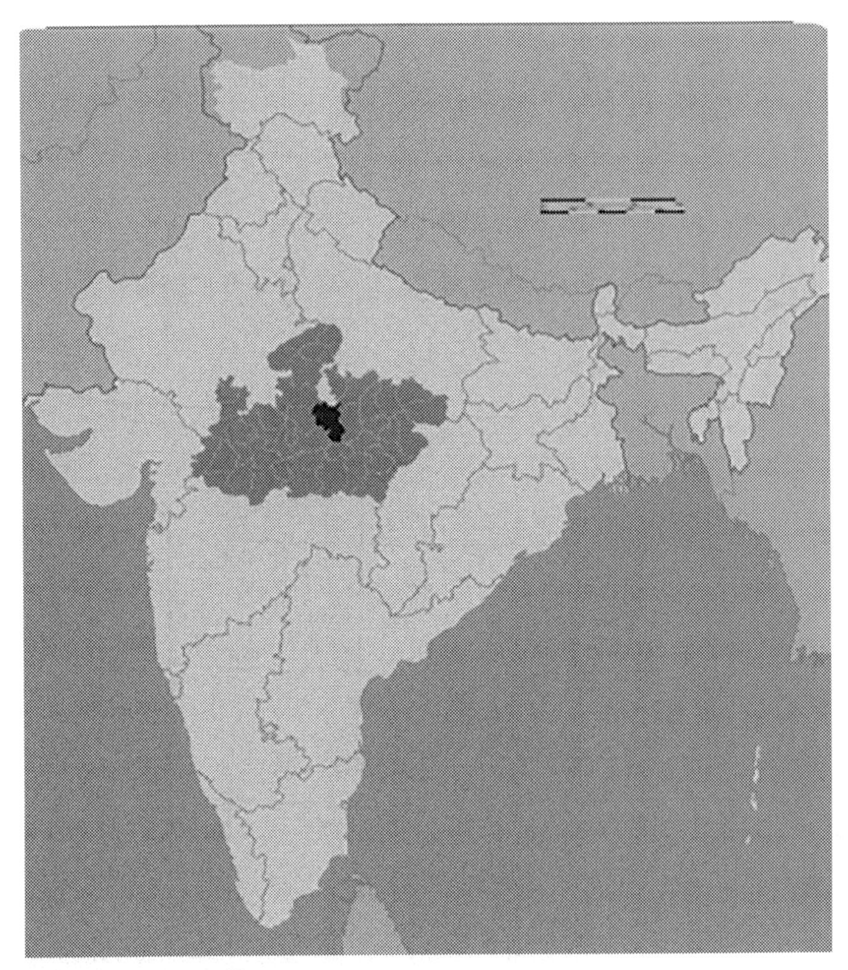

Figure 8.1 Location of Sagar.

weather data are important factors for prefeasibility analysis of renewable hybrid energy system for the study area [1,2,3,4].

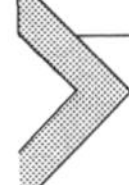

8.3 Solar radiation & wind velocity

8.3.1 Solar radiation

Solar radiation is a measure of how much solar power you are getting at your location. This irradiance varies throughout the year, depending on the seasons. It also varies throughout the day, depending on the position of the sun in the sky, and the weather. Solar insolation is a measure of solar irradiance over of a period of time—typically over the period of a single day.

Table 8.1 Clearness index of study area.

Month	Clearness index
January	0.651
February	0.669
March	0.668
April	0.664
May	0.640
June	0.542
July	0.440
August	0.403
September	0.541
October	0.680
November	0.673
December	0.660

HOMER uses the solar resource input to calculate the PV array to each hour of the year. Enter the latitude, and either an average daily radiation value or an average clearness index for each month. HOMER uses the latitude value to calculate the average daily radiation from the clearness index and vice versa. Table 8.1 shows month-wise clearness index of the study area. The clearness index can be defined on an instantaneous, hourly, or monthly basis. The clearness index values in HOMER's solar resource Inputs window are put in the form of monthly average values [5,6].

It is a measure of the clearness of the atmosphere and the fraction of the solar radiation that is transmitted through the atmosphere to strike the surface of the Earth. It is a dimensionless number between 0 and 1, defined as the surface radiation divided by the extraterrestrial radiation. The clearness index has a high value under clear, sunny conditions, and a low value under cloudy conditions. Fig. 8.2 shows the month-wise solar radiation of the study area, which shows maximum solar radiation is in the month of April and May and minimum is in the month of August. Fig. 8.3 shows month wise clearness index of the study area. On average the extraterrestrial irradiance is 1367 Watts/m^2 (W/m^2). This value varies by ±3% as the earth orbits the sun [7,8].

8.3.2 Wind velocity

Wind speed, or wind flow velocity, is a fundamental atmospheric rate. Wind speed is caused by air moving from high pressure to low pressure, usually due to changes in temperature. Wind speed data are measured at height above the ground is known as anemometer height. The mathematical modeling of wind energy conversion system depends on wind turbine dynamics and generator modeling. Nondimensional performance as a function of the tip

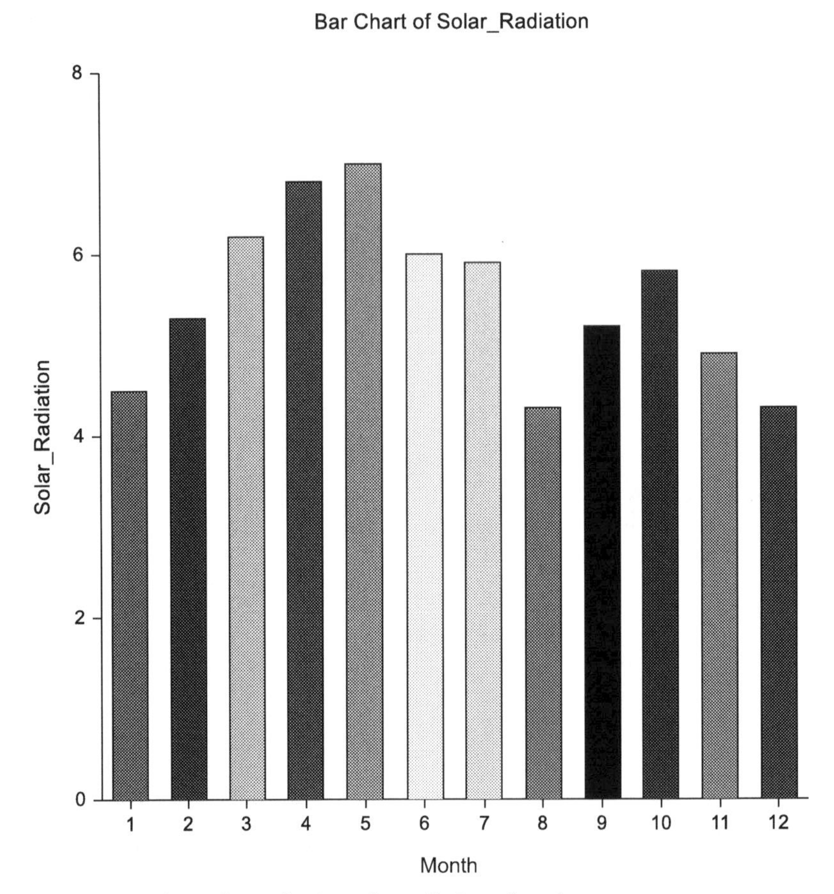

Figure 8.2 Bar chart of month wise solar radiation of study area.

speed ratio is a basic characteristic of wind turbines. Basically generated power is largely depending on the cube of the wind velocity. HOMER uses the wind resource input to calculate the wind turbine power to each hour of the year. Fig. 8.4 shows month-wise wind velocity in the study area.

8.4 Load profile of study area

Accurate study area energy consumption data are required for planning the optimal production capacity of a hybrid renewable power system. The data of electricity consumption are usually a sum of energy of numerous devices without detailed information about the events on each individual.

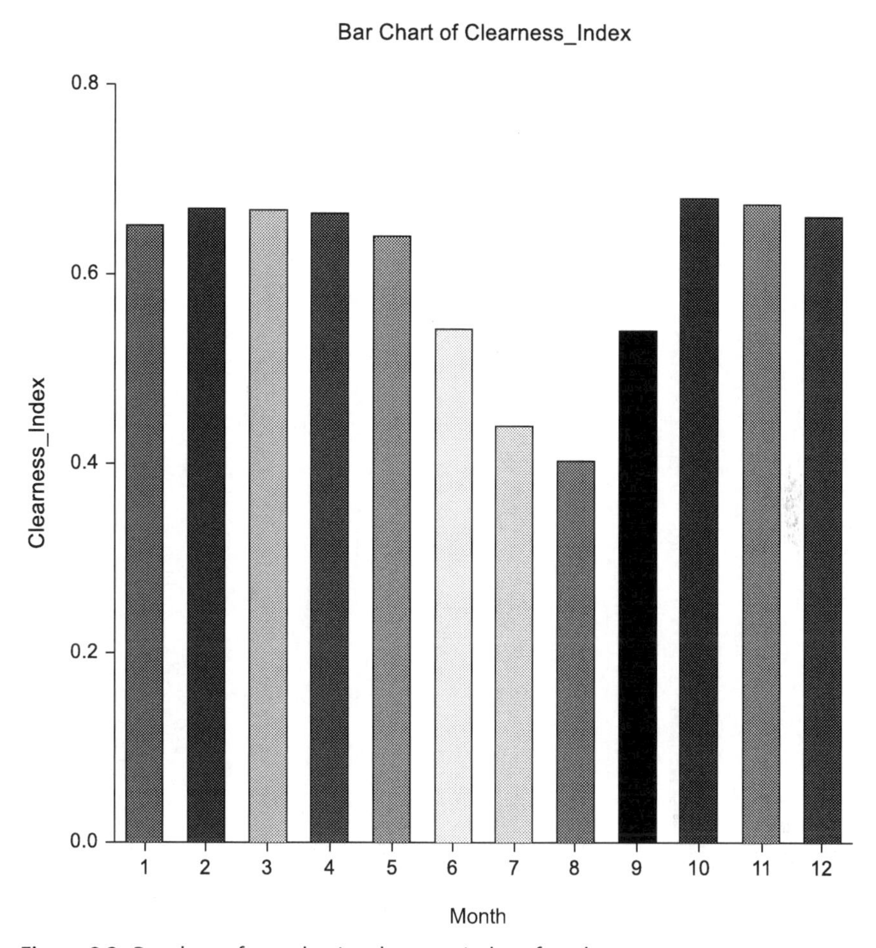

Figure 8.3 Bar chart of month wise clearness index of study area.

An ideal case is the one with a known consumption pattern and with details of various appliances. Yet another way is to consider statistical averages and sample data. Analyzing energy consumption data, we could identify the basic characteristics of load curves of devices that change on a periodical basis. A survey was conducted to identify energy consumption in a typical grid-connected police station in central India. Fig. 8.5 presents seasonal load profiles of the study area.

A set of energy consumption data for a typical grid-connected police station in Sagar was collected. This data was sampled every 1 hour for 365 days of a year. In a typical day, energy consumption is higher in the morning from 6 A.M. to 10 A.M. and in evening from 6 P.M. to 11 P.M.

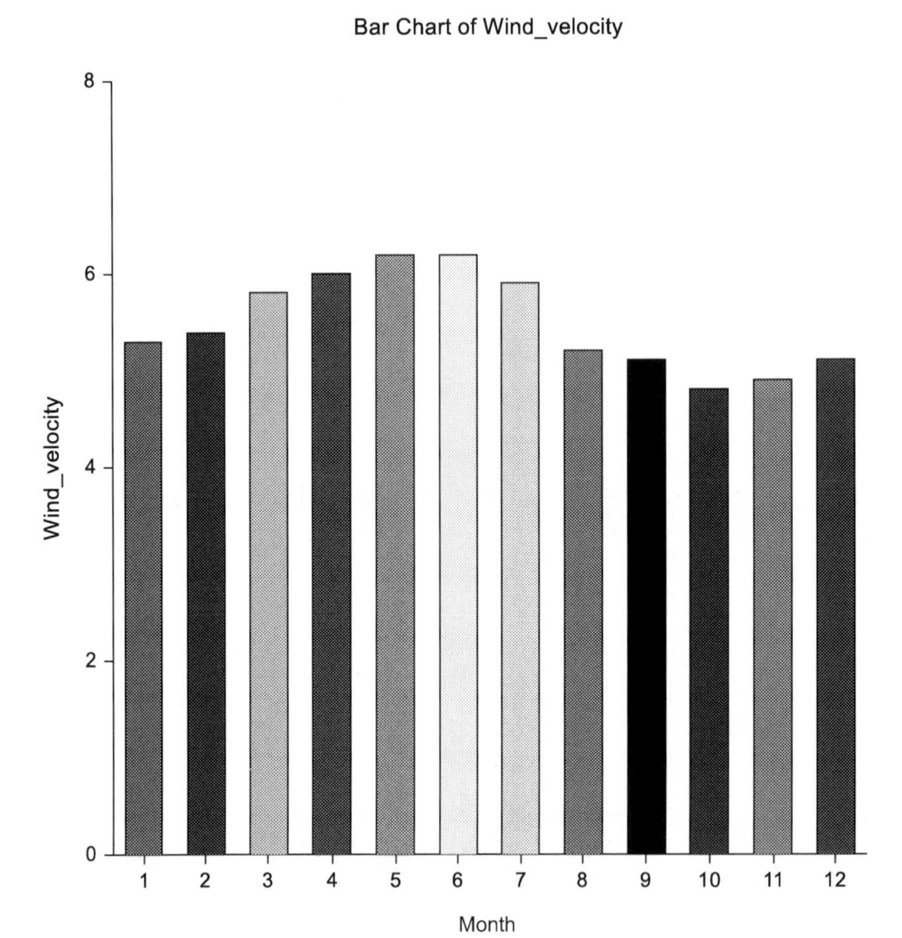

Figure 8.4 Month wise wind velocity of study area.

8.5 Statistical assessment of datasets

Statistics is the branch of mathematics that deals with the gathering, organizing, analyzing, interpreting, and presenting of data. It is customary to start with a statistical population or model to be researched when applying statistics to a scientific, industrial, or social problem. In this section, the statistical assessment of solar radiation, wind velocity, and load demand data [9,10]. Table 8.2 shows the statistical parameters of solar radiation & wind velocity and load demand, respectively. Table 8.3 shows a statistical analysis of load demand.

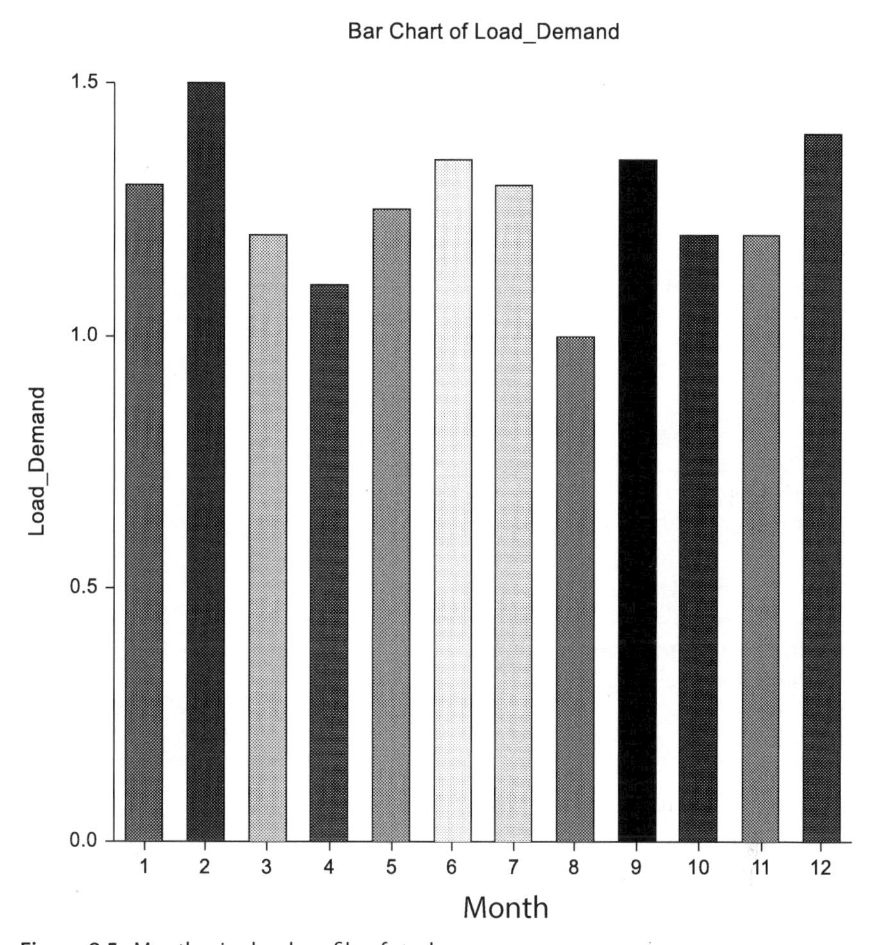

Figure 8.5 Month wise load profile of study area.

8.5.1 Correlation between solar radiation and load demand and between wind velocity and load demand

The correlation coefficient, often known as correlation, is a measure of the relationship between two variables that do not have a unit. This approach allows you to estimate three different forms of correlations: Pearson (product-moment) correlation, Spearman rank correlation, and Kendall's Tau correlation [11,12]. Table 8.4 shows the basic parameters of solar radiation and load demand. Table 8.5 shows the comparison between different parameters of load demand and solar radiation. Table 8.6 shows the Pearson correlation confidence interval section. The Pearson correlation coefficient is the most

Table 8.2 Statistical parameter of solar radiation and wind velocity.

Statistical parameters	Solar radiation	Wind velocity
Count	12	12
Sum	66.2	65.9
Mean	5.516667	5.491667
Standard deviation	0.9193212	0.503548
Standard error	0.2653852	0.1453618
Lower 95% CL mean	4.932558	5.171728
Upper 95% CL mean	6.100776	5.811606
Median	5.55	5.35
Minimum	4.3	4.8
Maximum	7	6.2
Range	2.7	1.4
Interquartile range	1.55	0.875
25th Percentile	4.6	5.1
75th Percentile	6.15	5.975
90th Percentile	6.94	6.2
Variance	0.8451515	0.2535606
Mean absolute deviation	0.7666667	0.4402778
Mean absolute deviation from median	0.7666667	0.425
Coefficient of variation	0.1666443	0.09169311
Coefficient of dispersion	0.1381381	0.07943925
Skewness	0.1172111	0.184977
Kurtosis	1.869111	1.564474

often used metric of correlation. The simple correlation coefficient, also known as the Pearson product-moment correlation, the sample correlation coefficient, or simple linear correlation, is a type of correlation coefficient. Table 8.7 shows the comparison between different parameters of load demand and wind velocity. Where X is the solar radiation and Y is the load demand. Fig. 8.6 shows the graph between load demand and solar radiation. Table 8.8 shows the Pearson coefficient between load demand and wind velocity. The estimated Pearson correlation is 0.0084. This report presents two sets of confidence limits: one based on the exact correlation distribution, and the other based on a Normal approximation using Fisher's Z-transformation. Fig. 8.7 shows the plot between wind velocity and load demand.

8.5.2 Forecasting of clearness index, solar radiation, wind velocity, and load demand

Forecasting is the technique of predicting the future based on historical and current data. These can then be compared to what really occurs. For

Table 8.3 Statistical analysis of load demand (kW).

Load demand	
Count	12
Sum	15.15
Mean	1.2625
Standard deviation	0.1350505
Standard error	0.03898572
Lower 95% CL mean	1.176693
Upper 95% CL mean	1.348307
Median	1.275
Minimum	1
Maximum	1.5
Range	0.5
Interquartile range	0.15
25th Percentile	1.2
75th Percentile	1.35
90th Percentile	1.47
Variance	0.01823864
Mean absolute deviation	0.1041667
Mean absolute deviation from median	0.1041667
Coefficient of variation	0.1069707
Coefficient of dispersion	0.08169935
Skewness	−0.2186451
Kurtosis	2.728361

Table 8.4 Basic parameter of solar radiation and load demand.

Parameter	Value	Parameter	Value
Y axis variable	Load demand	Rows processed	12
X axis variable	Solar radiation	Rows used in estimation	12
Frequency variable	None	Rows with X missing	0
Sum of frequencies	12	Rows with frequency missing	0

Table 8.5 Comparison between different parameters of load demand and solar radiation.

Variable	Count	Mean	Standard deviation	Minimum	Maximum
Load demand	12	1.26	0.14	1.00	1.50
Solar radiation	12	5.52	0.92	4.30	7.00

Table 8.6 Pearson correlation confidence interval section two-sided confidence interval of ρ.

R distribution Pearson correlation	Normal approximation 95% confidence limits count	95% confidence limits lower	Upper	Lower	Upper
−0.1153	12	0.4719	−0.6249	0.6464	0.4911

Table 8.7 Comparison between different parameters of load demand and wind velocity.

Standard variable	Count	Mean	Deviation	Minimum	Maximum
Load demand	12	1.26	0.14	1.00	1.50
Wind velocity	12	5.49	0.50	4.80	6.20

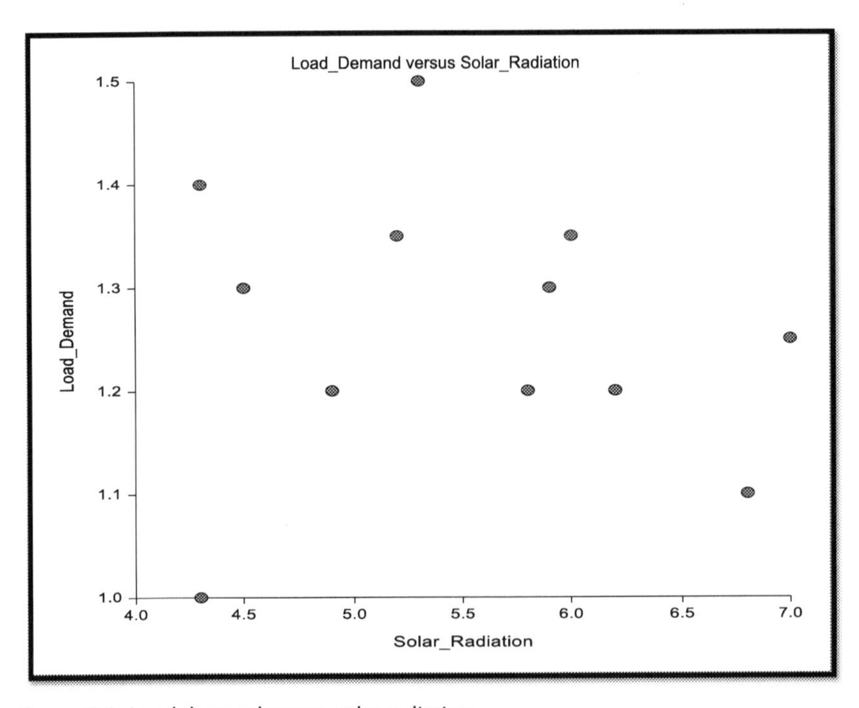

Figure 8.6 Load demand versus solar radiation.

Table 8.8 Pearson correlation confidence interval section two-sided confidence interval of ρ.

Pearson correlation	Count	R distribution 95% confidence limits		Normal approximation 95% confidence limits	
0.0084	12	−0.5474	0.5584	−0.5683	0.5795

example, a corporation might forecast sales for the coming year and then compare it to actual outcomes. Prediction is a related but broader phrase. In this case study, it is necessary to identify future value of solar radiation, wind velocity, and load demand to predict the performance of solar wind hybrid renewable energy systems [13,14]. Table 8.9 shows the forecast summary section of solar radiation. Table 8.10 shows that forecasting assessment, which

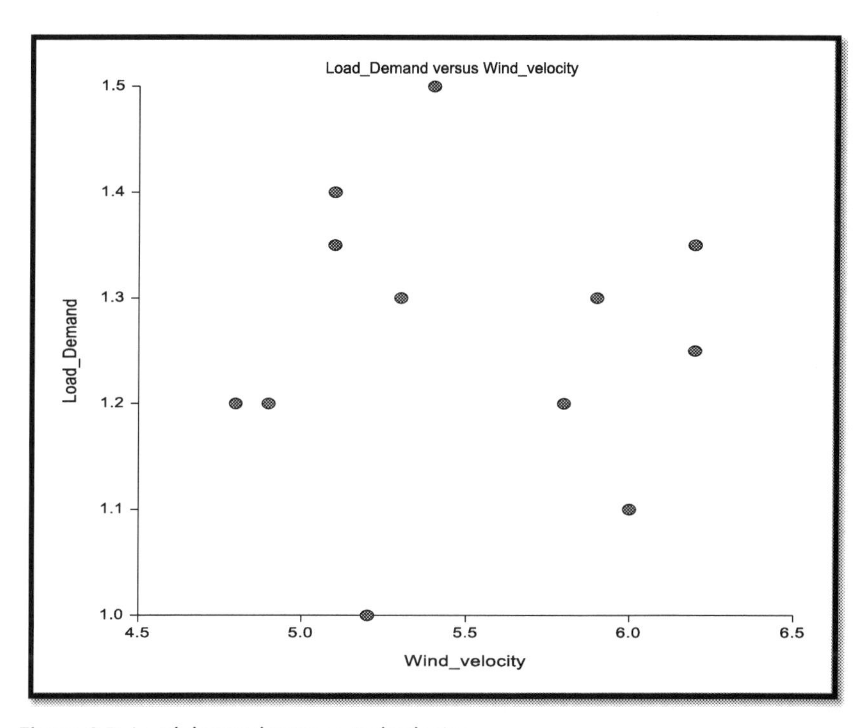

Figure 8.7 Load demand versus wind velocity.

Table 8.9 Forecast summary section.

Variable	**Solar radiation**
Number of rows	12
Mean	5.516667
Missing values	None
Pseudo R-Squared	0.000000
Mean square error	0.8843192
Mean |Error|	0.8138332
Mean |Percent Error|	15.56389
Forecast method	Double Smooth
Search criterion	Mean Square Error
Alpha	0.2205387
Intercept (A)	5.998731
Slope (B)	−0.0955909

is done through the double smooth techniques, and it is identified the mean value of solar radiation is 5.51,667 kwh/m^2/day. Figs 8.8 and 8.9 show the predicted and residual plot of solar radiation. In a similar way, Tables 8.11,

Table 8.10 Forecasts section.

No.	Solar radiation
13	4.756049
14	4.660458
15	4.564867
16	4.469276
17	4.373686
18	4.278095
19	4.182504
20	4.086913
21	3.991322
22	3.895731
23	3.80014
24	3.704549

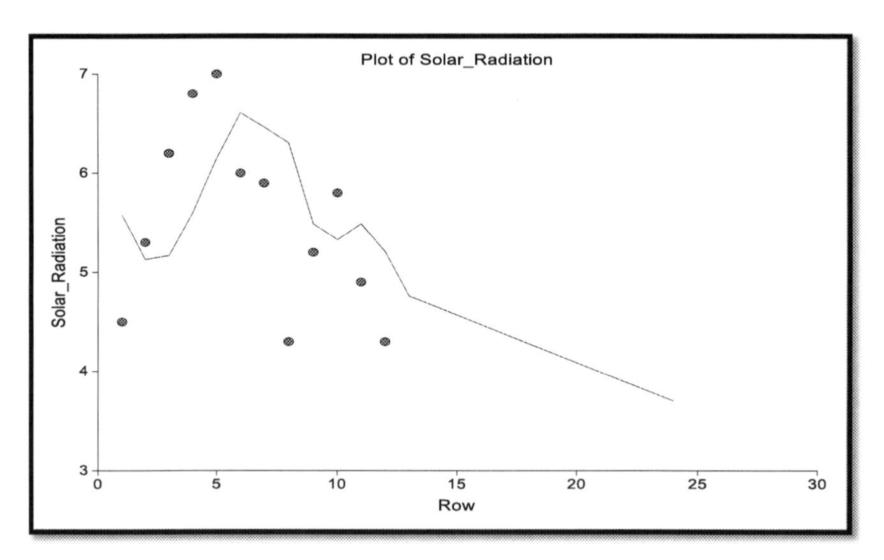

Figure 8.8 Forecasting plot of solar radiation.

8.13, and 8.15 show the forecast summary of wind velocity, load demand, and clearness index. Tables 8.12, 8.14, and 8.16 show the predicted value of the next 12 months for wind velocity, load demand, and clearness index. Figs 8.10, 8.12, and 8.14 show the predicted value of wind velocity, load demand, and clearness index, respectively. Figs 8.11, 8.13, and 8.15 show the residual plot of wind velocity, load demand, and clearness index, respectively.

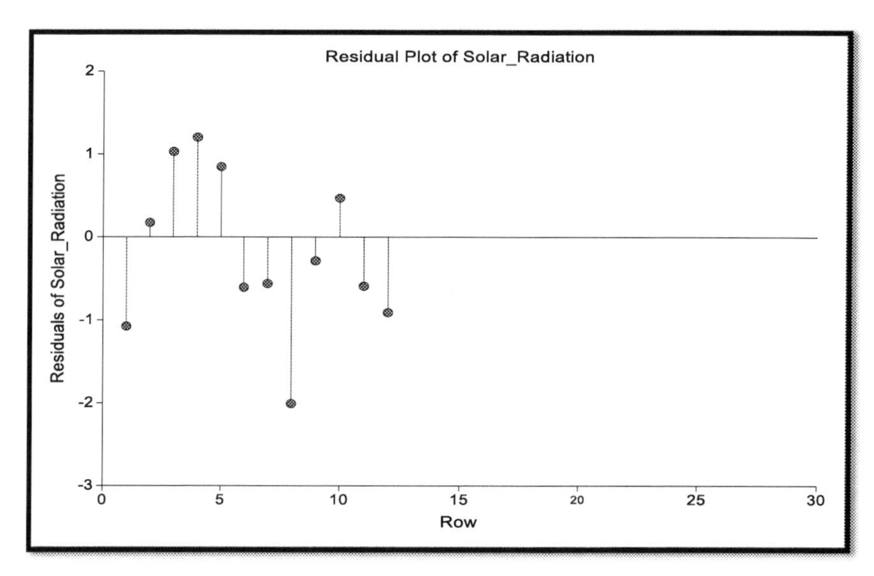

Figure 8.9 Residual plot of solar radiation.

Table 8.11 Forecast summary section of wind velocity.

Variable	Wind velocity
Number of rows	12
Mean	5.491667
Missing values	None
Pseudo R-Squared	0.656753
Mean square error	0.07978105
Mean error	0.2323239
Mean percent error	4.324657
Forecast method	Double smooth
Search criterion	Mean square error
Alpha	0.9093792
Intercept (A)	3.012773
Slope (B)	0.1738198

8.6 Modeling of solar–wind hybrid renewable energy system

Mathematical modeling is the initial phase in the design of any hybrid renewable energy system, and it provides a precise picture of the system. Wind energy, photovoltaics, diesel engines, and battery storage are all part

Table 8.12 Predicted value of wind velocity.

No.	Wind velocity
13	5.27243
14	5.44625
15	5.62007
16	5.79389
17	5.96771
18	6.141529
19	6.315349
20	6.489169
21	6.662989
22	6.836808
23	7.010628
24	7.184448

Table 8.13 Forecast summary section of load demand.

Variable	Load demand
Number of rows	12
Mean	1.2625
Missing values	None
Pseudo R-squared	0.000000
Mean square error	0.0183557
Mean error	0.1072382
Mean percent error	8.859279
Forecast method	Double smooth
Search criterion	Mean square error
Alpha	0.04689655
Intercept (A)	1.285854
Slope (B)	−0.002406881

Table 8.14 Predicted value of load demand.

No.	Load demand
13	1.254564
14	1.252157
15	1.249751
16	1.247344
17	1.244937
18	1.24253
19	1.240123
20	1.237716
21	1.235309
22	1.232902
23	1.230496
24	1.228089

Table 8.15 Forecast summary section.

Variable	Clearness index
Number of rows	12
Mean	0.6025833
Missing values	None
Pseudo R-squared	0.397331
Mean square error	0.005232417
Mean error	0.04425
Mean percent error	7.908377
Forecast method	Double smooth
Search criterion	Mean square error
Alpha	1
Intercept (A)	0.816
Slope (B)	−0.013

Table 8.16 Predicted value of clearness index.

No.	Clearness index
13	0.647
14	0.634
15	0.621
16	0.608
17	0.595
18	0.582
19	0.569
20	0.556
21	0.543
22	0.53
23	0.517
24	0.504

of the proposed system [15,16,17]. Fig. 8.16 represented a block diagram of a conventional PV–wind hybrid renewable energy system.

8.6.1 Modeling of PV system

According to this phenomena, a photovoltaic solar cell turns sunlight into an electric current by using the photovoltaic effect. A photovoltaic module is similar to a solar cell, but it also contains a diode equivalent; Fig. 8.17 depicts the diode equivalent, which includes five parameters. Because shunt resistance (Rsh) is substantially higher than series resistance, it is in parallel with the diode (Rs). The relationship between the output voltage and load

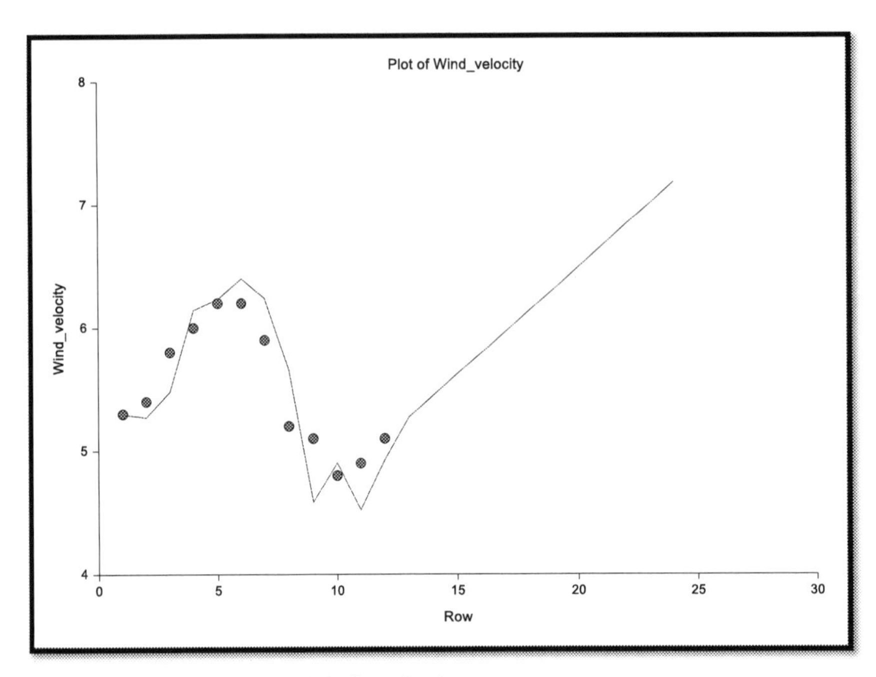

Figure 8.10 Forecasting model of wind velocity.

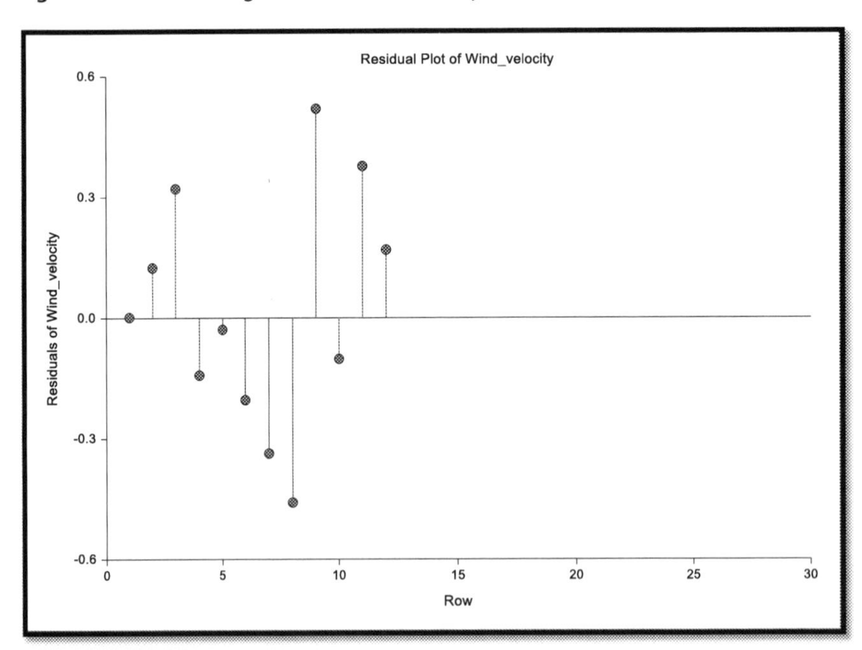

Figure 8.11 Residual plot of wind velocity.

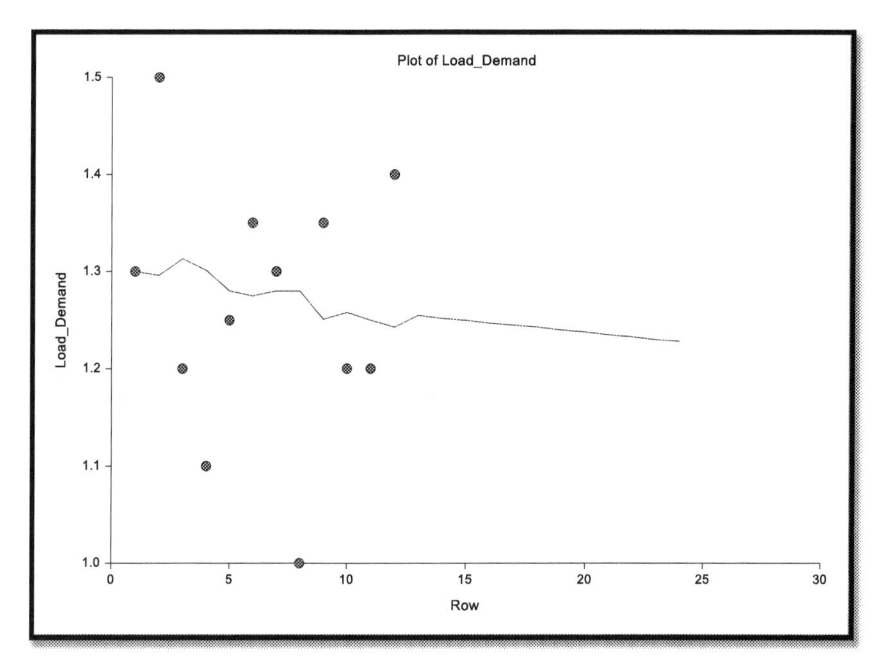

Figure 8.12 Forecasting plot of load demand.

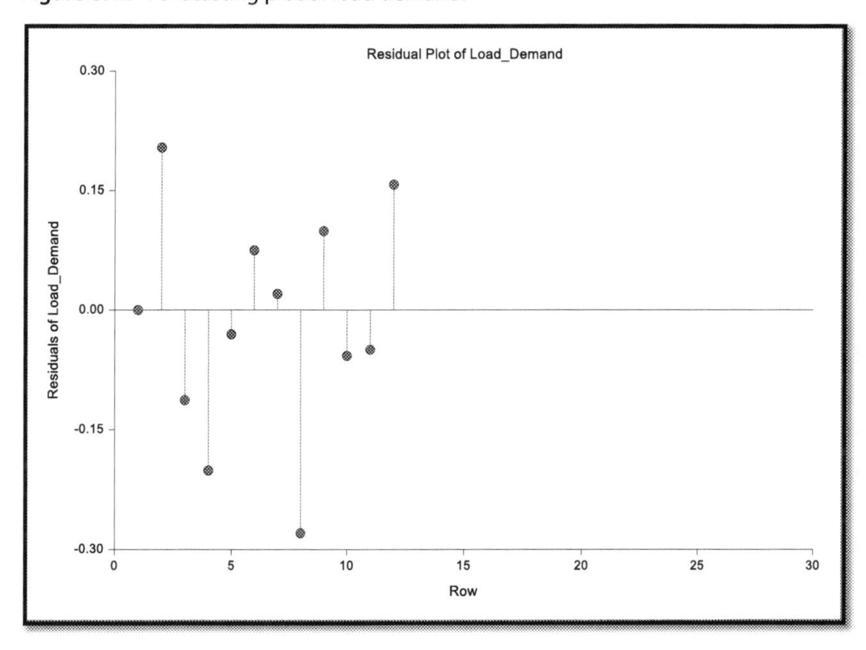

Figure 8.13 Residual plot of load demand.

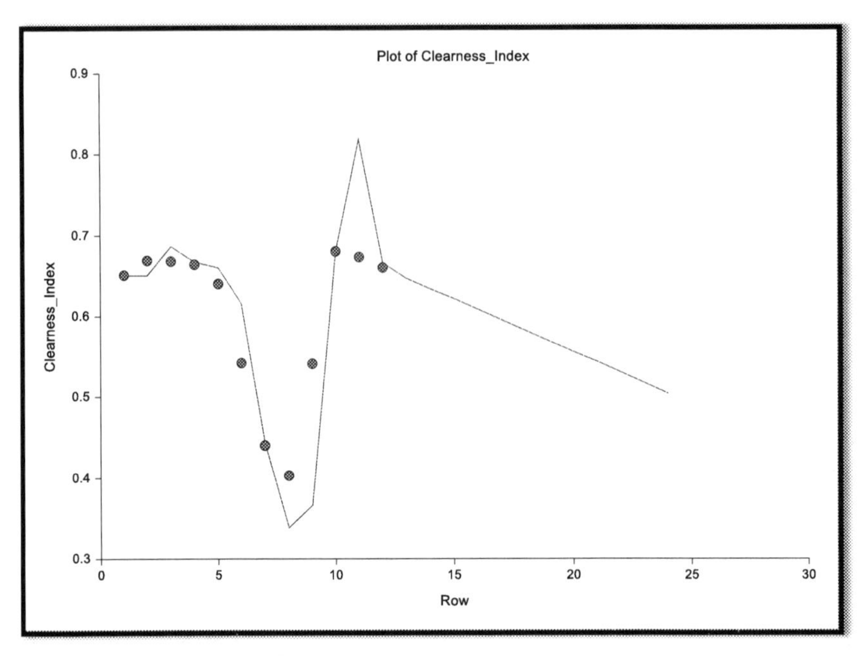

Figure 8.14 Forecasting of clearness index.

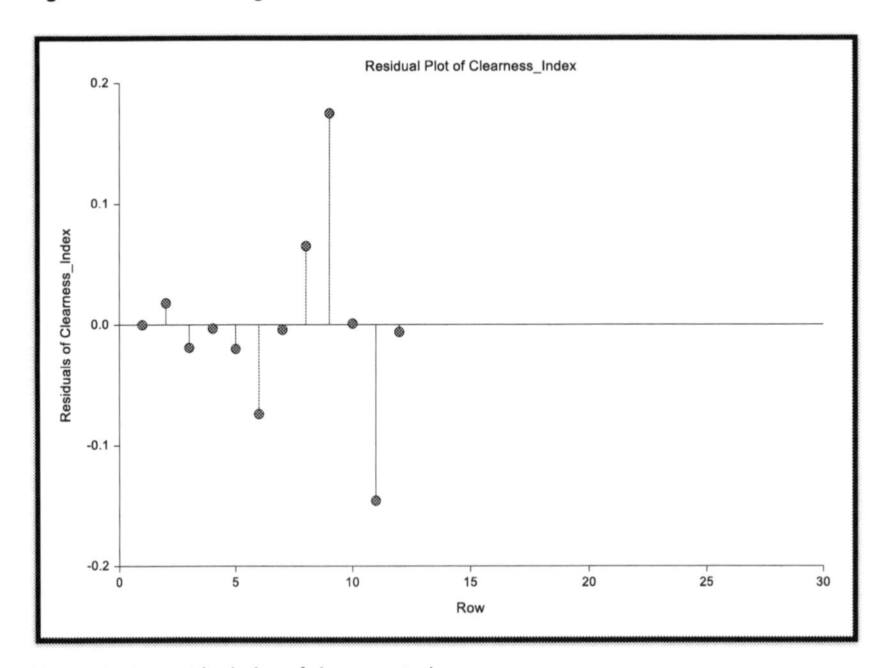

Figure 8.15 Residual plot of clearness index.

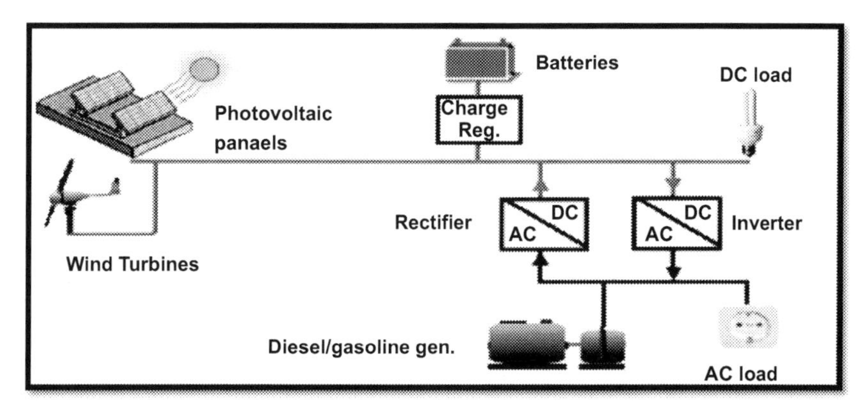

Figure 8.16 Block diagram of a conventional PV–wind–diesel–battery system.

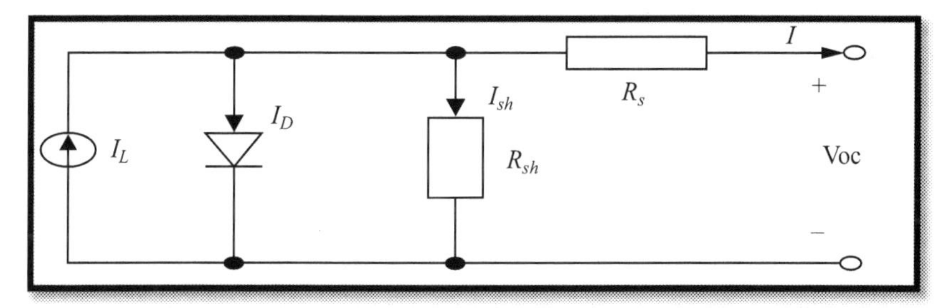

Figure 8.17 Equivalent circuit of photo voltaic system.

current, as shown below, is the most significant aspect of the circuit.

$$I = I_{SC}\left[1 - \exp\left(\frac{V - V_{OC} + IR_S}{V_t}\right)\right]$$

where I_{sc}, V_{oc}, R_S, V_t are short circuit current, open-circuit voltage, series resistance, thermal voltage, respectively. The voltage V_t equals 25 mVat 300 K for a typical monocrystalline silicon solar cell.

The power generated by the solar system is calculated as:

$$P_v(t) = N_{pv}.\ V_m(t).I_m(t)$$

where $P(t)$ is power generated by the solar system, N_{pv} is the number of modules used in solar system, $V_m(t)$ and $I_m(t)$ are the values of maximum voltage and current of module at time t, respectively.

8.6.2 Modeling of wind system

Anemometer height is the height at which wind speed data are measured above the ground. Wind turbine dynamics and generator modeling are part of the mathematical modeling of a wind energy conversion system.

A primary property of wind turbines is nondimensional performance as a function of tip speed ratio. Basically, the amount of power generated is proportional to the cube of the wind velocity. The output of mechanical power captured from the wind by a wind turbine can be formulated as:

$$W_t = -\frac{\left(C_p \lambda \rho A V^3\right)}{2}$$

And torque developed by a wind turbine can be expressed as:

$$T_t = {W_t}/{\omega_m}$$

where W_t is the output power, T_t is the torque developed by the wind turbine, C_p is the power coefficient, λ is the tip speed ratio, ρ is the air density in kg/m^3, A is the frontal area of wind turbine, V the wind speed. In wind energy system data, Weibull distribution factor which is the distribution of wind speed over a year is 1.961. Autocorrelation factor is 0.86 which present random behavior of wind speed.

8.6.3 Modeling of diesel generator

The selection of a diesel generator depends on the category and nature of the load. To determine the rated capability of the engine generator to be installed, the following two cases should be considered:

1. The rated capacity of the generator must be at least equal to the maximum load, then it is possible for the diesel generator is directly connected to a load.
2. If the diesel generator is works as a battery charger, then the current produced by the generator should not be greater than CAh/5 A, where CAh is the ampere-hour capacity of the battery.

   Overall η of diesel generator is given by

$$\eta_{\text{overall}} = \eta_{\text{break thermal}} \times \eta_{\text{generator}}$$

Here $\eta_{\text{break thermal}}$ is brake thermal efficiency of diesel-engine. Normally, diesel generators are modeled in the control of the hybrid power system in order to achieve the required autonomy.

8.6.4 Modeling of battery bank

The cumulative sum of the daily charge/discharge condition of the battery is the battery state of charge (SOC). The maximum battery capacity is reached when the battery is fully charged and holds the total quantity of energy. The status of the battery at any given hour is related to the prior SOC as well as the system's energy production and consumption situation from $t-1$

to t. During the charging process, when the total output of all generators exceeds the load demand, the available battery bank capacity at hour t can be described by:

$$B_{BAT}(t) = B_{BAT}(t-1) - B_{out}(t) \times \eta_{charging}$$

where B_{BAT} (t) is energy stored in the battery at hour t kWh, B_{BAT} $(t-1)$ is energy stored in the battery at hour t–1 kWh, B_{out} (t) is energized out of battery in time t, $\eta_{charging}$ is battery charging efficiency.

8.7 Standalone hybrid renewable energy system

A hybrid renewable energy system combines several renewable energy sources. The fundamental component of this suggested system is a solar and wind energy system. The battery serves as a storage medium, and the diesel generator serves as a backup. The HOMER model reflects the overall system. HOMER replicates a system's operation by calculating energy balances for each of the 8760 hours in a year. HOMER calculates the energy flows to and from each component of the system by comparing the hour's electric and thermal demand to the energy that the system can deliver in that hour. HOMER chooses how to operate the generators and whether to charge or discharge the batteries for each hour in systems using batteries or fuel-powered generators. For each system configuration you choose to investigate, HOMER performs these energy balance computations. It then analyses whether a configuration is practical, that is, whether it can meet the electricity demand under the specified parameters, and calculates the cost of installing and operating the system during the project's lifetime. Costs such as capital, replacement, operation and maintenance, fuel, and interest rate are all factored into the system cost calculations [18,19]. Fig. 8.18 shows the interconnection of components of a hybrid renewable energy system.

To use HOMER, we feed it with inputs that show equipment possibilities, component pricing, and the availability of wind and solar reserves. Homer uses these inputs to simulate various system configurations or component combinations, producing results that can be viewed as a list of feasible configurations sorted by net present cost. Homer also shows simulation results in a variety of tables and graphs, which may be used to compare and assess setups based on their economic and technical advantages.

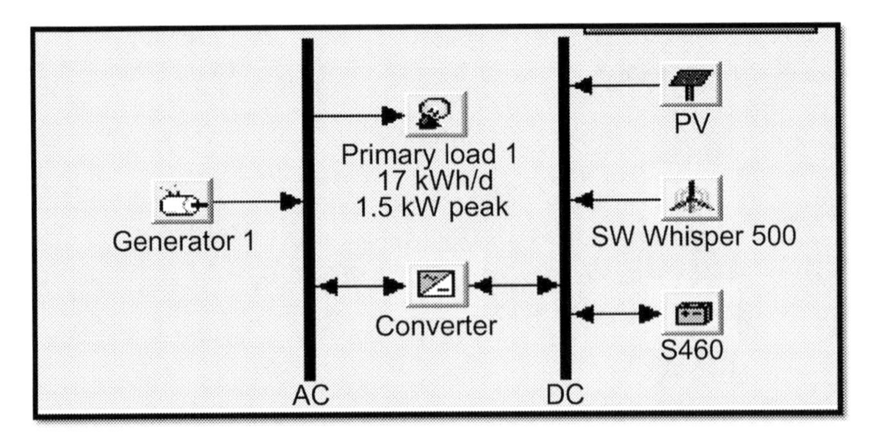

Figure 8.18 Hybrid system component.

8.7.1 Photo-voltaic system

In the solar resource input window of the HOMER simulation program, month-by-month average sun radiation statistics are required. When all of the relevant data have been entered, HOMER must calculate the solar radiation impinge on the PV array's surface for each time step; the temperature coefficient is an important aspect of this analysis. HOMER assumes that the temperature coefficient is 0 if the effect of temperature on the PV array is ignored. PV arrays might cost anywhere between Rs. 60 and Rs. 120 per watt at first. In a more optimistic scenario, the cost of installing a 3 KW solar energy system is Rs. 189,780, and the cost of replacing a component at the end of its lifetime is Rs. 170,760. Three distinct sizes are considered: 3KW, 6KW, and 8KW, with a lifetime of 20 years for the PV array. The study also takes into account the horizontal axis, monthly adjustments, and the effect of temperature. The specification of the PV system employed in the case study is shown in Table 8.17. Global horizontal irradiance is the total quantity of shortwave radiation received from above by a surface horizontal to the ground. Fig. 8.19 depicts changes in global solar radiation with frequency (percent) throughout 24 hours. This figure, which combines both direct normal irradiance and diffuse horizontal irradiance, is of special significance to photovoltaic installations (DIF).

8.7.2 Wind turbine

The wind reserve participation window is used to describe the available wind resources. HOMER uses this information to compute the production

Table 8.17 Photovoltaic specification.

PV	
Rated capacity	3kW
Mean output	0.519 kW
Mean output	11.65 kWh/day
Capacity factor	16.6%
Total production	4756 kWh/yr
Minimum output	0.00 kW
Maximum output	2.39 kW
PV Penetration	74.5%
Hours of operation	4267 hr/yr
Levelized cost	Rs. 3.6501/kWh

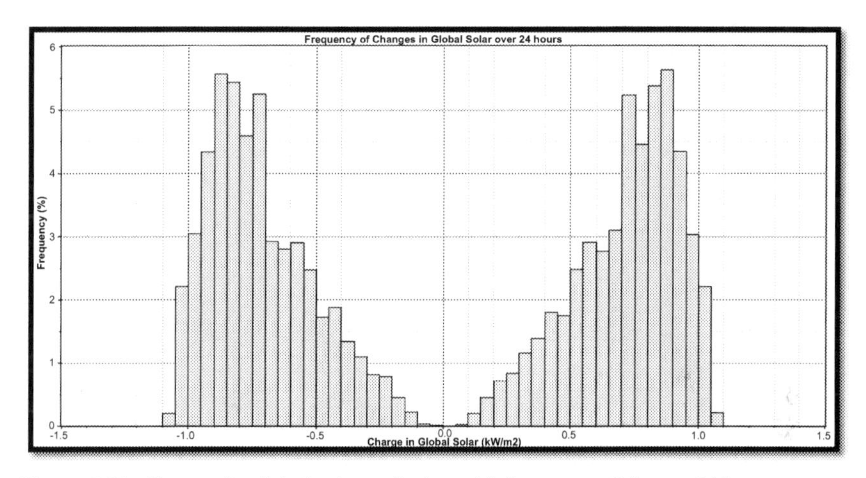

Figure 8.19 Change in global solar radiation with frequency (%) over 24 hour.

of the wind turbine each hour of the year. The most simplified models to simulate the power output of a wind turbine could be calculated from its power–speed curve which is shown in Fig. 8.20. This curve is given by the manufacturer and usually describes the real power transferred from WG to DC bus.

The available wind resources are desçribed in the wind resource input window. This information is used by HOMER to compute the wind turbine's output for each hour of the year. The baseline data consist of 8760 numbers that indicate the average wind speed in meters per second. In the wind resource table and graph, HOMER displays the monthly average generated from the baseline data for each hour of the year. The SW Whisper 500 model is used in this investigation. It has a 3KW rated capacity and a

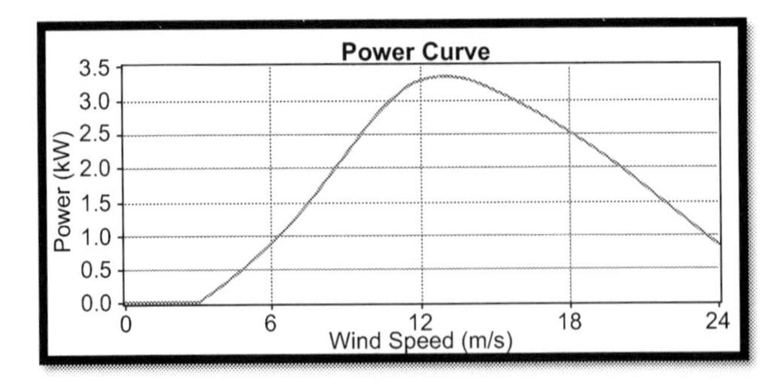

Figure 8.20 Power curve of wind system.

Table 8.18 Specification of wind turbine.
DC WIND TURBINE: SW-WHISPER 500

Rated capacity	3kW
Mean output	1.09kW
Capacity factor	35.78%
Total production	9579 kWh/yr
Minimum output	0 kW
Maximum output	3.63 kW
Wind Penetration	148%
Hours of operation	7147 hr/yr
Levelized cost	Rs. 9.36/kWh

4.5 m rotor diameter. In this analysis, the hub height is 25 m and the lifetime is 15 years. The cost of one unit is considered to be Rs. 780,000 and the cost of replacing a component at the end of its lifetime is Rs. 660,000. Table 8.18 shows the specification of wind turbine system which is used in the case study.

8.7.3 Battery

When the surplus energy generated by WGs and PVs exceeds the load requirement, the excess energy is stored in storage batteries for later use. If we want to expand energy storage capacity and improve energy security, we need to use multiple energy storage systems. On the other hand, if there is a shortfall in renewable energy generation, the stored energy will be used to meet the load. The system's reliability will be improved as a result of this. HOMER battery window displays the properties of the selected battery type. Surrette battery Rs. 27,600 (6V, 460Ah, 2.76kWh) is one of the commercially available models investigated for this investigation. One battery costs Rs. 21,000 in the capital. Because 12 batteries are required, the capital cost of 12 batteries is Rs. 252,000. Round trip efficiency is 80%, and the minimum SOC is 40%, respectively. The maximum charge current is 18A and the maximum charge rate is 1A/Ah. The DOD is the inverse of the SOC: when one rises, the other falls. DOD can use Ah units (e.g., 0 = full, 50 Ah = empty) or percent points (100% = empty; 0% = full) instead of percent points (100% = empty; 0% = full). As a battery may actually have higher capacity than its nominal rating, it is possible for the DOD value to exceed the full value (e.g., 55 Ah or 110%), something that is not possible when using SOC.

8.7.4 Diesel generator

The cost of a commercially available diesel generator may vary from Rs. 3600 to 57,000/KW. The per kW cost for larger units is lower and smaller cost more. The cost of 3.5 kW unit is considered to be Rs. 126,000 and the cost of replacing a component at the end of its lifetime is Rs. 78,000.

8.8 Objective function

The entire installed cost of a component at the start of the project is the initial capital cost of that component. Nonlinear programming techniques are used to optimize an equation given particular constraints and variables that must be minimized or maximized. An objective function is a mathematical representation of a business aim that can be used in decision analysis, operations research, or optimization studies. A freestanding hybrid renewable energy system's objective function is a mathematical equation that is used to reduce the system's cost using various variables and limitations. The total cost of the system is the sum of the capital, replacement, and operating and maintenance costs of the complete element employed in that hybrid.

8.8.1 Annual capital cost

Annual capital cost is mostly determined by the payback duration and interest rate. HOMER calculates the net present cost of an integrated system based on the capital costs of connected elements such as solar panels, wind turbines, batteries, and converters. Annual capital cost $C_{\text{ann}}^{\text{cap}}$ is the cost that needs to be recovered yearly for a payback period of "**y**" years and interest rate "**r.**" The capital recovery factor (CRF) is a ratio used to calculate the present value of an annuity. HOMER calculates the annualized capital cost of each component using the following equation:

$$C_{\text{ann}}^{\text{cap}} = C_{\text{capital}} \cdot \text{CRF}(r, y)$$

The equation for the CRF is expressed as:

$$\text{CRF}(r, y) = \frac{r(r+1)^y}{(r+1)^y - 1}$$

where C_{capital} is capital cost of component, r is interest rate, y is number of year.

8.8.2 Annual replacement cost

The amount of different elements required to replace an asset at this moment is referred to as replacement value. The net present cost of a system is primarily affected by the replacement cost of solar and wind systems when proposing a system. The annual replacement cost of a system component is the annualized value of all replacement expenses that occur over the project's lifetime. HOMER uses the following equation to calculate each component's annualized replacement cost:

$$C_{\text{replacement}} = C_{\text{replacement}} \cdot F_{\text{rep}} \cdot \text{SFF}(r, y)$$

where $C_{\text{replacement}}$, replacement cost of individual component F_{rep}, is a factor arising because the component lifetime can be different from the project lifetime.

Sinking fund factor is a ratio used to calculate the future value of a series of equal annual cash flows. It is given by

$$\text{SFF}(r, y) = \frac{r}{(r+1)^y - 1}$$

8.8.3 Annual O&M cost

The cost of operating and maintaining a component is referred to as the component's O&M cost. The sum of the O&M expenses of each system component is the total O&M cost of the system. Enter the O&M cost as an annual figure for most components. Miscellaneous annual costs, such as the system fixed O&M cost, as well as penalties, such as pollution penalties and capacity shortage penalties, are classified as additional O&M cost by HOMER. On the cost summary and cash flow tabs of the Simulation Results window, HOMER displays the O&M costs. HOMER calculates

O&M cost by the following equation:

$$C_{O\&M} = C_{O\&M,\text{fixed}} + C_{CS} + C_{\text{emission}}$$

where $C_{O\&M,\text{fixed}}$ is fixed operation and maintenance cost, $\boldsymbol{C_{CS}}$ is the penalty factor for capacity shortage in \$/year, and C_{emission} is the penalty for emission.

System total net present cost =

$$\sum_{i=1}^{N_{PV}} N_{PV}^{i} \times \left(C^{i}_{\text{capital,PV}} + C^{i}_{O\&M,PV} + C^{i}_{\text{Replacement,PV}}\right)$$

$$+ \sum_{(j=1)}^{(N_W T)} N^{j}_{WT} \times (C^{j}_{\text{capital,WT}} + C^{i}_{O\&M,WT} + C^{i}_{\text{Replacement,WT}})$$

$$\sum_{j=1}^{N_{WT}} N^{j}_{\text{Battery}} \times \left(C^{j}_{\text{capital, Battery}} + C^{i}_{O\&M,WT} + C^{i}_{\text{Replacement, Battery}}\right)$$

$$+ C_{\text{capital cost\_Generator}}$$

$$+ C_{O\&M\text{ cost\_Generator}} + C_{\text{fuel cost\_Generator}} - C_{\text{salvage value}}$$

where N_{PV}, N_{WT}, $N_{Battery}$ are number of PV panel, a number of wind turbine, and number of batteries, respectively.

8.9 Result and discussion

The major aspect of the HOMER simulation tool is economic issues and limits. We use a 6% annual real interest rate, a 1% annual capacity shortfall, and a 6.5% operating reserve as a proportion of hourly demand in this proposed system. The maximum annual capacity deficiency is calculated by dividing the total capacity shortage by total electrical load. This proposed independent system has a life expectancy of about 25 years, implying that the project has a 25-year lifespan. According to the simulation, overall electrical energy output is a mix of energy produced by solar panels (33%), wind turbines (67%), and diesel generators (12%). The hybrid renewable energy system's net present cost is Rs. 2,088,660, its annual operating cost is Rs. 55,620, and the levelized cost of energy is Rs. 26.4/kWh. Fig. 8.21 depicts the HOMER software's simulated result. Fig. 8.22 depicts monthly average electricity production, which includes solar, wind, and generator generation. The wind energy system produced the most electricity in the month of May. Fig. 8.23 depicts the relationship between incidence solar and solar powers. Fig. 8.24 shows the change of frequency with total renewable power output, where the renewable electricity production is the total amount of

	PV (kW)	W500	GENER (kW)	S460	Conv. (kW)	Initial Capital	Operating Cost ($/yr)	Total NPC	COE ($/kWh)	Ren. Frac.	Diesel (L)	GENER (hrs)	Batt. Lf. (yr)
	3	1	3.5	12	2	$ 22,663	950	$ 34,811	0.444	0.99	80	79	8.0
	8		3.5	14	2	$ 15,635	1,620	$ 36,345	0.464	0.92	651	605	4.9
		1	3.5	14	2	$ 20,200	1,637	$ 41,130	0.525	0.90	671	537	8.0
	6	1	3.5		2	$ 21,626	4,901	$ 84,283	1.075	0.80	3,348	4,157	
	8		3.5			$ 10,535	5,783	$ 84,458	1.077	0.66	4,383	5,437	
			3.5	12	1	$ 6,400	6,799	$ 93,314	1.190	0.00	5,177	4,190	4.5
		2	3.5		2	$ 28,300	5,656	$ 100,600	1.283	0.80	3,610	4,482	
			3.5			$ 2,100	9,182	$ 119,478	1.524	0.00	7,064	8,759	

Figure 8.21 Simulated result of HOMER software.

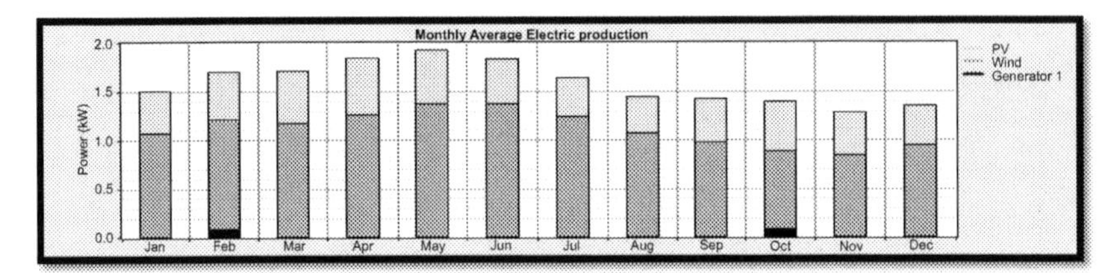

Figure 8.22 Monthly average electric production.

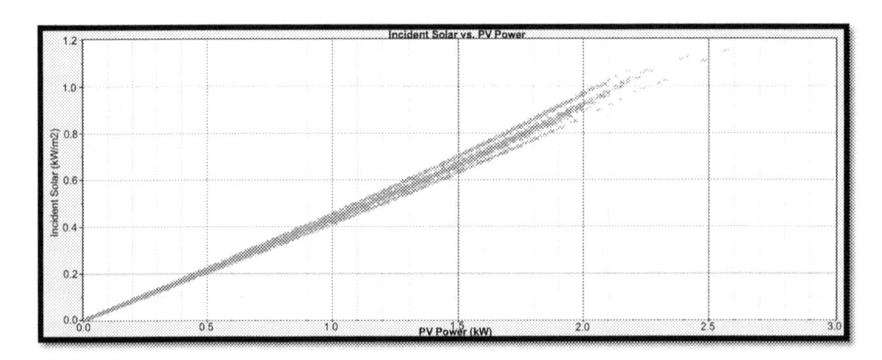

Figure 8.23 Incident solar versus solar power.

electrical energy produced annually by the renewable components of the power system. It is the sum of the electrical energy produced by the PV array, the wind energy system. Fig. 8.25 presents frequencies of changes with renewable penetration. HOMER measures renewable penetration at each time step, which is defined as the ratio of total renewable electrical power output [kW] to total electrical load served in this time step [kW]. Figs 8.26 and 8.27 show total renewable energy output as a function of solar radiation and wind speed, respectively. Renewable energy generation is

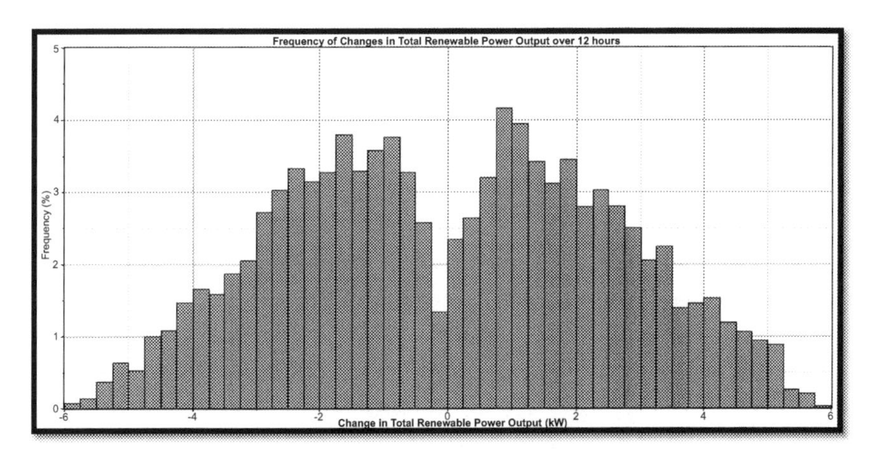

Figure 8.24 Change of frequency with total renewable power output.

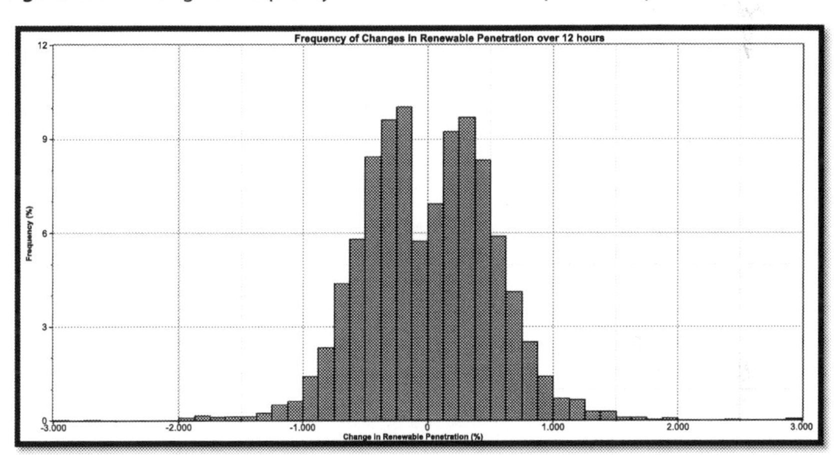

Figure 8.25 Frequency of changes with renewable penetration.

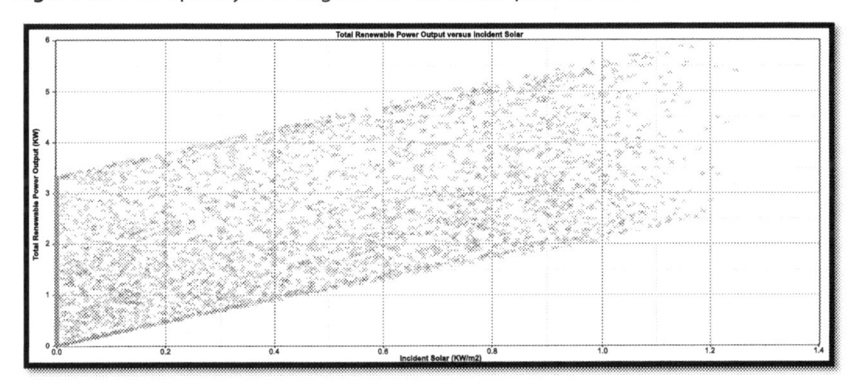

Figure 8.26 Total renewable power output versus incident solar.

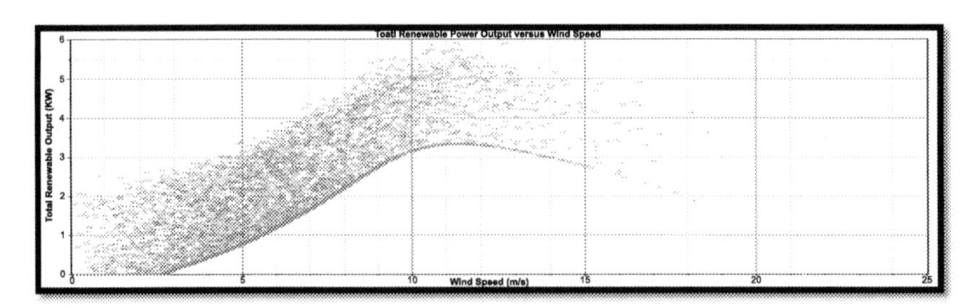

Figure 8.27 Total renewable power output versus wind speed.

Table 8.19 Electricity production through different energy source.

Production	kWh/yr	%
PV array	4,745	33
Wind turbine	9,667	67
Generator 1	87	1
Total	14,499	100

Table 8.20 Different parameter of HRES.

Quantity	kWh/yr	%
Excess electricity	7,479	51.6
Unmet electric load	0.00000703	0.0
Capacity shortage	0.00	0.0
AC primary load	6,132	100

directly proportional to the cube of wind velocity and is directly dependent on solar radiation. Table 8.19 depicts the production of electricity using various energy sources. Solar and wind systems create 4745 and 9667 kWh/yr, respectively, indicating that solar and wind systems contribute 33% and 67% of total electricity output, respectively. Excess electricity is surplus electrical energy that must be dumped because it cannot be used to serve a load or charge the batteries. Excess electricity occurs when there is a surplus of power being produced (either from a renewable source or by the generator when its minimum output exceeds the load) and the batteries are unable to absorb it all. Table 8.20 shows excess electricity is 7479kWh/yr which is 51.6%, AC primary load is 6132kWh/yr, and the capacity shortage is 0%. The capacity shortage fraction is equal to the total capacity shortage divided by the total electrical demand. HOMER considers a system feasible (or acceptable) only if the capacity shortage fraction is less than or equal to the maximum annual capacity shortage.

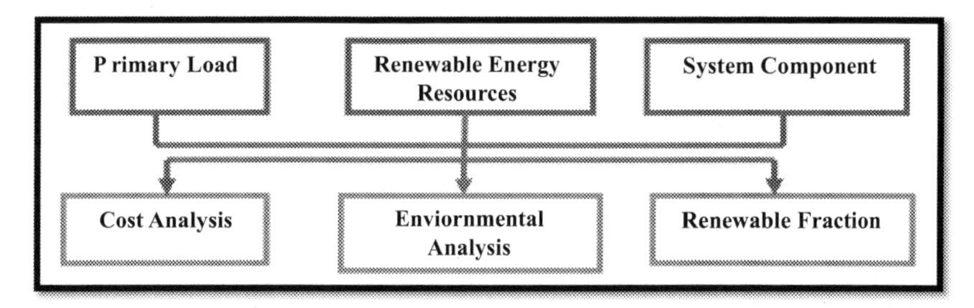

Figure 8.28 Process of life cycle analysis.

Table 8.21 Impact categories in streamlined LCA's.

Impact category	Inventory parameter	Characterization factor U	Reference
Global Warming	CO_2, CH_4, N_2O	Global warming potential	CO_2 equivalent
Acidification	SO_2, NO_X, NH_3, HCl	Acidification Potential	SO_2 equivalent
Eutrophication	NO_X, NH_3	Eutrophication Potential	PO_4^{3-} equivalent

8.10 Life cycle analysis

Individual components of a hybrid renewable energy system, such as solar panels, wind turbines, batteries, generators, and converters, are assessed as part of the life cycle evaluation. We developed a hybrid renewable energy system for the Sagar with the help of that component. From the search and supply of resources and fuels to the production and operation of the investigated objects to their disposal recycling, it studies the environmental impact system on products from cradle to grave across the whole life cycle. LCA is a technique for evaluating the environmental aspects and potential impacts of a product, process, or service by compiling an inventory of relevant energy and material inputs and environmental releases, evaluating the potential environmental impacts associated with identified inputs and releases, and interpreting the results to assist in making a more informed decision. The method of life cycle analysis is depicted in Fig. 8.28. The preliminary phase of the life cycle assessment is load measurement, prefeasibility of renewable energy resources, and system component analysis. After that, the cost and environmental analysis must be completed in the right order. Table 8.21 of life cycle analysis impact categories in terms of global warming, acidification, and eutrophication. The inventory parameter and characterization factor are used to analyze the impact category, with the characterization factor indicating the potential for global warming, acidification, and eutrophication. Emissions of nitrogen, phosphorus, and organic material are the inventory parameters that are commonly allocated to Eutrophication.

8.10.1 Component description

8.10.1.1 Solar panel

This system includes 3 KW of the solar panel that has been sourced as an appropriate supplier being able to provide 3.1 KW worth of power in a 14 solar panel array. These panels have a 10-year warranty and an 80% power output guarantee for 20 years, ensuring the panels will not have to be replaced during the lifetime operation of the system. In this project, a life cycle assessment of the electricity generated with PV power plant using an amorphous silicon module produced by Flexcell. The Flexcell module is evaluated per square meter. PV electricity is analyzed per kWh of Busbar. In compliance with the ecoinvent quality guidelines, life cycle inventories of 3 KW PV plants on a flat roof with Flexcell module are set up. The carbon footprint of Flexcell module is 19 Kg CO_2-eq per square meter. Major contribution stems from the emission of chlorofluorocarbons in the ETFE feedstock (52%), the CO_2 emission from the combustion of natural gas (23%), and the disposal of the module in municipal waste incineration (10%). Component and material for the production of a Flexcell PV module of 4.88 m^2, including production losses. Table 8.22 gave material requirement in solar panel system. The table shows material requirement for silica production is charcoal, low ash coal, cokes and wood scrap, and process is done in the form of silica production, high purity silica production, casting, wafering, texturing, and framing. In total, 56 kg quantity of material is required for design of solar panels. Several types of auxiliary material is also required for solar panel design, which is given in Table 8.23, approximate 20 kg of auxiliary material is required to design a 3KW solar panel.

8.10.1.2 Wind turbine

The SW Whisper 500 wind turbine was chosen as part of the hybrid system. The specifications of a whisper wind turbine are shown in Table 8.24. One of these turbines, with a power rating of 3 KW, is required for the system. With a diameter of 4.5 m and heights of 9.1 m, 12.8 m, and 21.3 m, this turbine was chosen because of its compact size in comparison to other turbines.

Table 8.22 Material requirement in solar panel.

Material	Process	Quantity (Kg)
Quartz	Si production	2.99
Charcoal	Silica reduction	0.41
Low ash coal	Silica reduction	0.59
Cokes	Silica reduction	0.47
Wood scrap	Silica reduction	1.32
High purity carbon	High purity Si production	0.89
HCl	High purity Si production	42
Na2co3	High purity Si production	0.70
Caco3	High purity Si production	1.39
Al2o3	High purity Si production	0.78
Argon gas	Casting	0.42
Mineral oil	Wafering	0.69
SiC	Wafering	0.92
NaOH	Etching/Texturing	0.59
H_2SO4	Etching/Texturing	0.49
KOH	Etching/Texturing	0.87
HNO3	Etching/Texturing	0.046
Al/Ag paste	Metallization	0.049
SiH4	Passivation	0.0019
EVA foil	Module assembly	0.046
Al	Framing	2.19
Total		56 Kg

Table 8.23 Auxiliary material used in the solar panel system.

Ethyl vinyl acetate	369 g
Copper foil	13.99 g
Acrylic polymer	3.47 g
Acrylate polymer	8.89 g
Polyester film	5.79 g
Rasin core solder	1.37 g
Acrylic tape	8.01 g
Nylon	1.94 g
Polycarbonate	49.99 g
Thermoplastic rubber	9.99 g
Copper	8.99 g
Silicon elastomer	11.99 g
Glass clear float	9756 g
Glass soda lime	9788 g
Polyester	5.91g
Corrugated box	569 g

Table 8.24 SW whisper wind turbine specification.

Model	SW-Whisper
Rotor diameter	4.5 m
Swept area	15.9 m^2
Tower height	9.1 m, 12.8 m, 21.3 m
Weight	80 Kg
Mount	127 mm diameter pipe mounted
Startup wind speed	3.1 m/s
Rated wind speed	12.5 m/s
Alternator	PM-3 Phase Alternator
Alternator Efficiency	85%
Magnet	Ceramic Magnet
Insulation class	'H'
Voltage configuration (L.V.Model)	24V/48V Nominal
Voltage configuration (H.V.Model)	96V/120V/240V
Rated power	3200W
Number of blades	2
Material of blades	Carbon fiber composite
Material of body	Power coated with marinization treatment
Survival wind speed	55 m/s
Overspeed protection	Furling, dump load, manual break switch
Controller	External regulator
Bearing	Low friction, Totally enclosed self-lubricated
Controller	Voltage options: 24V/48V/96V/240V

Because of its reduced size, the turbine may be installed sooner and is less likely to be damaged in inclement weather. The material requirements for a 3 kW wind turbine are shown in the table. Despite the fact that the SW whisper 500 wind turbine is made in the United States, it has been linked to an Indian company called KOTTURS. Wind turbines will also need enough wind exposure and room to operate. Over the course of 15 years, the output power of a wind turbine and a solar system has degraded. When compared to the warrantees (0.8–1%/yr on average), the actual degradation in output power for two lots of mono and one lot of polycrystalline silicon was found to be remarkably significant (3.9, 3.0, and 2%/yr). Surprisingly, the lot containing thin film Amorphous-Si modules had greater output power ratings than the nameplate. The modules have no evident flaws other than fading and discoloration. The increased rates of degradation can be linked to the module quality as well as the effects of extreme field weather conditions. The material requirements for a 2 KW rectifier and inverter are shown in the tables below. Figs 8.29 and 8.30 wind power and solar power degradation,

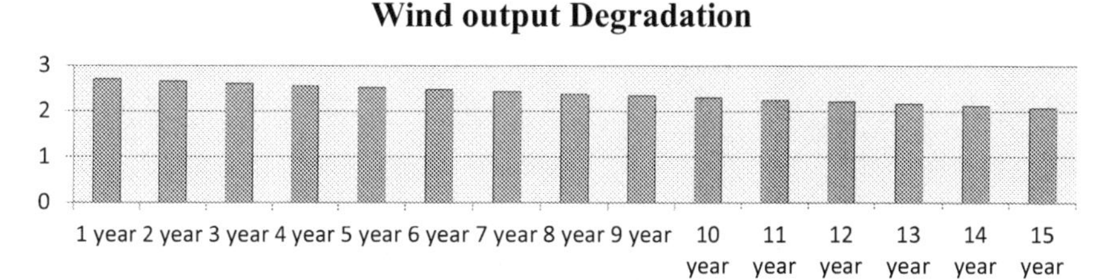

Figure 8.29 Wind output (kW) degradation.

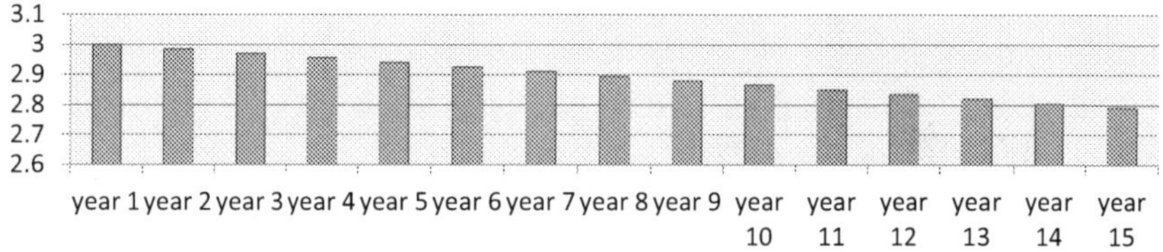

Figure 8.30 PV output (kW) degradation.

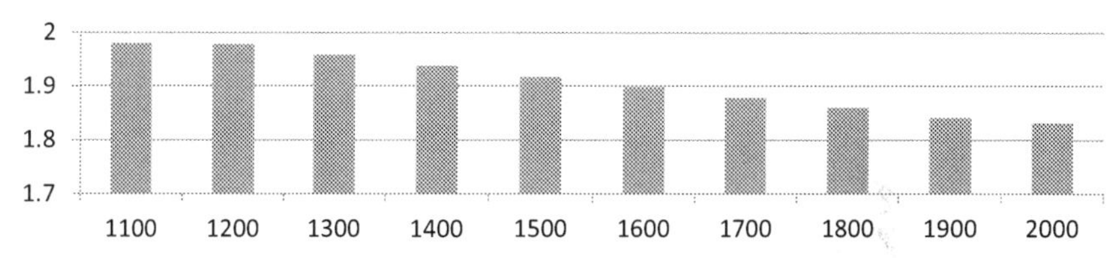

Figure 8.31 Power (kW) derating of inverter when altitude is over 1000 m.

respectively. Fig. 8.31 shows power derating of inverter over the altitude of 1000 m. Factors affecting the degree of degradation include the quality of materials used in manufacture, the manufacturing process, the quality of assembly and packaging. Regular maintenance and cleaning regimes may reduce degradation rates but the main impact is specific to the characteristics of the inverter being used. Table 8.25 shows the specification of wind turbine material. Tables 8.26 and 8.27 represent specification of rectifier material and inverter specification, respectively.

8.10.2 Wind system life cycle calculation

In the case of wind energy LCAs, there is significant variation among study scopes to conclude that a commonly accepted scope has not been found.

Table 8.25 Wind turbine material (3KW).

Component	Material	Total mass (Kg)	Manufacturing process
Tower structure	Low carbon steel	251	Forging rolling
Tower cathodic protection	Zinc alloy	0.3038	Casting
Nacelle, gears	Stainless steel	27.6	Forging rolling
Nacelle, generator core	Cast iron gray	12.9	Forging rolling
Nacelle, gen. conductor	Copper	1.499	Forging rolling
Nacelle, transformer core	Cast iron gray	8.99	Forging rolling
Nacelle, transformer conductor	Copper	2.99	Forging rolling
Nacelle, transformer conductor	Aluminum alloys	2.67	Forging rolling
Nacelle covers	GFRP, Epoxy matrix	5.99	Composite forming
Nacelle, main shaft	Cast iron ductile	17.98	Casting
Nacelle, forged component	Stainless steel	4.49	Forging rolling
Nacelle, cast Component	Cast Iron, ductile	5.83	Casting
Rotor blades	CFRP, Epoxy matrix	36	Composite forming
Rotor, iron component	Cast iron ductile	2.98	Casting
Rotor, spinner	GFRP, Epoxy matrix	4.89	Composite forming
Rotor, spinner	Cast iron ductile	3.39	Casting
Foundation, pile & platform	Concrete	1199	Construction
Foundations, steel	Low carbon steel	40.49	Forging rolling
Transmission, conductor	Copper	0.379	Forging rolling
Transmission, conductor	Aluminum alloys	0.109	Forging rolling
Transmission, insulation	Poly ethylene	2.67	Polymer extrusion

Table 8.26 Rectifier material.

Material	Quantity
Aluminium	3.19 kg
Polyvinyl Chloride	0.040 kg
Copper	0.025 kg
Cast iron, steel	19.98 kg
Connecting wire	0.69 kg
Transformer	11.98 kg
Coils	1.01 kg
IC's	0.079 kg
Transistor & Diode	0.17.98 kg
Capacitor & Resistor	1.89 kg

Table 8.27 Inverter specification.

2 KW Inverter	Specification
AC startup voltage	60–70V AC
Auto restart voltage	90V AC
Acceptable input voltage range	80–130V AC
Max. input current	30A
Nominal output voltage	101/110/120/127V AC
Output frequency	50Hz
Output Wave	Pure sine wave
Efficiency	90%
Nominal DC voltage	48V DC
Max. charging current	25A
Communication port	RS-232/USB
Operating temperature	0–40°C
Altitude	0~1000m

Table 8.28 Emission analysis of wind system.

GHG	Emission (kg/yr)
Carbon dioxide	59.89
Carbon monoxide	0.141
Unburnt hydrocarbon	0.0163
Particulate matter	0.01104
Sulfur dioxide	0.111
Nitrogen oxide	1.29

This variation can lead to differing energy intensity and air emission results. Table 8.28 presents emission analyses of wind system and it shows maximum emission is come due to carbon dioxide. Emission of carbon dioxide is 60 kg/yr which is 98% of total GHG emission from wind systems. Table 8.29 explores life cycle CO_2 emission from wind system. Life cycle CO_2 emission analysis is done through upstream, operational, and downstream processes. The emission of CO_2 in gCO_2/kWh is 6.28, 2.32, and 1.06 by upstream, operational, and downstream processes, respectively.

- **CO_2 emission rate [gCO_2/kWh]** = Total CO_2 emission during the life cycle [gCO_2]/annual power generation [kWh/yr] × Lifetime [Yr] = $(60 \times 1000 \times 20)/(6205 \times 20) = 9.67$ gCO_2/kWh.
- **Emission Intensity** = Total indirect emission [g]/lifetime electricity production [kW] = 923,898/93,075 = 9.92

Table 8.29 Life cycle CO_2 emission from wind system.

Life cycle stages	Life cycle process	CO_2 emission(gCO_2/kWh)
Upstream processes=65%	Raw material extraction Material production System manufacture System/plant component manufacture Installation/plant construction	$9.67 \times 0.65 = 6.28$
Operational processes 24%	Power generation System/plant O& M	$9.67 \times 0.24 = 2.32$
Downstream processes 11%	System/plant decommissioning Disposal	$9.67 \times 0.11 = 1.06$

- **Payback period** = System cost/annual energy delivered × cost of electricity = 780,000/9668 × 7= 11.5 years.
- **Saving to investment ratio** = Annual energy delivered × cost of electricity × payback period = 9668 × 7 × 11.5/780,000 = 0.99.
- **Electricity rate to achieve saving to investment ratio** = System cost/annual energy delivered × payback period = 780,000/9668 × 11.5 = Rs. 7.06.

8.10.3 Solar system life cycle calculation

Mono-crystalline, poly-crystalline, amorphous, CdTe/CIS, and other solar PV systems have had their energy requirements, EPBT, and GHG emissions calculated. Crystalline modules have a high conversion efficiency, but they need a lot of primary energy and produce a lot of EPBT and GHG emissions, whereas thin-film modules use less primary energy and produce less EPBT and GHG emissions, but they have a low conversion efficiency. The table below displays the greenhouse gas emissions of a solar system, with carbon dioxide accounting for the majority of the emissions (90kg/yr). Table 8.30 explores life cycle CO_2 emission from the solar system. The emission of CO_2 in gCO_2/kWh is 9.425, 3.48, and 1.6 by upstream, operational, and downstream processes, respectively. Table 8.31 shows life cycle stages of co_2 emission in gco_2/kwh.

- **CO_2 emission rate [gCO_2/kWh]** = Total CO_2 emission during the life cycle [gCO_2]/annual power generation [kWh/yr] × Lifetime [Yr] = (90 × 1000 × 20)/ (6205 × 20) = 14.5 gCO_2/kWh.

Table 8.30 Life cycle CO_2 emission from solar system.

GHG	Emission (kg/yr)
Carbon dioxide	90
Carbon monoxide	0.247
Unburnt hydrocarbon	0.0192
Particulate matter	0.01203
Sulfur dioxide	0.19
Nitrogen oxide	2.6

Table 8.31 Life cycle stages of CO_2 emission in gCO_2/kWh.

Life cycle stages	Life cycle process	CO_2 emission(gCO_2/kWh)
Upstream processes = 65%	Raw material extraction Material production Module manufacture System/plant component manufacture Installation/plant construction	$14.5 \times 0.65 = 9.425$
Operational processes 24%	Power generation System/plant O& M	$14.5 \times 0.24 = 3.48$
Downstream processes 11%	System/plant decommissioning Disposal	$14.5 \times 0.11 = 1.6$

- **Emission Intensity** = Total Indirect emission [g]/lifetime electricity production [kW] = 147,1023/93,075 = 15.8.
- **Payback period** = System cost/annual energy delivered × cost of electricity = 189,780/ 4133 × 7 = 6.5 years.
- **Saving to investment ration** = Annual energy delivered × cost of electricity × payback period = 4133 × 7 × 6.5/189,780 = 0.99.

This is the percent of money recovered from the system generation compared to the initial investment cost and ongoing O&M costs, with reference to the specific desired payback period. A value of 0.99 indicates that within the desired payback period, 99% of the investment cost was recovered.

- **Electricity rate to achieve saving to investment ratio** = System cost/annual energy delivered × payback period = 189,780/ 4133 × 6.5 = Rs. 7.06.

Table 8.32 Life cycle environment analysis of hybrid renewable energy system.

Emissions (Kg/Yr)	PV–wind–generator	Generator
Global warming potential(Kg CO_2- Equivalent)	199.99	13678
Carbon monoxide	0.455	33.7
Particulate matter	0.0373	3.73
Unburned hydrocarbon	0.0545	2.54
Acidification Potential (Kg SO_2- Equivalent)	0.403	27.4
Eutrophication Potential (Kg NO_X-Equivalent)	4.422	300

The annual average electricity rate required to recover the initial investment cost and ongoing O&M cost within the desired payback period. Table 8.32 presents life cycle environment analysis of a hybrid renewable energy system. Data show greenhouse gas emission from generator is much larger than PV–wind hybrid renewable energy system. So that PV–wind hybrid renewable energy system is much better from the environmental point of view.

8.11 Regression analysis

In statistical modeling, regression analysis is a statistical process for estimating the relationships among variables. It includes many techniques for modeling and analyzing several variables, when the focus is on the relationship between a dependent variable and one or more independent variables. For manual calculation of the relation, we have used the process of regression analysis as it is given by the exponential relation of the type:

$$y = AB^x$$

where A and B are constant.

Taking the log on both sides of the equation, we get

$$\log y = \log A + x \log B$$

Putting $\log y = Y$, $\log A = a$, $\log B = b$ in equation

$$Y = a + bx$$

From the equation, we get

$$\Sigma Y = \Sigma a + \Sigma bx$$

or

$$\Sigma Y = na + b\Sigma x$$

$$\Sigma xY = \Sigma ax + \Sigma bx^2$$

or

$$\Sigma xY = a\Sigma x + b\Sigma x^2$$

Table 8.33 Data set for regression analysis.

Month	Clearness index	Wind velocity (m/s)	Solar radiation (kwh/m²/day)	Load demand (kW)
January	0.651	5.3	4.5	1.3
February	0.669	5.4	5.3	1.5
March	0.668	5.8	6.2	1.2
April	0.664	6	6.8	1.1
May	0.640	6.2	7	1.25
June	0.542	6.2	6	1.35
July	0.440	5.9	5.9	1.3
August	0.403	5.2	4.3	1
September	0.541	5.1	5.2	1.35
October	0.680	4.8	5.8	1.2
November	0.673	4.9	4.9	1.2
December	0.660	5.1	4.3	1.4

8.11.1 Load demand and clearness Index

The equation of the straight line relating load demand and clearness Index is estimated as:

$$\text{Load Demand} = (1.0467) + (0.3581)\ \text{Clearness Index}$$

using the 12 observations in this dataset. The y-intercept, the estimated value of load demand when clearness index is zero, is 1.0467 with a standard error of 0.2585. The slope, the estimated change in load demand per unit change in clearness index, is 0.3581 with a standard error of 0.4240. The value of R-squared, the proportion of the variation in load demand that can be accounted for by variation in clearness index, is 0.0666. The correlation between load demand and clearness index is 0.2581. The whole assessment of the regression analysis is based on the data in Table 8.33.

A significance test that the slope is zero resulted in a *t*-value of 0.8447. The significance level of this *t*-test is 0.4180. Since $0.4180 > 0.0500$, the hypothesis that the slope is zero is not rejected. The estimated slope is 0.3581. The lower limit of the 95% confidence interval for the slope is −0.5865 and the upper limit is 1.3028. The estimated intercept is 1.0467. The lower limit of the 95% confidence interval for the intercept is 0.4707 and the upper limit is 1.6227. Table 8.34 shows the regression estimation of load demand and clearness index.

Table 8.34 shows the least-squares estimates of the intercept and slope followed by the corresponding standard errors, confidence intervals, and

Table 8.34 Regression estimation section.

	Intercept	Slope
Parameter	**B(0)**	**B(1)**
Regression coefficients	1.0467	0.3581
Lower 95% confidence limit	0.4707	−0.5865
Upper 95% confidence limit	1.6227	1.3028
Standard error	0.2585	0.4240
Standardized coefficient	0.0000	0.2581
T value	4.0490	0.8447
Prob level (*t*-test)	0.0023	0.4180
Reject HO (Alpha = 0.0500)	Yes	No
Power (Alpha = 0.0500)	0.9532	0.1195
Regression of Y on X	1.0467	0.3581
Inverse regression from X on Y	−1.9775	5.3769
Orthogonal regression of Y and X	−0.5009	2.9264

Y, load demand; X, clearness index.

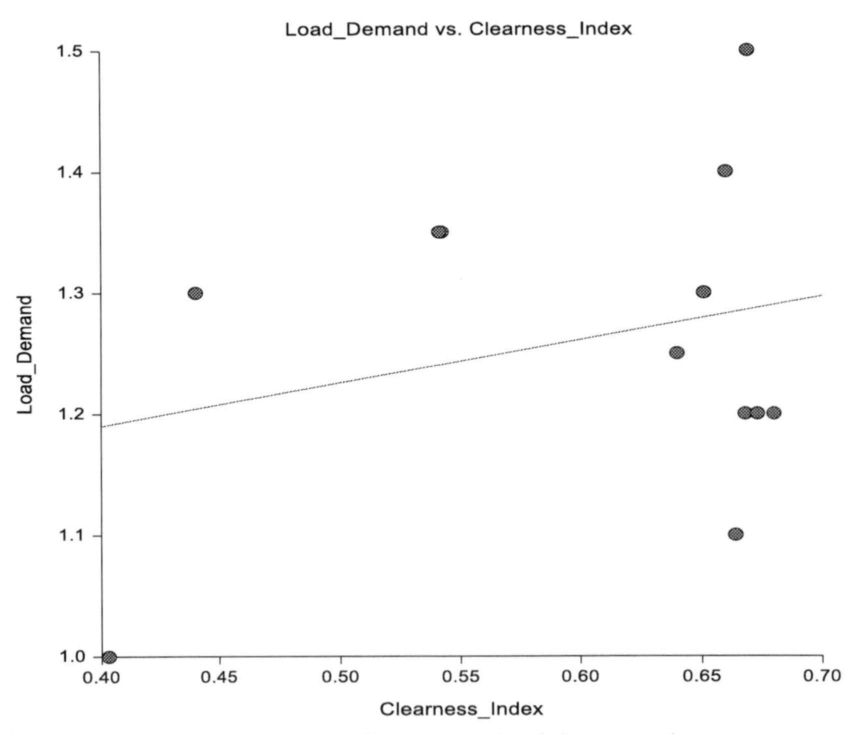

Figure 8.32 Linear regression plot of load demand and clearness index.

hypothesis tests. Note that these results are based on several assumptions that should be validated before they are used. Fig. 8.32 represents linear regression plot load demand and clearness index.

8.11.2 Load demand and wind velocity

The equation of the straight line relating load demand and wind velocity is estimated as:

$$\text{Load Demand} = (1.2502) + (0.0022)\ \text{Wind velocity}$$

using the 12 observations in this dataset. The y-intercept, the estimated value of load demand when wind velocity is zero, is 1.2502 with a standard error of 0.4675. The slope, the estimated change in load demand per unit change in wind velocity, is 0.0022 with a standard error of 0.0848. The value of R-squared, the proportion of the variation in load demand that can be accounted for by variation in wind velocity, is 0.0001. The correlation between load demand and wind velocity is 0.0084. A significance test that the slope is zero resulted in a *t*-value of 0.0264. The significance level of this *t*-test is 0.9794. Since $0.9794 > 0.0500$, the hypothesis that the slope is zero is not rejected. The estimated slope is 0.0022. The lower limit of the 95% confidence interval for the slope is −0.1867 and the upper limit is 0.1912. The estimated intercept is 1.2502. The lower limit of the 95% confidence interval for the intercept is 0.2085 and the upper limit is 2.2919. Fig. 8.33 represents linear regression plot load demand and wind velocity. Table 8.35 shows the regression estimation of load demand and wind velocity.

8.11.3 Load demand and solar radiation

The equation of the straight line relating load demand and solar radiation is estimated as:

$$\text{Load Demand} = (1.3560) + (-0.0169)\ \text{Solar Radiation}$$

using the 12 observations in this dataset. The y-intercept, the estimated value of load demand when solar radiation is zero, is 1.3560 with a standard error of 0.2578. The slope, the estimated change in load demand per unit change in solar radiation, is −0.0169 with a standard error of 0.0461. The value of R-squared, the proportion of the variation in load demand that can be accounted for by variation in Solar Radiation, is 0.0133. The correlation between load demand and solar radiation is −0.1153. A significance test that the slope is zero resulted in a *t*-value of −0.3671. The significance level of this *t*-test is 0.7212. Since $0.7212 > 0.0500$, the hypothesis that the slope is zero is not rejected. The estimated slope is −0.0169. The lower limit of the 95% confidence interval for the slope is −0.1198 and the upper limit

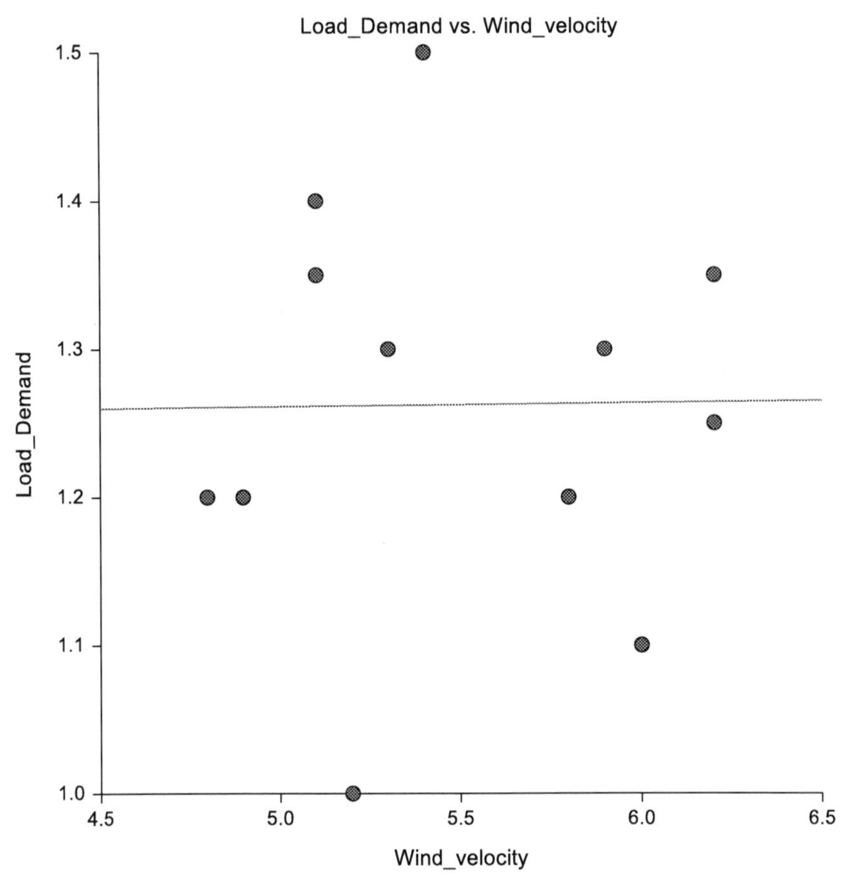

Figure 8.33 Regression plot of load demand and wind velocity.

Table 8.35 Regression estimation section.

	Intercept	Slope
Parameter	**B(0)**	**B(1)**
Regression coefficients	1.2502	0.0022
Lower 95% confidence limit	0.2085	−0.1867
Upper 95% confidence limit	2.2919	0.1912
Standard error	0.4675	0.0848
Standardized coefficient	0.0000	0.0084
T value	2.6740	0.0264
Prob level (*t*-test)	0.0233	0.9794
Reject HO (Alpha = 0.0500)	Yes	No
Power (Alpha = 0.0500)	0.6744	0.0501
Regression of Y on X	1.2502	0.0022
Inverse regression from X on Y	−175.0200	32.1000
Orthogonal regression of Y and X	1.2492	0.0024

Y, load demand; X, wind velocity (analysis through NCSS Tool).

Table 8.36 Regression estimation section.

	Intercept	Slope
Parameter	B(0)	B(1)
Regression coefficients	1.3560	−0.0169
Lower 95% confidence limit	0.7816	−0.1198
Upper 95% confidence limit	1.9303	0.0859
Standard error	0.2578	0.0461
Standardized coefficient	0.0000	−0.1153
T value	5.2601	−0.3671
Prob level (*t*-test)	0.0004	0.7212
Reject HO (Alpha = 0.0500)	Yes	No
Power (Alpha = 0.0500)	0.9970	0.0628
Regression of Y on X	1.3560	−0.0169
Inverse regression from X on Y	8.2897	−1.2738
Orthogonal regression of Y and X	1.3580	−0.0173

Y, load demand; X, solar radiation (analysis through NCSS Tool).

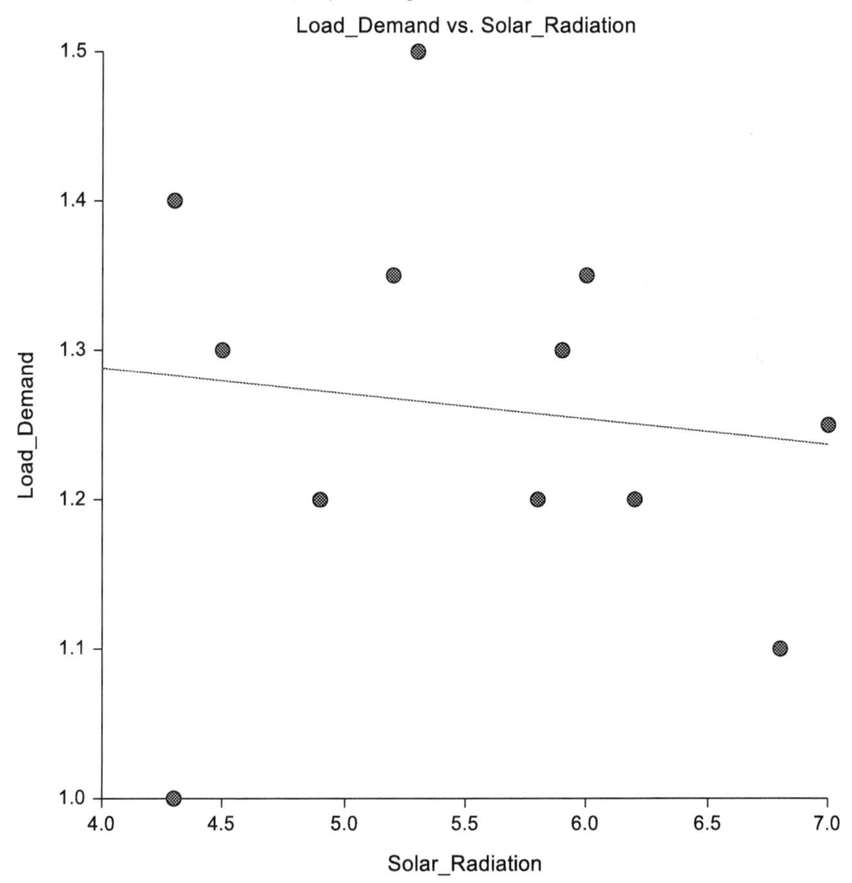

Figure 8.34 Regression plot of load demand and solar radiation.

is 0.0859. The estimated intercept is 1.3560. The lower limit of the 95% confidence interval for the intercept is 0.7816 and the upper limit is 1.9303. Table 8.36 shows the least-squares estimates of the intercept and slope followed by the corresponding standard errors, confidence intervals, and hypothesis tests. Note that these results are based on several assumptions that should be validated before they are used. Fig. 8.34 represents linear regression plot load demand and solar radiation.

8.12 Conclusion

Renewable energy-based hybrid energy systems are currently not economically competitive with fossil fuel-based stand-alone or grid-connected power sources. Solar- and wind-powered sites, whether in densely populated or distant places, benefit the environment as well as the operator's bottom line. This analysis took into account a variety of energy sources (wind, solar, and diesel generators) as well as storage devices. Because the base station is powered by a hybrid renewable energy system, it will minimize carbon and other harmful gas emissions in the environment by around 90%. The following conclusion could be drawn based on this analysis:

- When compared to wind energy, solar resources in Sagar (India) have a lot of potential, and wind energy may not be cost-effective in most circumstances.
- From an environmental standpoint, a PV–wind–diesel–battery system is the most viable option for a standalone application in Sagar.
- The solar–wind hybrid renewable energy system's net current cost is Rs. 20,70,960.
- The levelized cost of electricity is Rs. 26.4/kWh, and the annual running cost of a standalone system is Rs. 55,620.

References

[1] V. Khare, S. Nema, P. Baredar, Solar–wind hybrid renewable energy system: a review, Renewable Sustain. Energy Rev. 58 (2016) 23–33 Elsevier.
[2] V. Khare, S. Nema, P. Baredar, Status of solar wind renewable energy in India, Renewable Sustain. Energy Rev. 27 (2013) 1–10 Elsevier.
[3] V. Khare, S. Nema, P. Baredar, Optimization of hydrogen based hybrid renewable energy system using HOMER, BB-BC and GAMBIT, Int. J. Hydrogen Energy 41 (38) (2016) 16743–16751 Elsevier.
[4] V. Khare, S. Nema, P. Barear, Optimisation of the hybrid renewable energy system by HOMER, PSO and CPSO for the study area, Int. J. Sustain. Energy 36 (4) (2017) 326–343 Taylor and Francis.

[5] V. Khare, Prediction, investigation, and assessment of novel tidal–solar hybrid renewable energy system in India by different techniques, Int. J. Sustainable Energy 38 (5) (2019) 447–468 Taylor and Francis.
[6] V. Khare, Status of tidal energy system in India, J. Marine Eng. Technol. 20 (5) (2021) 289–298 Taylor and Francis.
[7] V. Khare, C. Khare, S. Nema, P. Baredar, Renewable energy system paradigm change from trending technology: a review, Int. J. Sustain. Energy 40 (7) (2021) 697–718 Taylor and Francis.
[8] V. Khare, S. Nema, P. Baredar, C.J. Khare, Game theory-based framework of solar–wind renewable energy system, J. Institution Engineers (India): Series B 100 (6) (2019) 575–587.
[9] R. Belfkira, G. Barakat, C. Nichita, Sizing optimization of a stand-alone hybrid power supply unit: wind/PV system with battery storage, Int. Review Electrical Eng. 3 (2008) 820–828.
[10] A.G. Bhave, Hybrid solar–wind domestic power generating system—case study, Renewable Energy 17 (1999) 355–358.
[11] B.S. Borowy, Z.M. Salameh, Methodology for optimally sizing the combination of a battery bank and PV array in a wind/PV hybrid system, IEEE Trans. Energy Convers. 11 (1996) 367–375.
[12] B.S. Borowy, Z.M. Salameh, Dynamic response of stand-alone wind energy conversation system with battery energy storage to a wind gust, IEEE Trans. Energy Conversation 12 (1997) 73–78.
[13] M.J. Khan, M.T. Iqbal, Pre-feasibility study of stand-alone hybrid energy systems for applications in new found land, Renewable Energy 30 (2004) 835–854.
[14] J. Lagorse, D. Paire, A. Miraoui, Hybrid stand alone power supply using PEMFC, PV and battery, modelling and optimization, in: IEEE Conference on Clean Electrical Power (ICCEP), Italy, Capri, 2009.
[15] D.K. Lal, B.B. Dash, A.K. Akella, Optimization of PV/wind/micro-hydro/diesel hybrid power system in HOMER for the study area, Int. J. Electrical Eng. Informatics 3 (2011) 1–10.
[16] R. Madlener, C.H. Antunes, L.C. Dias, Assessing the performance of biogas plants with multi-criteria and data envelopment analysis, European J. Operations Res. 197 (2006) 1084–1094.
[17] A. Mellit, S.A. Kalogirou, L. Hontoria, S. Shaari, Artificial intelligence techniques for sizing photovoltaic systems: a review, Renewable Sustain. Energy Rev. 13 (2009) 406–419.
[18] S.M. Moghaddas-Tafreshi, S.M. Hakimi, Optimal sizing of a stand-alone hybrid power system via particle swarm optimization for Kahnouj Area in Southeast of Iran, Renewable Energy 34 (2009) 1855–1862.
[19] M. Mohammadi, Review of simulation and optimization of autonomous and grid connected hybrid renewable energy system at microgrid, ISESCO J. Sci. Technol. 16 (2013) 60–67.

CHAPTER NINE

Data analysis of solar energy system with Python

Learning objectives

- Understand the basics of Python programming in the field of solar energy system.
- Know how regression analysis with Python programming can apply on the data analysis of solar energy system.
- Learn application of random forest and Naïve Bayes with Python programming on the data analysis of solar energy system.
- Understand the data analysis of solar energy system with decision tree and support vector machine.

9.1 Introduction

Various difficulties confront the energy system today as society evolves fast. The need for renewable energy supplies has risen as a result of growing awareness of pollution and global warming. Because of its availability, diversity, and environmental friendliness, solar energy is the most promising renewable energy source. The evolution and usage of this energy not only give a means of utilizing these resources but also generate an efficient evaluation that allows resources to be better adjusted in order to overcome energy resource shortages [1,2]. Solar energy is the most beneficial renewable energy source. It may be feasible to improve the efficiency of this energy source by using data processing and computer algorithms. Because of their availability, renewable energy resources are quickly becoming one of the most important sources of energy. Python is one of the most promising languages for proper data assessment of the solar energy system. Python is a scripting language that is high-level, interpreted, interactive, and object-oriented. Python is intended to be a very understandable language. It typically employs English terms instead of punctuation, and it has fewer syntactic structures than other languages. The new oil is data. This statement demonstrates how data capture, storage, and analysis are at the heart of every contemporary IT system [3,4]. It does not matter if you are making a business choice, forecasting the weather, investigating protein structures in

Decision Science and Operations Management of Solar Energy Systems.
DOI: https://doi.org/10.1016/B978-0-323-85761-1.00008-1

biology, or creating a marketing strategy. All of these circumstances call for a multidisciplinary approach that includes the use of mathematical models, statistics, graphs, databases, and, of course, the commercial or scientific rationale that underpins the data analysis. As a result, we need a programming language that can handle all of these different data science requirements. This chapter mentions data analysis of solar energy systems with Python programming. Python is a general-purpose programming language with a high level of abstraction. Its design philosophy prioritizes code readability and makes extensive use of indentation. Its language elements and object-oriented approach are aimed at assisting programmers in writing clear, logical code for both small and big projects [5].

The following platform can be used to run the command of Python programming.

1. **Jupyter:** JupyterLab is the most recent interactive development environment for notebooks, code, and data that are available on the web. Users may create and arrange workflows in data science, scientific computing, computational journalism, and machine learning using its versatile interface. Extensions to enhance and enrich functionality are encouraged by a modular architecture. The purpose of Project Jupyter is to "create open-source software, open standards, and services enabling interactive computing spanning dozens of programming languages." Fernando Pérez and Brian Granger split it out from IPython in 2014. The name Project Jupyter is a play on the three primary programming languages supported by Jupyter: Julia, Python, and R, as well as a nod to Galileo's notebooks documenting the discovery of Jupiter's moons.
2. **Google Colab:** Google Research's Colaboratory, or "Colab" for short, is a product. Colab is a web-based Python editor that allows anybody to create and run arbitrary Python code. It is notably useful for machine learning, data analysis, and teaching. Colab, in more technical terms, is a hosted Jupyter notebook service that requires no installation and provides free access to computer resources, including GPUs.
3. **Anaconda:** Anaconda is a Python and R programming language distribution aimed for simplifying package management and deployment in scientific computing (data science, machine learning applications, large-scale data processing, predictive analytics, and so on). Data-science software for Windows, Linux, and macOS is included in the release. Anaconda, Inc., created by Peter Wang and Travis Oliphant in 2012, is responsible for its development and maintenance. Anaconda Distribution or Anaconda Individual Edition is other Anaconda, Inc. goods, whereas

```
# more than one dimensions of solar radiation or load demand
import numpy as np
a = np.array([[2, 3], [3, 4]])
print (a)
Output: [[2 3]
    [3 4]]
```

Figure 9.1 Python programming with NumPy with output of solar radiation.

```
pandas. Series( data, index, dtype, copy)
```

Figure 9.2 Syntax of pandas library.

Anaconda Team Edition and Anaconda Enterprise Edition, both of which are not free, are other Anaconda, Inc. products.

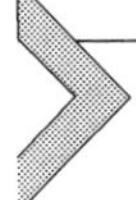

9.2 First level data analysis of solar energy data with Python library

9.2.1 Solar Data Operations in Numpy

The most significant object in NumPy is the ndarray type of N-dimensional array. It refers to a group of goods that are all of the same sort. A zero-based index may be used to find items in the collection. Different array construction procedures discussed later in the course can be used to create an instance of the ndarray class. The following is how to make a basic ndarray in NumPy using an array function:

numpy.array

Now we can use the numpy as np is used for more than one dimension of solar radiation or load demand. Fig. 9.1 shows the Python programming with numpy with output of solar radiation or load demand.

9.2.2 Pandas data operations

Pandas work with data in three ways: Series, Data Frame, and Panel. Pandas Series is a one-dimensional labeled array that may contain any form of data (integer, string, float, Python objects, etc.). The axis labels are referred to as indexes. The function Object () {[native code]} below may be used to create a pandas Series. Fig. 9.2 shows the syntax of the pandas library. Fig. 9.3 shows the data gathering and their output of month-wise average solar radiation through the pandas library. Fig. 9.4 shows the syntax of pandas DataFrame, and Fig. 9.5 shows the DataFrame for solar energy system parameters.

```
#import the pandas library and aliasing as pd for average solar radiation for 12 Month
import pandas as pd
import numpy as np
data = np.array(['4.5','6.2','7.3','5.9','5.8','8.1','5.3','4.9','8.9','3.2','9.3','4.1'])
s = pd.series(data)
print (s)

Output:
0   4.5
1   6.2
2   7.3
3   5.9
4   5.8
5   8.1
6   5.3
7   4.9
8   8.9
9   3.2
10  9.3
11  4.1
```

Figure 9.3 Import pandas library for average solar radiation.

```
pandas.DataFrame( data, index, columns, dtype, copy)
```

Figure 9.4 Command for pandas DataFrame.

```
# DataFrame for solar energy system parameters
import pandas as pd
data = {'solar system parameter':['Solar Radiation', 'Wind Velocity', 'Temperature', 'Load Deman
d'],
'Value':[5.6,12.1,40,0.87]}
df = pd.DataFrame(data, index=['Location1','Location2','Location3','Location4'])
print (df)

Output:

          solar system parameter  Value
Location1        Solar Radiation   5.60
Location2          Wind Velocity  12.10
Location3            Temperature  40.00
Location4            Load Demand   0.87
```

Figure 9.5 DataFrame for solar energy system parameter.

9.2.3 Solar energy system data input as CSV file

The csv file is a text file in which the values in the columns are separated by a comma. Let us consider the following data present in the file named

```
import pandas as pd
data = pd.read_csv('solar energy.csv')
print (data)
```

Figure 9.6 Import pandas library for solar energy system.

```
pd.merge(left, right, how='inner', on=None, left_on=None, right_on=None,
left_index=False, right_index=False, sort=True)
```

Figure 9.7 Command for merge operation.

```
import pandas as pd
left = pd.DataFrame({
        'Solar Radiation':[5.8,6.3,8.2,7.9,4.9],
        'Location': ['Indore', 'Bhopal', 'Dubai', 'Bombay', 'Delhi'],
        'Load demand in KW':['50','70','30','25','50']})
right = pd.DataFrame(
        {'Solar Radiation':[6.8,7.3,8.2,6.9,5.9],
        'Location': ['Chennai', 'Sagar', 'Jaipur', 'Patna', 'Jaipur'],
        'Load demand in KW':['45','72','65','12','18']})
print (left)
print (right)

Output:
  Solar Radiation Location Load demand in KW
0           5.8  Indore            50
1           6.3  Bhopal             70
2           8.2   Dubai            30
3           7.9  Bombay              25
4           4.9   Delhi            50
   Solar Radiation Location Load demand in KW
0           6.8  Chennai             45
1           7.3   Sagar            72
2           8.2  Jaipur            65
3           6.9   Patna            12
4           5.9  Jaipur            18
```

Figure 9.8 Merge operation for solar energy system.

input.csv. Fig. 9.6 shows the command to upload the solar energy data in the csv file.

The Pandas Python package includes a single merge function that may be used to perform all typical database join operations between DataFrame objects. Fig. 9.7 shows the join operation with DataFrame objects. Fig. 9.8 shows the data frame of solar energy parameters.

```
import pandas as pd
solar_data = {'location': ['Bhopal', 'Bhopal', 'Delhi', 'Delhi', 'Jaipur',
    'Jaipur', 'Jaipur', 'Jaipur', 'Bhopal', 'Bombay', 'Bombay', 'Bhopal'],
    'number of solar plant ': [1, 2, 2, 3, 3,4 ,1 ,1,2 , 4,1,2],
    'Year of commisioning': [2014,2015,2014,2015,2014,2015,2016,2017,2016,2014,2015,
2017],
    'load demand in KW':[876,789,863,673,741,812,756,788,694,701,804,690]}
df = pd.DataFrame(solar_data)

grouped = df.groupby('Year of commisioning')
print (grouped.get_group(2014))

Output:

 location  number of solar plant  Year of commissioning  load demand in KW
                  0  Bhopal              1              2014           876
                  2   Delhi              2              2014           863
                  4  Jaipur              3              2014           741
                  9  Bombay              4              2014           701
```

Figure 9.9 Grouping operation of solar energy system.

9.2.4 Grouping of solar energy data

In data analysis, grouping datasets is a common need when the outcome must be expressed in terms of the many groups included in the dataset. Panadas provides built-in mechanisms for grouping data into several categories. Fig. 9.9 shows the grouping of solar energy data, where the different attributes are location, number of solar plant, year of commissioning, and load demand in KW.

9.2.5 Solar energy data concatenation

Pandas has a number of features that make it simple to combine Series, DataFrame, and Panel objects. The concat function is used to conduct concatenation operations along an axis in the example below. Let us make some distinct items and concatenate them. Fig. 9.10 shows the concatenation of different parameters of solar energy data.

9.3 Second level data analysis of solar energy data with Python library

In the first level of data analysis, the solar energy data analyze with Numpy and Pandas library, with the function of grouping, merging, and

```
import pandas as pd
one = pd.DataFrame({
        'location': ['Bhopal', 'Delhi', 'Jaipur', 'Jodhpur', 'Sagar'],
        'location_id':['loc1','loc2','loc4','loc6','loc5'],
        'Solar_radiation':[7.5,6.3,8.1,6.2,7.8]},
        index=[1,2,3,4,5])
two = pd.DataFrame({
        'location': ['Bombay', 'Chennai', 'Rewa', 'Indore', 'Patna'],
        'location_id':['loc2','loc4','loc3','loc6','loc5'],
        'Solar_radiation':[8.9,8.0,7.9,9.7,8.8]},
        index=[1,2,3,4,5])
print (pd.concat([one,two]))
```

Output:

```
   location  location_id  Solar radiation
1  Bhopal     loc1         7.5
2  Delhi      loc2         6.3
3  Jaipur     loc4         8.1
4  Jodhpur    loc6         6.2
5  Sagar      loc5         7.8
1  Bombay     loc2         8.9
2  Chennai    loc4         8.0
3  Rewa       loc3         7.9
4  Indore     loc6         9.7
5  Patna      loc5         8.8
```

Figure 9.10 Solar energy data concatenation.

```
import pandas as pd
solar = pd.read_csv("SOLAR RADIATION DATA.csv")
solar.head()
```

Figure 9.11 Import pandas library for solar radiation data.

concatenation. Now this section elaborates data analysis of solar energy system through different types of techniques with Python programming. Table 9.1 shows the dataset of solar radiation in different cities in India. This dataset is used in Python programming. Following is the step-by-step process of data analysis of solar energy system by Python.

9.3.1 Load and representation of solar data

To load the solar energy data through Python programming, the first data are load in the CSV file (or through Excel File) from the command, which is shown in Fig. 9.11.

In the above command pandas as a library and SOLAR RADIATION DATA is the name of the CSV file. When we are type command solar.head (),

Table 9.1 Dataset of solar radiation of different city of India.

Patna	Bhopal	Mumbai	Chennai	Delhi	Goa	Jaipur	Kolkata	Lucknow	Nagpur
8.34	9.1	7.01	8.13	7.42	7.22	3.12	4.121	8.12	3.14
8.33	9.02	7.02	8.12	7.34	7.23	3.13	4.122	8.145	3.15
8.32	8.94	7.03	8.11	7.26	7.24	3.14	4.123	8.17	3.16
8.31	8.86	7.04	8.1	7.18	7.25	3.15	4.124	8.195	3.17
8.3	8.78	7.05	8.09	7.1	7.26	3.16	4.125	8.22	3.18
8.29	8.7	7.06	8.08	7.02	7.27	3.17	4.126	8.245	3.19
8.28	8.62	7.07	8.07	6.94	7.28	3.18	4.127	8.27	3.2
8.27	8.54	7.08	8.06	6.86	7.29	3.19	4.128	8.295	3.21
8.26	8.46	7.09	8.05	6.78	7.3	3.2	4.129	8.32	3.22
8.25	8.38	7.1	8.04	6.7	7.31	3.21	4.13	8.345	3.23
8.24	8.3	7.11	8.03	6.62	7.32	3.22	4.131	8.37	3.24
8.23	8.22	7.12	8.02	6.54	7.33	3.23	4.132	8.395	3.25
8.22	8.14	7.13	8.01	6.46	7.34	3.24	4.133	8.42	3.26
8.21	8.06	7.14	8	6.38	7.35	3.25	4.134	8.445	3.27
8.2	7.98	7.15	7.99	6.3	7.36	3.26	4.135	8.47	3.28
8.19	7.9	7.16	7.98	6.22	7.37	3.27	4.136	8.495	3.29
8.18	7.82	7.17	7.97	6.14	7.38	3.28	4.137	8.52	3.3
8.17	7.74	7.18	7.96	6.06	7.39	3.29	4.138	8.545	3.31
8.16	7.66	7.19	7.95	5.98	7.4	3.3	4.139	8.57	3.32
8.15	7.58	7.2	7.94	5.9	7.41	3.31	4.14	8.595	3.33
8.14	7.5	7.21	7.93	5.82	7.42	3.32	4.141	8.62	3.34
8.13	7.42	7.22	7.92	5.74	7.43	3.33	4.142	8.645	3.35
8.12	7.34	7.23	7.91	5.66	7.44	3.34	4.143	8.67	3.36
8.11	7.26	7.24	7.9	5.58	7.45	3.35	4.12	8.695	3.37
8.1	7.18	7.25	7.89	5.5	7.46	3.36	4.13	8.72	3.38
8.09	7.1	7.26	7.88	5.42	7.47	3.37	4.14	8.745	3.39
8.08	7.02	7.27	7.87	5.34	7.48	3.38	4.15	8.77	3.4
8.07	6.94	7.28	7.86	5.26	7.49	3.39	4.16	8.795	3.41
8.06	6.86	7.29	7.85	5.18	7.5	3.4	4.17	8.82	3.42
8.05	6.78	7.3	7.84	5.1	7.51	3.41	4.18	8.845	3.43
8.04	6.7	7.31	7.83	5.02	7.52	3.42	4.19	8.87	3.44
8.03	6.62	7.32	7.82	4.94	7.53	3.43	4.2	8.895	3.45
8.02	6.54	7.33	7.81	4.86	7.54	3.44	4.21	8.92	3.46
8.01	6.46	7.34	7.8	4.78	7.55	3.45	4.22	8.945	3.47
8	6.38	7.35	7.79	4.7	7.56	3.46	4.23	8.97	3.48
7.99	6.3	7.36	7.78	4.62	7.57	3.47	4.24	8.995	3.49
7.98	6.22	7.37	7.77	4.54	7.58	3.48	4.25	9.02	3.5
7.97	6.14	7.38	7.76	4.46	7.59	3.49	4.26	9.045	3.51
7.96	6.06	7.39	7.75	4.38	7.6	3.5	4.27	9.07	3.52
7.95	5.98	7.4	7.74	4.3	7.61	3.51	4.28	9.095	3.53
7.94	5.9	7.41	7.73	4.22	7.62	3.52	4.29	9.12	3.54
7.93	5.82	7.42	7.72	4.14	7.63	3.53	4.3	9.145	3.55

(continued on next page)

Table 9.1 Dataset of solar radiation of different city of India—cont'd

Patna	Bhopal	Mumbai	Chennai	Delhi	Goa	Jaipur	Kolkata	Lucknow	Nagpur
7.92	5.74	7.43	7.71	4.06	7.64	3.54	4.31	9.17	3.56
7.91	5.66	7.44	7.7	3.98	7.65	3.55	4.32	9.195	3.57
7.9	5.58	7.45	7.69	3.9	7.66	3.56	4.33	9.22	3.58
7.89	5.5	7.46	7.68	3.82	7.67	3.57	4.34	9.245	3.59
7.88	5.42	7.47	7.67	3.74	7.68	3.58	4.35	9.27	3.6
7.87	5.34	7.48	7.66	3.66	7.69	3.59	4.36	9.295	3.61
7.86	5.26	7.49	7.65	3.58	7.7	3.6	4.37	9.32	3.62
7.85	5.18	7.5	7.64	3.5	7.71	3.61	4.38	9.345	3.63
7.84	5.1	7.51	7.63	3.42	7.72	3.62	4.39	9.37	3.64
7.83	5.02	7.52	7.62	3.34	7.73	3.63	4.4	9.395	3.65
7.82	4.94	7.53	7.61	3.26	7.74	3.64	4.41	9.42	3.66
7.81	4.86	7.54	7.6	3.18	7.75	3.65	4.42	9.445	3.67
7.8	4.78	7.55	7.59	3.1	7.76	3.66	4.43	9.47	3.68
7.79	4.7	7.56	7.58	3.02	7.77	3.67	4.44	9.495	3.69
7.78	4.62	7.57	7.57	2.94	7.78	3.68	4.45	9.52	3.7
7.77	4.54	7.58	7.56	2.86	7.79	3.69	4.46	9.545	3.71
7.76	4.46	7.59	7.55	2.78	7.8	3.7	4.47	9.57	3.72
7.75	4.38	7.6	7.54	2.7	7.81	3.71	4.48	9.595	3.73
7.74	4.3	7.61	7.53	2.62	7.82	3.72	4.49	9.62	3.74
7.73	4.22	7.62	7.52	2.54	7.83	3.73	4.5	9.645	3.75
7.72	4.14	7.63	7.51	2.46	7.84	3.74	4.51	9.67	3.76
7.71	4.06	7.64	7.5	2.38	7.89	3.75	4.52	9.695	3.77
7.7	3.98	7.65	7.49	2.3	7.88	3.76	4.53	9.72	3.78
7.69	3.9	7.66	7.48	2.22	7.87	3.77	4.54	9.745	3.79
7.68	3.82	7.67	7.47	2.14	7.86	3.78	4.55	9.77	3.8
7.67	3.74	7.68	7.46	2.06	7.85	3.79	4.56	9.795	3.81
7.66	3.66	7.69	7.45	1.98	7.84	3.8	4.57	9.82	3.82
7.65	3.58	7.7	7.44	1.9	7.83	3.81	4.58	9.845	3.83
7.64	3.5	7.71	7.43	1.82	7.82	3.82	4.59	9.87	3.84
7.63	3.42	7.72	7.42	1.74	7.81	3.83	4.6	9.895	3.85
7.62	3.34	7.73	7.41	1.66	7.8	3.84	4.61	9.92	3.86
7.61	3.26	7.74	7.4	1.58	7.79	3.85	4.62	9.945	3.87
7.6	3.18	7.75	7.39	1.5	7.78	3.86	4.63	9.97	3.88
7.59	3.1	7.76	7.38	1.42	7.77	3.87	4.64	9.995	3.89
7.58	3.02	7.77	7.37	1.34	7.76	3.88	4.65	10.02	3.9
7.57	2.94	7.78	7.36	1.26	7.75	3.89	4.66	10.045	3.91
7.56	2.86	7.79	3.45	1.18	7.74	3.9	4.67	10.07	3.92
7.55	2.78	7.8	3.46	1.1	7.73	3.91	4.68	10.095	3.93
7.54	2.7	7.81	3.47	1.02	7.72	3.92	4.69	10.12	3.94
7.53	2.62	7.82	3.48	0.94	7.71	3.93	4.7	10.145	3.95
7.52	2.54	7.83	3.49	0.86	7.7	3.94	4.71	10.17	3.96

(continued on next page)

Table 9.1 Dataset of solar radiation of different city of India—cont'd

Patna	Bhopal	Mumbai	Chennai	Delhi	Goa	Jaipur	Kolkata	Lucknow	Nagpur
7.51	2.46	7.84	3.5	0.78	7.69	3.95	4.72	10.195	3.97
7.5	2.38	7.89	3.51	0.7	7.68	3.96	4.73	10.22	3.98
7.49	2.3	7.88	3.52	0.62	7.67	3.97	4.74	10.245	3.99
7.48	2.22	7.87	3.53	0.54	7.66	3.98	4.75	10.27	4
7.47	2.14	7.86	3.54	0.52	7.65	3.99	4.76	10.295	4.01
7.46	2.06	7.85	3.55	0.53	7.64	4	4.77	10.32	4.02
7.45	1.98	7.84	3.56	0.54	7.63	4.01	4.78	10.345	4.03
7.44	1.9	7.83	3.57	0.55	7.62	4.02	4.79	10.37	4.04
7.43	1.82	7.82	3.58	0.56	7.61	4.03	4.8	10.395	4.05
7.42	1.74	7.81	3.59	0.57	7.6	4.04	4.81	10.42	4.06
7.41	1.66	7.8	3.6	0.58	7.59	4.05	4.82	10.445	4.07
7.4	1.58	7.79	3.61	0.59	7.58	4.06	4.83	10.47	4.08
7.39	1.5	7.78	3.62	0.6	7.57	4.07	4.84	10.495	4.09
7.38	1.42	7.77	3.63	0.61	7.56	4.08	4.85	10.52	4.1
7.37	1.34	7.76	3.64	0.62	7.55	4.09	4.86	10.545	4.11
7.36	1.26	7.75	3.65	0.63	7.54	4.1	4.87	10.57	4.12
3.45	1.18	7.74	3.66	0.64	7.53	4.11	4.88	10.595	4.13
3.46	1.1	7.73	3.67	0.65	7.52	4.12	4.89	10.62	4.14
3.47	1.02	7.72	3.68	0.66	7.51	4.13	4.9	10.645	4.15
3.48	0.94	7.71	3.69	0.67	7.5	4.14	4.91	10.67	4.16
3.49	0.86	7.7	3.7	0.68	7.49	4.15	4.92	10.695	4.17
3.5	0.78	7.69	3.71	0.69	7.48	4.16	4.93	10.72	4.18
3.51	0.7	7.68	3.72	0.7	7.47	4.17	4.94	10.745	4.19
3.52	0.62	7.67	3.73	0.71	7.46	4.18	4.95	10.77	4.2
3.53	0.54	7.66	3.74	0.72	7.45	4.19	4.96	10.795	4.21
3.54	0.52	7.65	3.75	0.73	7.44	4.2	4.97	10.82	4.22
3.55	0.53	7.64	3.76	0.74	7.43	4.21	4.98	10.845	4.23
3.56	0.54	7.63	3.77	0.75	7.42	4.22	4.99	10.87	4.24
3.57	0.55	7.62	3.78	0.76	7.41	4.23	5	10.895	4.25
3.58	0.56	7.61	3.79	0.77	7.4	4.24	5.01	10.92	4.26
3.59	0.57	7.6	3.8	0.78	7.39	4.25	5.02	10.945	4.27
3.6	0.58	7.59	3.81	0.79	7.38	4.26	5.03	10.97	4.28
3.61	0.59	7.58	3.82	0.8	7.37	4.27	5.04	10.995	4.29
3.62	0.6	7.57	3.83	0.81	7.36	4.28	5.05	11.02	4.3
3.63	0.61	7.56	3.84	0.82	7.35	4.29	5.06	11.045	4.31
3.64	0.62	7.55	3.85	0.83	7.34	4.3	5.07	11.07	4.32
3.65	0.63	7.54	3.86	0.84	7.33	4.31	5.08	11.095	4.33
3.66	0.64	7.53	3.87	0.85	7.32	4.32	5.09	11.12	4.34
3.67	0.65	7.52	3.88	0.86	7.31	4.33	5.1	11.145	4.35
3.68	0.66	7.51	3.89	0.87	7.3	4.34	5.11	11.17	4.36
3.69	0.67	7.5	3.9	0.88	7.29	4.35	5.12	11.195	4.37
3.7	0.68	7.49	3.91	0.89	7.28	4.36	5.13	11.22	4.38

(continued on next page)

Table 9.1 Dataset of solar radiation of different city of India—cont'd

Patna	Bhopal	Mumbai	Chennai	Delhi	Goa	Jaipur	Kolkata	Lucknow	Nagpur
3.71	0.69	7.48	3.92	0.9	7.27	4.37	5.14	11.245	4.39
3.72	0.7	7.47	3.93	0.91	7.26	4.38	5.15	11.27	4.4
3.73	0.71	7.46	3.94	0.92	7.25	4.39	5.16	11.295	4.41
3.74	0.72	7.45	3.95	0.93	7.24	4.4	5.17	11.32	4.42
3.75	0.73	7.44	3.96	0.94	7.23	4.41	5.18	11.345	4.43
3.76	0.74	7.43	3.97	0.95	7.22	4.42	5.19	11.37	4.44
3.77	0.75	7.42	3.98	0.96	7.21	4.43	5.2	11.395	4.45
3.78	0.76	7.41	3.99	0.97	7.2	4.44	5.21	11.42	4.46
3.79	0.77	7.4	4	0.98	7.19	4.45	5.22	11.445	4.47
3.8	0.78	7.39	4.01	0.99	7.18	4.46	5.23	11.47	4.48
3.81	0.79	7.38	4.02	1	7.17	4.47	5.24	11.495	4.49
3.82	0.8	7.37	4.03	1.01	7.16	4.48	5.25	11.52	4.5
3.83	0.81	7.36	4.04	1.02	7.15	4.49	5.26	11.545	4.51
3.84	0.82	7.35	4.05	1.03	7.14	4.5	5.27	11.57	4.52
3.85	0.83	7.34	4.06	1.04	7.13	4.51	5.28	11.595	4.53
3.86	0.84	7.33	4.07	1.05	7.12	4.52	5.29	11.62	4.54
3.87	0.85	7.32	4.08	1.06	7.11	4.53	5.3	11.645	4.55
3.88	0.86	7.31	4.09	1.07	7.1	4.54	5.31	11.67	4.56
3.89	0.87	7.3	4.1	1.08	7.09	4.55	5.32	11.695	4.57
3.9	0.88	7.29	4.11	1.09	7.08	4.56	5.33	11.72	4.58
3.91	0.89	7.28	4.12	1.1	7.07	4.57	5.34	11.745	4.59
3.92	0.9	7.27	4.13	1.11	7.06	4.58	5.35	11.77	4.6
3.93	0.91	7.26	4.14	1.12	7.05	4.59	5.36	11.795	4.61
3.94	0.92	7.25	4.15	1.13	7.04	4.6	5.37	11.82	4.62
3.95	0.93	7.24	4.16	1.14	7.03	4.61	5.38	11.845	4.63
3.96	0.94	7.23	4.17	1.15	7.02	4.62	5.39	11.87	4.64
3.97	0.95	7.22	4.18	1.16	7.01	4.63	5.4	11.895	4.65
3.98	0.96	7.21	4.19	1.17	7	4.64	5.41	11.92	4.66
3.99	0.97	7.2	4.2	1.18	6.99	4.65	5.42	11.945	4.67
4	0.98	7.19	4.21	1.19	6.98	4.66	5.43	11.97	4.68
4.01	0.99	7.18	4.22	1.2	6.97	4.67	5.44	11.995	4.69
4.02	1	7.17	4.23	1.21	6.96	4.68	5.45	12.02	4.7
4.03	1.01	7.16	4.24	1.22	6.95	4.69	5.46	12.045	4.71
4.04	1.02	7.15	4.25	1.23	6.94	4.7	5.47	12.07	4.72
4.05	1.03	7.14	4.26	1.24	6.93	4.71	5.48	12.095	4.73
4.06	1.04	7.13	4.27	1.25	6.92	4.72	5.49	12.12	4.74
4.07	1.05	7.12	4.28	1.26	6.91	4.73	5.5	12.145	4.75
4.08	1.06	7.11	4.29	1.27	6.9	4.74	5.51	12.17	4.76
4.09	1.07	7.1	4.3	1.28	6.89	4.75	5.52	12.195	4.77
4.1	1.08	7.09	4.31	1.29	6.88	4.76	5.53	12.22	4.78
4.11	1.09	7.08	4.32	1.3	6.87	4.77	5.54	12.245	4.79
4.12	1.1	7.07	4.33	1.31	6.86	4.78	5.55	12.27	4.8

(continued on next page)

Table 9.1 Dataset of solar radiation of different city of India—cont'd

Patna	Bhopal	Mumbai	Chennai	Delhi	Goa	Jaipur	Kolkata	Lucknow	Nagpur
4.13	1.11	7.06	4.34	1.32	6.85	4.79	5.56	12.295	4.81
4.14	1.12	7.05	4.35	1.33	6.84	4.8	5.57	12.32	4.82
4.15	1.13	7.04	4.36	1.34	6.83	4.81	5.58	12.345	4.83
4.16	1.14	7.03	4.37	1.35	6.82	4.82	5.59	12.37	4.84
4.17	1.15	7.02	4.38	1.36	6.81	4.83	5.6	12.395	4.85
4.18	1.16	7.01	4.39	1.37	6.8	4.84	5.61	12.42	4.86
4.19	1.17	7	4.4	1.38	6.79	4.85	5.62	12.445	4.87
4.2	1.18	6.99	4.41	1.39	6.78	4.86	5.63	12.47	4.88
4.21	1.19	6.98	4.42	1.4	6.77	4.87	5.64	12.495	4.89
4.22	1.2	6.97	4.43	1.41	6.76	4.88	5.65	12.52	4.9
4.23	1.21	6.96	4.44	1.42	6.75	4.89	5.66	12.545	4.91
4.24	1.22	6.95	4.45	1.43	6.74	4.9	5.67	12.57	4.92
4.25	1.23	6.94	4.46	1.44	6.73	4.91	5.68	12.595	4.93
4.26	1.24	6.93	4.47	1.45	6.72	4.92	5.69	12.62	4.94
4.27	1.25	6.92	4.48	1.46	6.71	4.93	5.7	12.645	4.95
4.28	1.26	6.91	4.49	1.47	6.7	4.94	5.71	12.67	4.96
4.29	1.27	6.9	4.5	1.48	6.69	4.95	5.72	12.695	4.97
4.3	1.28	6.89	4.51	1.49	6.68	4.96	5.73	12.72	4.98
4.31	1.29	6.88	4.52	1.5	6.67	4.97	5.74	12.745	4.99
4.32	1.3	6.87	4.53	1.51	6.66	4.98	5.75	3.11	5
4.33	1.31	6.86	4.54	1.52	6.65	4.99	5.76	3.12	5.01
4.34	1.32	6.85	4.55	1.53	6.64	5	5.77	3.13	5.02
4.35	1.33	6.84	4.56	1.54	6.63	5.01	5.78	3.14	5.03
4.36	1.34	6.83	4.57	1.55	6.62	5.02	5.79	3.15	5.04
4.37	1.35	6.82	4.58	1.56	6.61	5.03	5.8	3.16	5.05
4.38	1.36	6.81	4.59	1.57	6.6	5.04	5.81	3.17	5.06
4.39	1.37	6.8	4.6	1.58	6.59	5.05	5.82	3.18	5.07
4.4	1.38	6.79	4.61	1.59	6.58	5.06	5.83	3.19	5.08
4.41	1.39	6.78	4.62	1.6	6.57	5.07	5.84	3.2	5.09
4.42	1.4	6.77	4.63	1.61	6.56	5.08	5.85	3.21	5.1
4.43	1.41	6.76	4.64	1.62	6.55	5.09	5.86	3.22	5.11
4.44	1.42	6.75	4.65	1.63	6.54	5.1	5.87	3.23	5.12

so first five rows will be shown in the result window, which is shown in Table 9.2.

When we are using the command solar.head (20) (Fig. 9.12), so in that case first 20 rows are visible on the screen, which is shown in Table 9.3.

To identify the last five rows of the given dataset of solar radiation, we can use the command solar.tail () (Fig. 9.13) and result are shown in Table 9.4.

Table 9.2 Results of Fig. 9.11.

	Patna	Bhopal	Mumbai	Chennai	Delhi	Goa	Jaipur	Kolkata	Lucknow	Nagpur
0	8.34	9.10	7.01	8.13	7.42	7.22	3.12	4.121	8.120	3.14
1	8.33	9.02	7.02	8.12	7.34	7.23	3.13	4.122	8.145	3.15
2	8.32	8.94	7.03	8.11	7.26	7.24	3.14	4.123	8.170	3.16
3	8.31	8.86	7.04	8.10	7.18	7.25	3.15	4.124	8.195	3.17
4	8.30	8.78	7.05	8.09	7.10	7.26	3.16	4.125	8.220	3.18

```
Solar.head(20)
```

Figure 9.12 Command for first 20 rows of solar radiation data.

Table 9.3 Results of Fig. 9.12.

	Patna	Bhopal	Mumbai	Chennai	Delhi	Goa	Jaipur	Kolkata	Lucknow	Nagpur
0	8.34	9.10	7.01	8.13	7.42	7.22	3.12	4.121	8.120	3.14
1	8.33	9.02	7.02	8.12	7.34	7.23	3.13	4.122	8.145	3.15
2	8.32	8.94	7.03	8.11	7.26	7.24	3.14	4.123	8.170	3.16
3	8.31	8.86	7.04	8.10	7.18	7.25	3.15	4.124	8.195	3.17
4	8.30	8.78	7.05	8.09	7.10	7.26	3.16	4.125	8.220	3.18
5	8.29	8.70	7.06	8.08	7.02	7.27	3.17	4.126	8.245	3.19
6	8.28	8.62	7.07	8.07	6.94	7.28	3.18	4.127	8.270	3.20
7	8.27	8.54	7.08	8.06	6.86	7.29	3.19	4.128	8.295	3.21
8	8.26	8.46	7.09	8.05	6.78	7.30	3.20	4.129	8.320	3.22
9	8.25	8.38	7.10	8.04	6.70	7.31	3.21	4.130	8.345	3.23
10	8.24	8.30	7.11	8.03	6.62	7.32	3.22	4.131	8.370	3.24
11	8.23	8.22	7.12	8.02	6.54	7.33	3.23	4.132	8.395	3.25
12	8.22	8.14	7.13	8.01	6.46	7.34	3.24	4.133	8.420	3.26
13	8.21	8.06	7.14	8.00	6.38	7.35	3.25	4.134	8.445	3.27
14	8.20	7.98	7.15	7.99	6.30	7.36	3.26	4.135	8.470	3.28
15	8.19	7.90	7.16	7.98	6.22	7.37	3.27	4.136	8.495	3.29
16	8.18	7.82	7.17	7.97	6.14	7.38	3.28	4.137	8.520	3.30
17	8.17	7.74	7.18	7.96	6.06	7.39	3.29	4.138	8.545	3.31
18	8.16	7.66	7.19	7.95	5.98	7.40	3.30	4.139	8.570	3.32
19	8.15	7.58	7.20	7.94	5.90	7.41	3.31	4.140	8.595	3.33

```
Solar.tail()
```

Figure 9.13 Command for last 5 rows of solar radiation data.

To identify the last 10 rows of the given dataset of solar radiation, we can use the command solar.tail(10) (Fig. 9.14) and result is looking like in Table 9.5.

To the descriptive analysis of solar radiation data, following command will be used (Fig. 9.15).

Table 9.4 Results of Fig. 9.13.

	Patna	Bhopal	Mumbai	Chennai	Delhi	Goa	Jaipur	Kolkata	Lucknow	Nagpur
194	4.40	1.38	6.79	4.61	1.59	6.58	5.06	5.83	3.19	5.08
195	4.41	1.39	6.78	4.62	1.60	6.57	5.07	5.84	3.20	5.09
196	4.42	1.40	6.77	4.63	1.61	6.56	5.08	5.85	3.21	5.10
197	4.43	1.41	6.76	4.64	1.62	6.55	5.09	5.86	3.22	5.11
198	4.44	1.42	6.75	4.65	1.63	6.54	5.10	5.87	3.23	5.12

```
Solar.tail(10)
```

Figure 9.14 Command for last 10 rows of solar radiation data.

Table 9.5 Results of Fig. 9.14.

	Patna	Bhopal	Mumbai	Chennai	Delhi	Goa	Jaipur	Kolkata	Lucknow	Nagpur
189	4.35	1.33	6.84	4.56	1.54	6.63	5.01	5.78	3.14	5.03
190	4.36	1.34	6.83	4.57	1.55	6.62	5.02	5.79	3.15	5.04
191	4.37	1.35	6.82	4.58	1.56	6.61	5.03	5.80	3.16	5.05
192	4.38	1.36	6.81	4.59	1.57	6.60	5.04	5.81	3.17	5.06
193	4.39	1.37	6.80	4.60	1.58	6.59	5.05	5.82	3.18	5.07
194	4.40	1.38	6.79	4.61	1.59	6.58	5.06	5.83	3.19	5.08
195	4.41	1.39	6.78	4.62	1.60	6.57	5.07	5.84	3.20	5.09
196	4.42	1.40	6.77	4.63	1.61	6.56	5.08	5.85	3.21	5.10
197	4.43	1.41	6.76	4.64	1.62	6.55	5.09	5.86	3.22	5.11
198	4.44	1.42	6.75	4.65	1.63	6.54	5.10	5.87	3.23	5.12

```
solar.describe()
```

Figure 9.15 Command for descriptive analysis of solar energy data.

With the help of the above command, result shown (Table 9.6) in the terms of number of counts, minimum value, maximum value, mean value, standard deviation, and 25, 50, and 75 percentile of solar radiation data.

9.3.2 Data visualization of solar energy data

The graphical depiction of information and data is known as data visualization. Data visualization tools make it easy to examine and comprehend trends, outliers, and patterns in data by employing visual components like charts, graphs, and maps. This representation may be thought of as a mapping between the original data and visual elements from an academic standpoint. The mapping specifies how the characteristics of these items change when the data changes. In this sense, a bar chart is a mapping of a variable's magnitude to the length of a bar. Mapping is a basic component of data

Table 9.6 Results of Fig. 9.15.

	Patna	Bhopal	Mumbai	Chennai	Delhi	Goa	Jaipur	Kolkata	Lucknow	Nagpur
count	199.0	199.0	199.0	199.0	199.0	199.0	199.0	199.0	199.0	199.0
mean	5.887688	3.059447	7.364322	5.498291	2.345025	7.314724	4.110000	4.895256	9.958065	4.130000
std	1.978467	2.668660	0.302737	1.834352	1.979794	0.371432	0.575905	0.553140	2.220241	0.575905
min	3.450000	0.520000	6.750000	3.450000	0.520000	6.540000	3.120000	4.120000	3.110000	3.140000
25%	3.945000	0.955000	7.125000	3.945000	0.955000	7.035000	3.615000	4.385000	9.032500	3.635000
50%	4.440000	1.400000	7.370000	4.440000	1.400000	7.370000	4.110000	4.880000	10.270000	4.130000
75%	7.845000	5.140000	7.620000	7.635000	3.460000	7.620000	4.605000	5.375000	11.507500	4.625000
max	8.340000	9.100000	7.890000	8.130000	7.420000	7.890000	5.100000	5.870000	12.745000	5.120000

```
import pandas as pd
import numpy as np
import matplotlib.pyplot as plt
solar = pd.read_csv("SOLAR RADIATION DATA.csv")
sd = pd.DataFrame(solar)
sd.plot.bar(x = 'Patna',color = 'orange')
plt.xlabel('solar radiation of Patna')
plt.ylabel('Count')
plt.title('solar radiation data of Patna')
```

Figure 9.16 Command for bar plot of solar radiation of Patna.

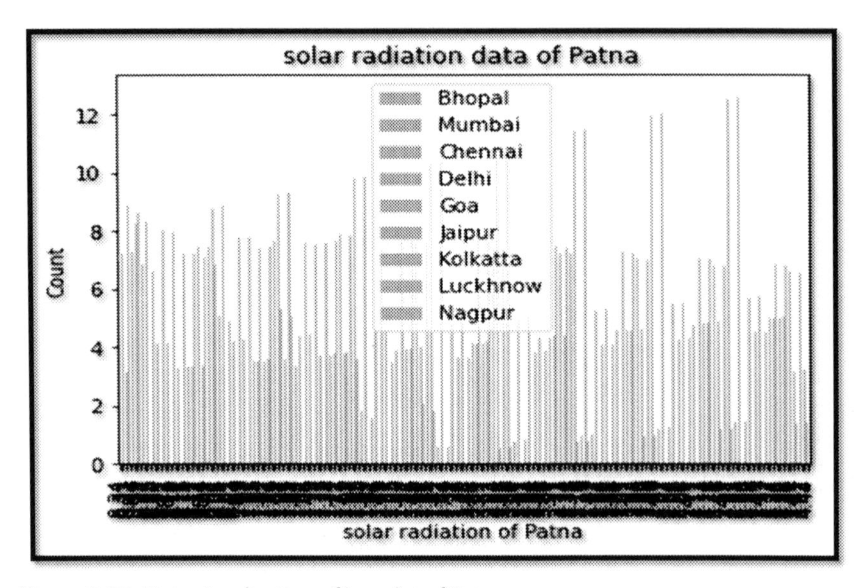

Figure 9.17 Data visualization of bar plot of Patna.

visualization since the graphic design of the mapping can impact the reading of a chart. Fig. 9.16 shows the Python command for bar plot of solar radiation of Patna (for data refer to the table). Fig. 9.17 shows data visualization of the bar plot of solar radiation of Patna. Figs 9.18 and 9.19 show the Python program and visualization of histogram of solar radiation data of Patna, respectively. Figs 9.20 and 9.21 show the Python program and visualization of scatter plot of solar radiation data of Bhopal and Patna, respectively.

Figs 9.22 and 9.23 show the Python program and visualization of box plot of solar radiation data of Bhopal and Patna, respectively. Figs 9.24 and 9.25 show the Python program and visualization of area plot of solar radiation data of Bhopal and Patna, respectively. Figs 9.26 and 9.27 show the Python program and visualization of line plot of solar radiation data of

```
import pandas as pd
import numpy as np
import matplotlib.pyplot as plt
solar = pd.read_csv("SOLAR RADIATION DATA.csv")
sd = pd.DataFrame(solar)
sd.plot.hist(x = 'Patna',color = 'orange')
plt.xlabel('solar radiation of Patna')
plt.ylabel('Count')
plt.title('solar radiation data of Patna')
```

Figure 9.18 Command for histogram plot of solar radiation of Patna.

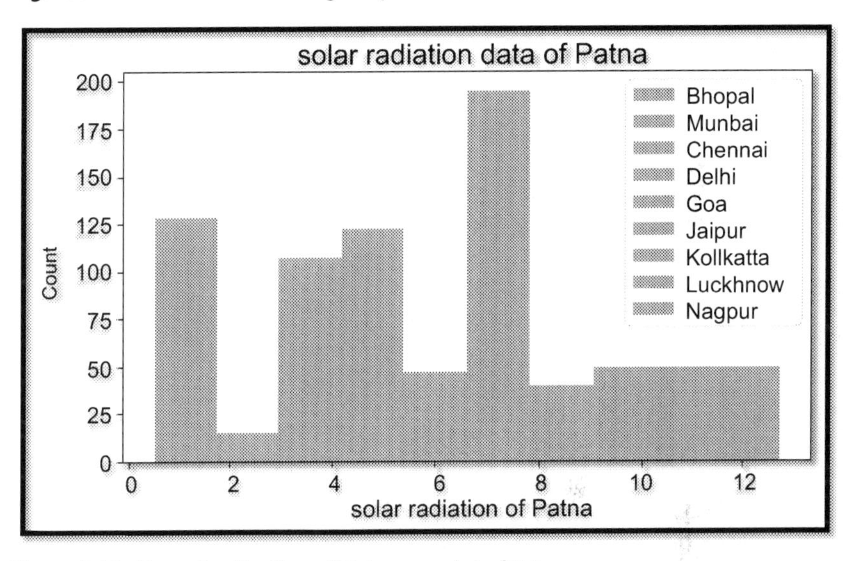

Figure 9.19 Data visualization of histogram plot of Patna.

```
import pandas as pd
import numpy as np
import matplotlib.pyplot as plt
solar = pd.read_csv("SOLAR RADIATION DATA.csv")
sd = pd.DataFrame(solar)
sd.plot.scatter(x = 'Patna',y = 'Bhopal',color = 'orange')
plt.xlabel('solar radiation of Patna')
plt.ylabel('solar radiation of Bhopal')
plt.title('solar radiation data of Patna and Bhopal')
```

Figure 9.20 Command for scatter plot of solar radiation of Patna and Bhopal.

Bhopal and Patna, respectively. Figs 9.28 and 2.29 show the Python program with seaborn command and visualization of line plot of solar radiation data of Bhopal and Patna, respectively. Figs 9.30 and 2.31 show the Python program with seaborn command and visualization of scatter plot of solar radiation data of Bhopal and Patna, respectively.

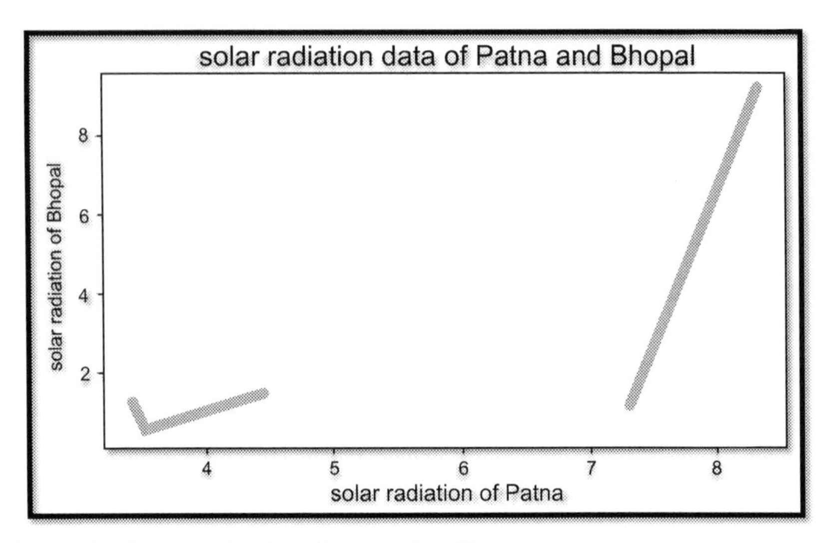

Figure 9.21 Data visualization of scatter plot of Patna.

```
import pandas as pd
import numpy as np
import matplotlib.pyplot as plt
solar = pd.read_csv("SOLAR RADIATION DATA.csv")
sd = pd.DataFrame(solar)
sd.plot.box(x = 'Patna',y = 'Bhopal',color = 'orange')
plt.xlabel('solar radiation of Patna')
plt.ylabel('solar radiation of Bhopal')
plt.title('solar radiation data of Patna and Bhopal')
```

Figure 9.22 Command for box plot of solar radiation of Patna and Bhopal.

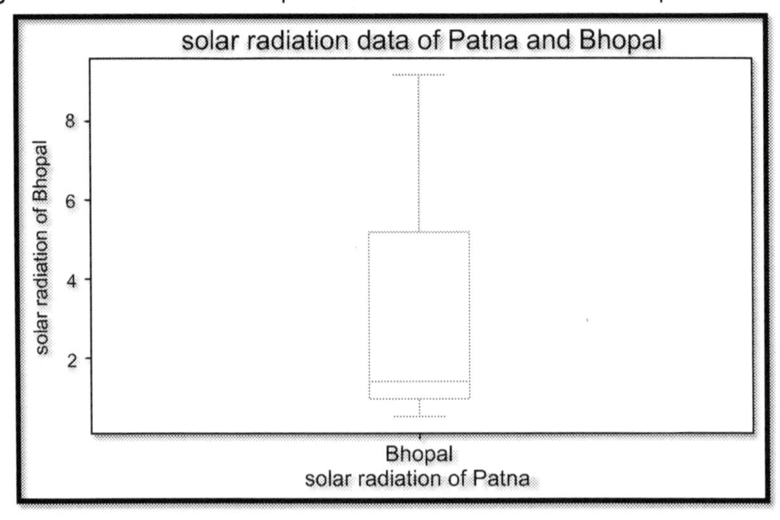

Figure 9.23 Data visualization of Box plot of Patna.

```
import pandas as pd
import numpy as np
import matplotlib.pyplot as plt
solar = pd.read_csv("SOLAR RADIATION DATA.csv")
sd = pd.DataFrame(solar)
sd.plot.area(x = 'Patna',y = 'Bhopal',color = 'orange')
plt.xlabel('solar radiation of Patna')
plt.ylabel('solar radiation of Bhopal')
plt.title('solar radiation data of Patna and Bhopal')
```

Figure 9.24 Command for area plot of solar radiation of Patna and Bhopal.

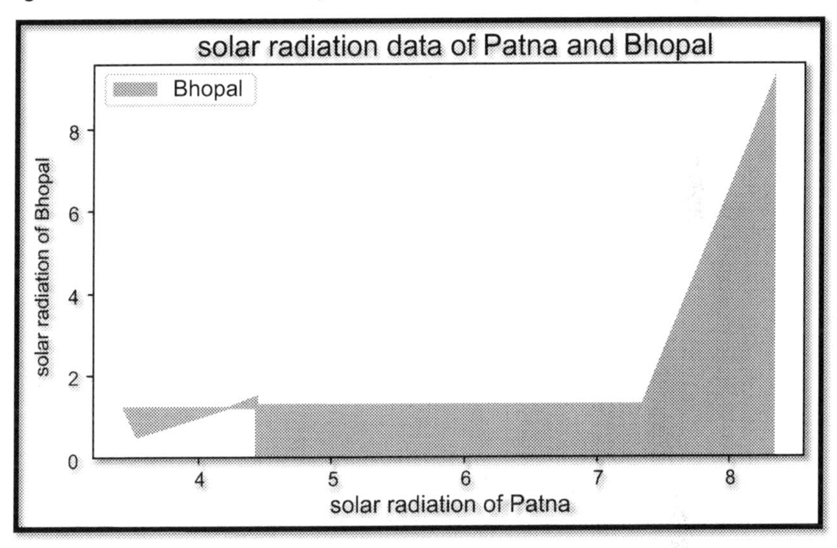

Figure 9.25 Data visualization of area plot of Patna.

```
import pandas as pd
import numpy as np
import matplotlib.pyplot as plt
solar = pd.read_csv("SOLAR RADIATION DATA.csv")
sd = pd.DataFrame(solar)
sd.plot.line(x = 'Patna',y = 'Bhopal',color = 'orange')
plt.xlabel('solar radiation of Patna')
plt.ylabel('solar radiation of Bhopal')
plt.title('solar radiation data of Patna and Bhopal')
```

Figure 9.26 Command for line plot of solar radiation of Patna and Bhopal.

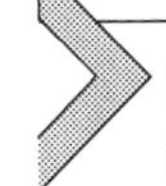

9.4 Data assessment of solar radiation by linear regression analysis

Linear regression analysis is a statistical technique for predicting the value of one variable based on the value of another. The dependent variable

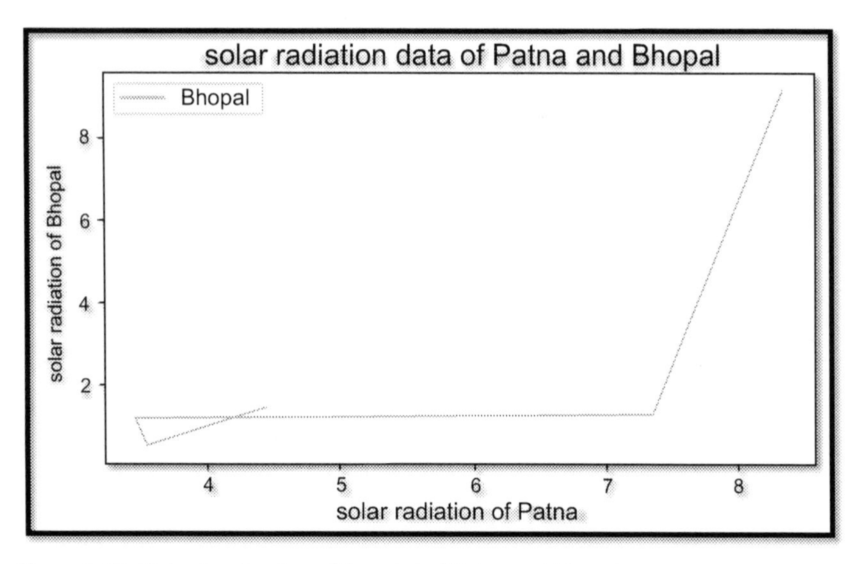

Figure 9.27 Data visualization of line plot of Patna and Bhopal.

```
import seaborn as sns
import matplotlib.pyplot as plt
import pandas as pd
# reading the database
data = pd.read_csv("SOLAR RADIATION DATA.csv")
# draw lineplot
sns.lineplot(x="Bhopal", y="Patna", data=data)
# setting the title using Matplotlib
plt.title('Solar Radiation Data')
plt.show()
```

Figure 9.28 Command for line plot of solar radiation of Patna and Bhopal with seaborn.

is the variable you wish to forecast. The independent variable is the one you are using to forecast the value of the other variable. This type of analysis involves one or more independent variables that best predict the value of the dependent variable in order to estimate the coefficients of the linear equation. Linear regression creates a straight line or surface that reduces the difference between expected and actual output values. Simple linear regression calculators that employ the "least squares" approach to get the best-fit line for a set of paired data are available. The value of X (dependent variable) is then estimated using Y. For linear regression analysis, first load the data by the command, which is shown in Fig. 9.32. In the command pandas as a library and SOLAR RADIATION DATA is the name of the

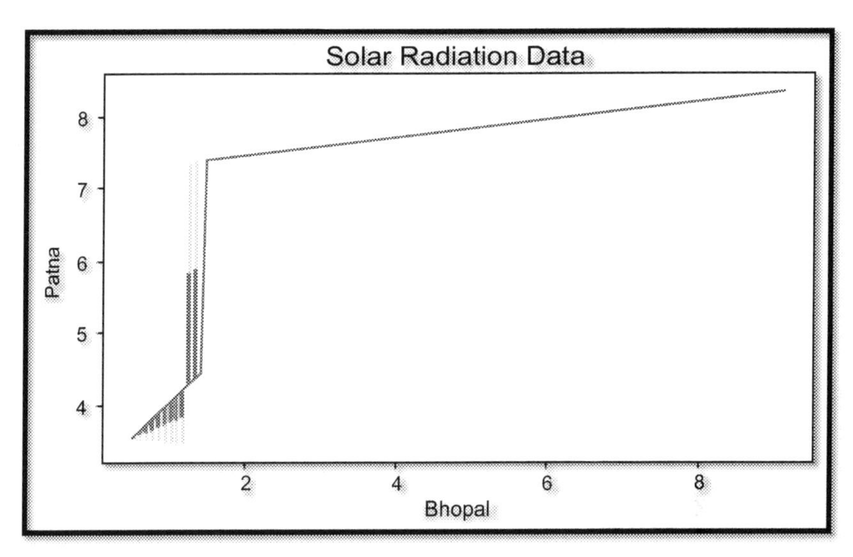

Figure 9.29 Data visualization of line plot of Patna and Bhopal with seaborn.

```
import seaborn as sns
import matplotlib.pyplot as plt
import pandas as pd
# reading the database
data = pd.read_csv("SOLAR RADIATION DATA.csv")
sns.scatterplot(x='Bhopal', y='Patna', data=data,)
plt.show()
```

Figure 9.30 Command for line plot of solar radiation of Patna and Bhopal.

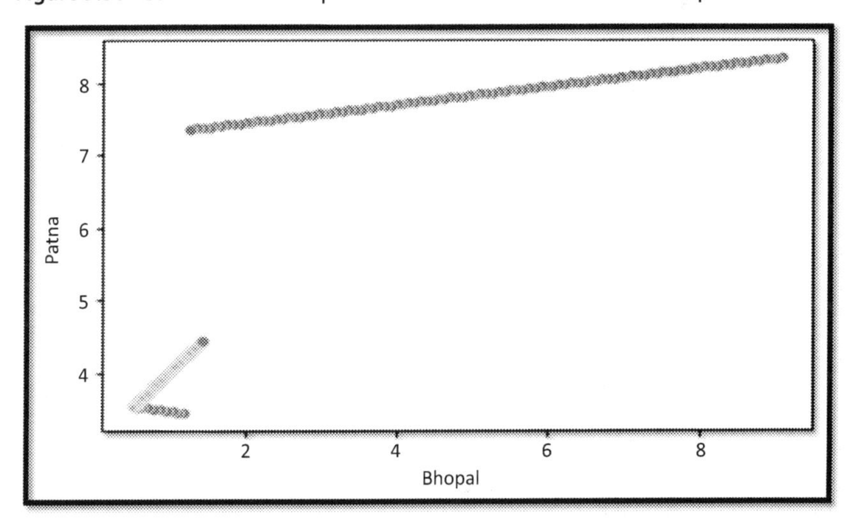

Figure 9.31 Data visualization of scatter plot of Patna and Bhopal with Seaborn.

```
import pandas as pd
solar = pd.read_csv("SOLAR RADIATION DATA.csv")
solar.head()
```

Figure 9.32 Command for pandas library for solar radiation.

Table 9.7 Results of Fig. 9.32.

	Patna	Bhopal	Mumbai	Chennai	Delhi	Goa	Jaipur	Kolkata	Lucknow	Nagpur
0	8.34	9.10	7.01	8.13	7.42	7.22	3.12	4.121	8.120	3.14
1	8.33	9.02	7.02	8.12	7.34	7.23	3.13	4.122	8.145	3.15
2	8.32	8.94	7.03	8.11	7.26	7.24	3.14	4.123	8.170	3.16
3	8.31	8.86	7.04	8.10	7.18	7.25	3.15	4.124	8.195	3.17
4	8.30	8.78	7.05	8.09	7.10	7.26	3.16	4.125	8.220	3.18

```
# For independent parameter
x = solar['Chennai']
x
```

Output:

0	8.13
1	8.12
2	8.11
3	8.10
4	8.09
	...
194	4.61
195	4.62
196	4.63
197	4.64
198	4.65

Figure 9.33 Command and their output for independent parameter.

CSV file. When we are type command solar.head (), so the first five rows will be shown in the result window, which is shown in Table 9.7.

In this regression analysis, we create a relationship between the solar radiation of Chennai and Goa, where solar radiation of Chennai is independent parameter and solar radiation of Goa is considered as dependent parameters. Figs 9.33 and 9.34 show the command and their output for independent and dependent parameters, respectively.

In the linear regression of solar energy system, command which is shown in Fig. 9.35 describe through the sk_learn model and utilize to create the train and test dataset of solar radiation data. Table 9.8 shows the output of training dataset of independent variable.

From the following command of linear regression, predict the value of dependent parameter (Fig. 9.36), and output shown in Table 9.9.

```
# For dependent parameter
y = solar['Goa']
y
```

Output:

0	7.22
1	7.23
2	7.24
3	7.25
4	7.26
	...
194	6.58
195	6.57
196	6.56
197	6.55
198	6.54

Figure 9.34 Command and their output for dependent parameter.

```
from sklearn.model_selection import train_test_split

x_train,x_test,y_train,y_test = train_test_split(x,y,test_size= 0.33, random_state =42)

x_train_reshaped= x_train.values.reshape(-1,1)
y_train_reshaped= y_train.values.reshape(-1,1)
x_test_reshaped= x_test.values.reshape(-1,1)
y_test_reshaped= y_test.values.reshape(-1,1

x_train_reshaped
```

Figure 9.35 Command for sk_learn model to create the train and test dataset of solar radiation data.

Table 9.8 Results of Fig. 9.35.

array([[7.4],	[7.67],	[4.53],	[3.86],	[3.46],	[4.51],	[3.98],
[4.03],	[3.79],	[8.09],	[7.49],	[3.51],	[7.64],	[7.96],
[4.02],	[4.52],	[7.81],	[7.69],	[7.74],	[3.47],	[7.41],
[7.62],	[4.05],	[3.67],	[4.48],	[4.34],	[7.79],	[4.41],
[3.71],	[7.77],	[4.11],	[7.85],	[7.66],	[8.06],	[4.01],
[4.28],	[7.52],	[4.61],	[7.73],	[3.61],	[3.77],	[4.33],
[7.72],	[7.91],	[4.1],	[3.92],	[4.2],	[3.58],	[4.49],
[3.81],	[4.29],	[8.03],	[4.21],	[7.7],	[3.5],	[7.5],
[3.76],	[3.65],	[7.51],	[4.22],	[4.04],	[4.42],	[7.59],
[7.87],	[7.8],	[3.84],	[7.88],	[8.1],	[4.47],	[3.74],
[3.99],	[8.02],	[4.12],	[7.9],	[3.72],	[3.56],	[7.63],
[3.64],	[4.6],	[8.13],	[4.17],	[7.6],	[8.05],	[4.62],
[4.32],	[4.25],	[4.45],	[4.13],	[4.00],	[8.00],	[7.55],
[8.11],	[8.07],	[4.19],	[3.48],	[4.64],	[7.54],	[7.65],
[7.36],	[7.86],	[7.43],	[3.75],	[4.23],	[4.37],	[3.55],

```
from sklearn.linear_model import LinearRegression

lin_reg = LinearRegression().fit(x_train_reshaped,y_train_reshaped)

y_pred = lin_reg.predict(x_test_reshaped)

y_pred
```

Figure 9.36 sk_learn linear model for linear regression.

Table 9.9 Results of Fig. 9.36.

array([[7.13183965],	**[7.13449068],**	**[7.22727679],**	**[7.51623925],**
[7.52861073],	[7.49326364],	[7.50210041],	[7.52507602],
[7.15746629],	[7.48884526],	[7.22904414],	[7.20871957],
[7.21578898],	[7.13890906],	[7.20430118],	[7.17160512],
[7.47470642],	[7.19988279],	[7.14862951],	[7.16365203],
[7.2034175],	[7.52772706],	[7.48089216],	[7.49237997],
[7.48177584],	[7.16895409],	[7.17072144],	[7.53744751],
[7.48265952],	[7.14421113],	[7.16807041],	[7.16718674],
[7.16541938],	[7.2113706],	[7.4755901],	[7.15923364],
[7.21225428],	[7.48354319],	[7.128830494],	[7.1415601],
[7.21490531],	[7.22639311],	[7.22109105],	[7.18220925],
[7.19016234],	[7.18927867],	[7.50828616],	[7.14332745],
[7.48442687],	[7.5339128],	[7.52065764],	[7.21048692],
[7.51535557],	[7.5259597],	[7.18486028],	[7.23257885],
[7.13537435],	[7.1724888],	[7.1839766],	[7.20783589],

```
y_pred1= lin_reg.predict(y_test_reshaped)

y_pred1
```

Figure 9.37 Predicted value of dependent variable.

Command shown in Fig. 9.37 shows the prediction of solar radiation of Goa, with the help of test dataset of solar radiation of Goa. Table 9.10 shows the output of predicted value of solar radiation of Goa.

Command shown in Fig. 9.38 shows the prediction of training dataset of solar radiation of Goa, with the help of training dataset of solar radiation of Chennai. Table 9.11 shows the output of predicted value of training dataset of solar radiation in Goa.

Command shown in Fig. 9.39 shows the regression coefficient and regression intercept between the solar radiation of Goa and Chennai. Fig. 9.40 shows the R^2 error of the training dataset of predicted value of solar radiation in Goa.

Table 9.10 Results of Fig. 9.37

array([[7.50386777],	**[7.50121674],**	**[7.40843063],**	**[7.4870779],**
[7.47470642],	[7.51005351],	[7.50121674],	[7.47824113],
[7.47824113],	[7.5144719],	[7.40666327],	[7.42698785],
[7.41991843],	[7.49679835],	[7.43140623],	[7.46410229],
[7.50916983],	[7.43582462],	[7.4870779],	[7.47205539],
[7.43228991],	[7.4755901],	[7.51535557],	[7.51093719],
[7.51623925],	[7.46675332],	[7.46498597],	[7.46586965],
[7.51712293],	[7.49149629],	[7.467637],	[7.46852068],
[7.47028803],	[7.42433682],	[7.51005351],	[7.47647377],
[7.42345314],	[7.51800661],	[7.50740248],	[7.49414732],
[7.42080211],	[7.4093143],	[7.41461637],	[7.45349817],
[7.44554507],	[7.44642875],	[7.495031],	[7.49237997],
[7.51889028],	[7.46940436],	[7.48265952],	[7.42522049],
[7.48796158],	[7.47735745],	[7.45084713],	[7.40312856],
[7.50033306],	[7.46321862],	[7.45173081],	[7.42787153],

```
y_train_pred = lin_reg.predict(x_train_reshaped)
y_train_pred
```

Figure 9.38 Command for predicted value of training dataset of dependent parameter.

9.5 Data assessment of solar energy system by logistic regression analysis

The (binary) logistic model (or logit model) in statistics is a statistical model that predicts the chance of one event (out of two options) occurring by making the event's log-odds (logarithm of the odds) a linear combination of one or more independent variables ("predictors"). Logistic regression (or logit regression) is a method of estimating the parameters of a logistic model in regression analysis (the coefficients in the linear combination). In binary logistic regression, there is a single binary dependent variable, coded by an indicator variable, with two values labeled "0" and "1," and the independent variables can be either binary variables (two classes, coded by an indicator variable) or continuous variables. The corresponding likelihood of the value labeled "1" might fluctuate between 0 (definitely the value "0") and 1 (definitely the value "1"), hence the labeling; the logistic function, which transforms log-odds to probability, is named after it. The logit, or logistic unit, is the unit of measurement for the log-odds scale, thus the different names. Table 9.12 shows the dataset of solar energy system, where temperature, clearness index, and solar radiation are the input attributes and power generation is the outcome; where power generation = "Yes" = "1" and power generation = "No" = "0."

Table 9.11 Results of Fig. 9.38.

array([[7.47735745],	[7.48796158],	[7.18751131],	[7.12918861],	[7.15658261],	[7.49768203],	[7.16188467],
[7.17955822],	[7.52242499],	[7.54186589],	[7.133607],	[7.13979274],	[7.23169518],	[7.13625803],
[7.17867454],	[7.20253383],	[7.21667266],	[7.50740248],	[7.13272332],	[7.49061261],	[7.47647377],
[7.49679835],	[7.14597848],	[7.19369705],	[7.20695221],	[7.21402163],	[7.49944938],	[7.16630306],
[7.15128055],	[7.51270454],	[7.48000848],	[7.50033306],	[7.21844002],	[7.13714171],	[7.23434621],
[7.20165015],	[7.53214544],	[7.1645357],	[7.14244377],	[7.13802539],	[7.52330867],	[7.52419235],
[7.50563512],	[7.22992782],	[7.48531055],	[7.19458073],	[7.53479647],	[7.49149629],	[7.22550943],
[7.16011732],	[7.19899912],	[7.50298409],	[7.50386777],	[7.53037809],	[7.20076647],	[7.47912481],
[7.15569893],	[7.53656383],	[7.21932369],	[7.18044189],	[7.48972893],	[7.22816047],	[7.1530479],
[7.51889028],	[7.51800661],	[7.51712293],	[7.53921486],	[7.20960324],	[7.17337248],	[7.52949441],
[7.17602351],	[7.22374208],	[7.5065188],	[7.15216422],	[7.17513983],	[7.50916983],	[7.14067642],
[7.1450948],	[7.53833118],	[7.16983777],	[7.495031],	[7.52684338],	[7.19811544],	[7.21755634],
[7.20518486],	[7.51358822],	[7.19546441],	[7.17690719],	[7.47824113],	[7.22462576],	[7.14951319]])
[7.54009854],	[7.14774584],	[7.19634809],	[7.23346253],	[7.21313795],	[7.54098221],	
[7.47382274],	[7.18662764],	[7.51977396],	[7.19723176],	[7.17779086],	[7.49591467],	
[7.50121674],	[7.2308115],	[7.52154131],	[7.22197473],	[7.20606854],	[7.19104602],	
[7.15834996],	[7.18574396],	[7.1919297],	[7.49856571],	[7.22020737],	[7.17425615],	
[7.2228584],	[7.53302912],	[7.18839499],	[7.13007229],	[7.48619422],	[7.19281338],	
[7.18132557],	[7.4870779],	[7.13095597],	[7.51182086],	[7.49414732],	[7.15039687],	
[7.51005351],	[7.16276835],	[7.15481525],	[7.53568015],	[7.15393158],	[7.14686216],	

```
print(lin_reg.coef_)
Output: [[0.08836773]]

print(lin_reg.intercept_)
Output: [6.82343628]
```

Figure 9.39 Command and their output of linear regression coefficient and intercept.

```
from sklearn.metrics import r2_score
print(r2_score(y_train_reshaped,y_train_pred))

Output: 0.19322642697253523
```

Figure 9.40 Command and their output of linear regression R^2 error.

Table 9.12 Dataset for logistic regression.

Temperature	Clearness index	Solar radiation	Power generation
17	0.4	5	0
21	0.6	6	0
25	0.7	7	1
29	0.72	8	1
35	0.8	10	1
37	0.8	12	1
41	0.83	12	1
18	0.6	7	0
17	0.63	6	0
19	0.7	6	0
22	0.7	8	1
42	0.83	10	1
43	0.85	11	1
41	0.83	12	1
37	0.79	10	1
20	0.55	6	0
38	0.71	9	1
18	0.4	7	0
36	0.81	11	1

For logistic regression analysis, first load the data by the command, which is shown in Fig. 9.41. In the command pandas as a library and SOLAR1 DATA is the name of the CSV file. When we are type command solar.head (), so first five rows will be shown in the result window, which is shown in the table.

Command shown in Fig. 9.42 is used to identify the outcome in terms of power generation and output shown in Table 9.13.

```
import pandas as pd
sd2 = pd.read_csv("solar1.csv")
sd2.head()
```

Output:

	Temperature	Clearness Index	Solar Radiation	Power generation
0	17	0.40	5	0
1	21	0.60	6	0
2	25	0.70	7	1
3	29	0.72	8	1
4	35	0.80	10	1

Figure 9.41 Import file for logistic regression with pandas library.

```
y = sd2["Power generation"]
y
```

Figure 9.42 Command for the outcome as a power generation.

Table 9.13 Results of Fig. 9.42.

0	0	10	1
1	0	11	1
2	1	12	1
3	1	13	1
4	1	14	1
5	1	15	0
6	1	16	1
7	0	17	0
8	0	18	1
9	0		

```
# extract data from every column except Outcome column in a variable named X
x = sd2.iloc[:,:-1] # all row and only left last column
x
```

Figure 9.43 Extract data from every column except outcome column in a variable named *x*.

Command shown in Fig. 9.43 is used to extract data from every column except the outcome column in a variable named *x* and Table 9.14 shows the output.

Command shown in Fig. 9.44 represents dividing the dataset into train and test, where 70% data are used for training and 30% used for testing. Sk_learn model is used to split the data into train and test dataset. Sk_learn model is also used to import model of logistic regression. Fig. 9.45 shows the predicted and test value of outcome in the form of power generation.

Command shown in Fig. 9.46 is used to find out the performance by calculating confusion matrix and accuracy score of the model. The confusion matrix shows the true positive, true negative, false positive, and false negative value through the actual and predicted value of power generation.

Table 9.14 Results of Fig. 9.43.

	Temperature	Clearness index	Solar radiation
0	17	0.40	5
1	21	0.60	6
2	25	0.70	7
3	29	0.72	8
4	35	0.80	10
5	37	0.80	12
6	41	0.83	12
7	18	0.60	7
8	17	0.63	6
9	19	0.70	6
10	22	0.70	8
11	42	0.83	10
12	43	0.85	11
13	41	0.83	12
14	37	0.79	10
15	20	0.55	6
16	38	0.71	9
17	18	0.40	7
18	36	0.81	11

```
# divide the dataset into train and test 70 and 30 percent
from sklearn.model_selection import train_test_split

x_train,x_test,y_train,y_test = train_test_split(x,y,train_size= 0.7, random_state =10)

# Create and Train Logistic Regression model on training set
from sklearn.linear_model import LogisticRegression

log_model = LogisticRegression().fit(x_train,y_train)
```

Figure 9.44 Command for divide the dataset into train and test 70% and 30%.

9.6 Data assessment of solar energy system by Naïve Bayes analysis

The Bayes' Theorem is used to create a set of classification algorithms known as Naive Bayes classifiers. It is a family of algorithms that share a similar idea, namely that each pair of characteristics being categorized is independent of the others.

The fundamental Naive Bayes assumption is that each feature makes an:

- Independent.
- Equal.
- Contribution to the outcome.

```
# make prediction based on the testing set using the trained model
y_pred = log_model.predict(x_test)
y_pred

Output: array([1, 0, 0, 1, 1, 1])

y_test

Output:
```

3	1
7	0
17	0
5	1
6	1
2	1

Figure 9.45 Command for make prediction based on the testing set using the trained model.

```
# Check the performance by calculating the confusion matrix and accuracy score of the model

from sklearn.metrics import accuracy_score, confusion_matrix
accuracy_score(y_test,y_pred)

Output: 1.0

confusion_matrix(y_test,y_pred)

Output: array([[2, 0],
    [0, 4]])
```

Figure 9.46 Command for calculating the confusion matrix and accuracy score of the model.

With relation to our dataset, this concept can be understood as:

This idea may be interpreted in respect to our dataset as follows:

We presume that no two attributes are mutually exclusive. The fact that the temperature is "Hot" has no bearing on the humidity, and the fact that the forecast is "Rainy" has no bearing on the winds. As a result, the characteristics are presumed to be independent.

Second, each feature is equally weighted (or importance). Knowing merely the temperature and humidity, for example, is insufficient to reliably forecast the outcome. None of the characteristics are unimportant, and they are all expected to have an equal impact on the outcome.

For Naïve Bayes analysis first load the data by the command, which is shown in Fig. 9.47. In the command pandas as a library and solar1 data with

```
import pandas as pd
df = pd.read_csv('solar1.csv')
df.head()
```

Output:

	Temperature	Clearness Index	Solar Radiation	Power generation
0	17	0.40	5	0
1	21	0.60	6	0
2	25	0.70	7	1
3	29	0.72	8	1
4	35	0.80	10	1

Figure 9.47 Command for load the data for Naïve Bayes analysis and their output.

```
y= df['Power generation']
y
```

Figure 9.48 Command for outcome in terms of power generation.

Table 9.15 Results of Fig. 9.48.

0	0	10	1
1	0	11	1
2	1	12	1
3	1	13	1
4	1	14	1
5	1	15	0
6	1	16	1
7	0	17	0
8	0	18	1
9 0			

the name of the CSV file. When we are type command solar.head (), so the first five rows will be shown in the output window.

Command shown in Fig. 9.48 is used to identify the outcome in terms of power generation and output shown in Table 9.15.

Command shown in Fig. 9.49 is used to extract data from every column except the outcome column in a variable named x and Table 9.16 shows the output.

Command shown in Fig. 9.50 represents dividing the dataset into train and test, where 70% data are used for training and 30% used for testing.

```
x = df.iloc[:,:-1]
x
```

Figure 9.49 Command for extract data from every column except outcome in a variable named *x*.

Table 9.16 Results of Fig. 9.49.

	Temperature	Clearness index	Solar radiation
0	17	0.40	5
1	21	0.60	6
2	25	0.70	7
3	29	0.72	8
4	35	0.80	10
5	37	. 0.80	12
6	41	0.83	12
7	18	0.60	7
8	17	0.63	6
9	19	0.70	6
10	22	0.70	8
11	42	0.83	10
12	43	0.85	11
13	41	0.83	12
14	37	0.79	10
15	20	0.55	6
16	38	0.71	9
17	18	0.40	7
18	36	0.81	11

```
# data set into two parts
from sklearn.model_selection import train_test_split
x_train,x_test,y_train,y_test = train_test_split(x,y,test_size = 0.3, random_state =1)

# Create and Train Naive Bayes Model
from sklearn.naive_bayes import GaussianNB
model = GaussianNB()
model.fit(x_train,y_train)

y_pred = model.predict(x_test)
y_pred
```

Figure 9.50 Command for data split into two parts for Naïve Bayes model.

Sk_learn model is used to split the data into train and test dataset. Sk_learn model was also used to import model of Naïve Bayes.

Command shown in Fig. 9.51 is used to find out the performance by calculating confusion matrix and accuracy score of the model. The confusion

```
print('Accuracy:%d',(model.score(x_test,y_test)))

from sklearn.metrics import confusion_matrix
confusion_matrix = confusion_matrix(y_test,y_pred)
print(confusion_matrix)
```

Figure 9.51 Command for accuracy and confusion matrix for Naïve Bayes model.

```
import pandas as pd
df = pd.read_csv('solar1.csv')
df.describe()
```

Figure 9.52 Command for load the data for random forest analysis and their descriptive analysis.

matrix shows the true positive, true negative, false positive, and false negative value through the actual and predicted value of power generation.

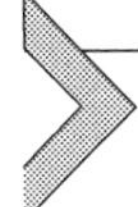

9.7 Data assessment of solar energy system by random forest

Random forest is a well-known machine learning algorithm that uses the supervised learning method. In machine learning, it may be utilized for both classification and regression issues. It is based on ensemble learning, which is a method of integrating several classifiers to solve a complicated issue and increase the model's performance.

"Random Forest is a classifier that contains a number of decision trees on various subsets of a given dataset and takes the average to enhance the predicted accuracy of that dataset," according to the name. Instead of depending on a single decision tree, the random forest collects the forecasts from each tree and predicts the final output based on the majority votes of predictions. For random forest analysis, first load the data by the command, which is shown in Fig. 9.52. In the command pandas as a library and solar1 data with the name of the CSV file. When we have type command solar.describe (), then descriptive statistics of all the parameters are shown in Table 9.17.

Command shown in Fig. 9.53 is used to identify the outcome in terms of power generation and output, as shown in Table 9.18.

Command shown in Fig. 9.54 is used to extract data from every column except outcome column in a variable named x and Table 9.19 shows the output.

Command shown in Fig. 9.55 represents divide the dataset into train and test, where 70% data are used for training and 30% used for testing. Sk_learn

Table 9.17 Descriptive statistics of solar energy system data.

	Temperature	Clearness index	Solar radiation	Power generation
count	19.000000	19.000000	19.000000	19.000000
mean	29.263158	0.697368	8.578947	0.631579
std	9.960155	0.137673	2.364454	0.495595
min	17.000000	0.400000	5.000000	0.000000
25%	19.500000	0.615000	6.500000	0.000000
50%	29.000000	0.710000	8.000000	1.000000
75%	37.500000	0.805000	10.500000	1.000000
max	43.000000	0.850000	12.000000	1.000000

```
y = df['Power generation']
y
```

Figure 9.53 Command for outcome in terms of power generation for random forest.

Table 9.18 Results of Fig. 9.53.

0	0	10	1
1	0	11	1
2	1	12	1
3	1	13	1
4	1	14	1
5	1	15	0
6	1	16	1
7	0	17	0
8	0	18	1
9	0		

```
x = df.iloc[:,:-1]
x
```

Figure 9.54 Command for extract data from every column except outcome in a variable named *x* for random forest.

model is used to split the data into train and test dataset. Sk_learn model was also used to import model of random forest classifier.

Command shown in Fig. 9.56 is used to find out the performance by calculating confusion matrix and accuracy score of the model. The confusion matrix shows the true positive, true negative, false positive, and false negative values through the actual and predicted value of power generation.

Table 9.19 Results of Fig. 9.54.

	Temperature	Clearness index	Solar radiation
0	17	0.40	5
1	21	0.60	6
2	25	0.70	7
3	29	0.72	8
4	35	0.80	10
5	37	0.80	12
6	41	0.83	12
7	18	0.60	7
8	17	0.63	6
9	19	0.70	6
10	22	0.70	8
11	42	0.83	10
12	43	0.85	11
13	41	0.83	12
14	37	0.79	10
15	20	0.55	6
16	38	0.71	9
17	18	0.40	7
18	36	0.81	11

```
# divide the data set into testing and training 70% and 30%
from sklearn.model_selection import train_test_split
x_train,x_test,y_train,y_test = train_test_split(x,y,test_size=0.3,random_state =1)


# create and train random forest model
from sklearn.ensemble import RandomForestClassifier
clf = RandomForestClassifier()
clf.fit(x_train,y_train)


Output: RandomForestClassifier()


y_pred = clf.predict(x_test)
y_pred
Output: array([1, 0, 1, 0, 0, 1])
```

Figure 9.55 Command for splitting the dataset in random forest.

```
print('Accuracy:%d',(clf.score(x_test,y_test)))
Output: Accuracy:%d 0.6666666666666666

from sklearn.metrics import confusion_matrix
confusion_matrix = confusion_matrix(y_test,y_pred)
print(confusion_matrix)

Results: [[1 0]
          [2 3]]
```

Figure 9.56 Accuracy and confusion matrix for random forest.

9.8 Data assessment of solar energy system by decision tree

A decision tree is a decision-making aid that employs a tree-like model of decisions and their probable outcomes, such as chance event outcomes, resource costs, and utility. It is one approach to showing an algorithm using simply conditional control statements. A decision tree is a flowchart-like structure in which each internal node represents a "test" on an attribute (e.g., whether a coin flip will come up heads or tails), each branch reflects the test's conclusion, and each leaf node represents a class label (decision taken after computing all attributes). Figures show different libraries are required for decision tree implementation. PyDotPlus is an improved version of the old pydot project that provides a Python interface to Graphviz's Dot language. Sk_learn also import decision tree classifier for decision tree implementation. Matplotlib.pyplot is a collection of command-style functions that make matplotlib work like MATLAB. Each pyplot function makes some change to a figure, for example, creates a figure, creates a plotting area in a figure, plots some lines in a plotting area, decorates the plot with labels, etc. The image module in matplotlib library is used for working with images in Python. The image module also includes two useful methods that are imread which is used to read images and imshow which is used to display the image. For decision tree analysis first load the data by the command, which is shown in Fig. 9.57. In the command pandas as a library and solar1 data with the name of the CSV file. Input data are shown in Table 9.20. When we are type command solar.describe (), then descriptive statistics of all the parameters are shown in the table (Table 9.21).

Command shown in Fig. 9.58 represents the independent and dependent parameters, where independent parameters are temperature, clearness index, and solar radiation and outcome as a dependent parameters in the terms

```
import pandas
from sklearn import tree
import pydotplus
from sklearn.tree import DecisionTreeClassifier
import matplotlib.pyplot as plt
import matplotlib.image as pltimg
df = pandas.read_csv("solar1.csv")
print(df)

df.describe()
```

Figure 9.57 Load the data through pandas and sk_learn command.

Table 9.20 Results of Table 57.

	Temperature	Clearness index	Solar radiation	Power generation
0	17	0.40	5	0
1	21	0.60	6	0
2	25	0.70	7	1
3	29	0.72	8	1
4	35	0.80	10	1
5	37	0.80	12	1
6	41	0.83	12	1
7	18	0.60	7	0
8	17	0.63	6	0
9	19	0.70	6	0
10	22	0.70	8	1
11	42	0.83	10	1
12	43	0.85	11	1
13	41	0.83	12	1
14	37	0.79	10	1
15	20	0.55	6	0
16	38	0.71	9	1
17	18	0.40	7	0
18	36	0.81	11	1

Table 9.21 Descriptive statistics of the load data for decision tree classifier.

	Temperature	Clearness index	Solar radiation	Power generation
count	19.000000	19.000000	19.000000	19.000000
mean	29.263158	0.697368	8.578947	0.631579
std	9.960155	0.137673	2.364454	0.495595
min	17.000000	0.400000	5.000000	0.000000
25%	19.500000	0.615000	6.500000	0.000000
50%	29.000000	0.710000	8.000000	1.000000
75%	37.500000	0.805000	10.500000	1.000000
max	43.000000	0.850000	12.000000	1.000000

```
features = ['Temperature', 'Clearness Index', 'Solar Radiation']
x = df[features]
y = df['Power generation']
print(x)
print(y)
```

Figure 9.58 Command for independent and dependent parameters for decision tree.

```
dtree = DecisionTreeClassifier()
dtree = dtree.fit(X, y)
data = tree.export_graphviz(dtree, out_file=None, feature_names=features)
graph = pydotplus.graph_from_dot_data(data)
graph.write_png('mydecisiontree.png')
img=pltimg.imread('mydecisiontree.png')
imgplot = plt.imshow(img)
plt.show()
```

Figure 9.59 Command for decision tree classifier.

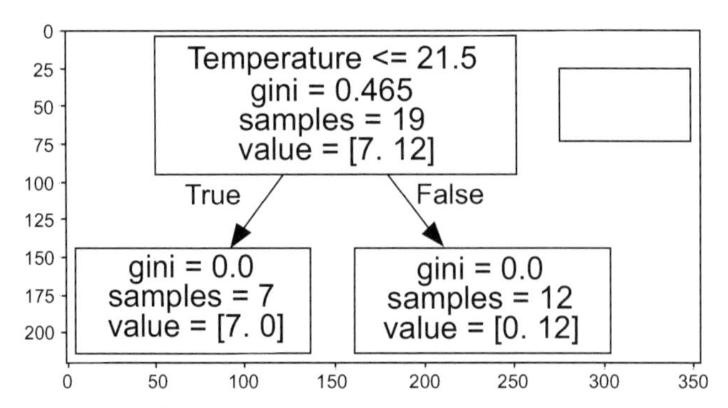

Figure 9.60 Gini index for temperature.

of power generation. Fig. 9.59 shows the command for the decision tree classifier, with png command. Fig. 9.60 shows the Gini index of temperature with true and false value.

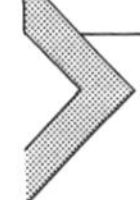

9.9 Data analysis of solar energy system by support vector machine

The support vector machine, or SVM, is a common supervised learning technique that may be used to solve both classification and regression issues. However, it is mostly utilized in machine learning for classification difficulties. The SVM algorithm's purpose is to find the optimum line or decision boundary for categorizing n-dimensional space into classes so that

```
import pandas as pd
data = pd.read_csv("SVM.csv")
data.head(12)
```

Output:

	Clearness Index	Solar Radiation	Outlook
0	0.89	8.7	Sunny
1	0.34	2.9	Mild
2	0.45	3.5	Mild
3	0.78	7.6	Sunny
4	0.81	7.9	Sunny
5	0.49	3.6	Mild
6	0.34	3.1	Mild
7	0.37	3.0	Mild
8	0.76	7.5	Sunny
9	0.85	8.3	Sunny
10	0.88	7.9	Sunny
11	0.91	8.6	Sunny

Figure 9.61 Command for load the data and their output for SVM.

additional data points may be readily placed in the proper category in the future. A hyperplane is the name for the optimal choice boundary. The extreme points/vectors that assist create the hyperplane are chosen via SVM. Support vectors are the extreme instances, and the method is called a support vector machine. For the support vector machine, first load the data and their output by the command, which is shown in Fig. 9.61. In the command pandas as a library and SVM.csv data with the name of the CSV file. The dataset is the 12 values of clearness index and solar radiation as an input parameter and outlook as an output parameter, where sunny and mild is the variable of the outlook. When we are type command data.head(), then descriptive statistics of all the parameters are shown in Fig. 9.62.

Command shown in Fig. 9.63 represents, splitting the dataset into training and test samples and also classifying predicted and target values. Fig. 9.64 shows the initializing support vector machine and fitting the

```
data.describe()

Output:
```

	Clearness Index	Solar Radiation
count	12.000000	12.000000
mean	0.655833	6.050000
std	0.235197	2.529283
min	0.340000	2.900000
25%	0.430000	3.400000
50%	0.770000	7.550000
75%	0.857500	8.000000
max	0.910000	8.700000

Figure 9.62 Command for descriptive statistics and their output for SVM.

```
from sklearn.model_selection import train_test_split
training_set, test_set = train_test_split(data, test_size = 0.2, random_state = 1)

X_train = training_set.iloc[:,0:2].values
Y_train = training_set.iloc[:,2].values
X_test = test_set.iloc[:,0:2].values
Y_test = test_set.iloc[:,2].values
```

Figure 9.63 Command for split the data and their output.

```
from sklearn.svm import SVC
classifier = SVC(kernel='rbf', random_state = 1)
classifier.fit(X_train,Y_train)
Output:
SVC(random_state=1)

Y_pred = classifier.predict(X_test)

test_set["Predictions"] = Y_pred
```

Figure 9.64 Command for classifier for SVM.

training dataset with the classifier. Predicting the classes for the test set is shown in Fig. 9.65. Command for accuracy for the dataset of support vector machine is shown in Fig. 9.66. Visualization of the classifier is shown in Fig. 9.67 and command for data visualization of clearness index and solar radiation are shown in Fig. 9.68.

```
# predicting the classes for test set
Y_pred = classifier.predict(X_test)

# Attaching the predictions to test set for comparing
test_set["Predictions"] = Y_pred
```

Figure 9.65 Command for predicting the classes for test set.

```
# calculating the accuracy of the predictions


from sklearn.metrics import confusion_matrix
cm = confusion_matrix(Y_test,Y_pred)
accuracy = float(cm.diagonal().sum())/len(Y_test)
print("\nAccuracy Of SVM For The Given Dataset : ", accuracy)
```

Output:
Accuracy Of SVM For The Given Dataset : 1.0

Figure 9.66 Command for accuracy of the prediction.

```
# visualizing the classifier


from sklearn.preprocessing import LabelEncoder
le = LabelEncoder()
Y_train = le.fit_transform(Y_train)
```

```
from sklearn.svm import SVC
classifier = SVC(kernel='rbf', random_state = 1)
classifier.fit(X_train,Y_train)
```

Output
SVC(random_state=1)

Figure 9.67 Visualizing the classifier.

```
import numpy as np
import matplotlib.pyplot as plt
from matplotlib.colors import ListedColormap
plt.figure(figsize = (7,7))
X_set, y_set = X_train, Y_train
X1, X2 = np.meshgrid(np.arange(start = X_set[:, 0].min() -
 1, stop = X_set[:, 0].max() + 1, step = 0.01), np.arange(start = X_set[:, 1].min() -
 1, stop = X_set[:, 1].max() + 1, step = 0.01))
plt.contourf(X1, X2, classifier.predict(np.array([X1.ravel(), X2.ravel()]).T).reshape(X1.shape
), alpha = 0.75, cmap = ListedColormap(('black', 'white')))
plt.xlim(X1.min(), X1.max())
plt.ylim(X2.min(), X2.max())
for i, j in enumerate(np.unique(y_set))
plt.scatter(X_set[y_set == j, 0], X_set[y_set == j, 1], c = ListedColormap(('red', 'orange'))(i), l
abel = j)
plt.title('Clearness Index Vs Solar Radiation ')
plt.xlabel('Unitless')
plt.ylabel('kwh/m2/day')
plt.legend()
plt.show()
```

Figure 9.68 Command for data visualization of clearness index and solar radiation.

9.10 Conclusion

According to the above assessment, it is find out, Python programming play a very vital role in the data analysis of different parameters of the solar energy system. Highly efficient Python's clean object-oriented design provides enhanced process control, and the language is equipped with excellent text processing and integration capabilities, as well as its own unit testing framework, which makes it a more efficient analysis of solar energy data.

9.11 Exercise/questions

1. Apply the Python programming for the data visualization of dataset in Table 9.22 in the following way.

Table 9.22 Solar radiation data of different city.

London	Perris	Delhi	Sagar	Bhopal	Ahmedabad	Indore	Pune	Perth	Bangalore
7.53	6.54	7.56	8.13	8.11	7.22	3.12	4.121	8.12	3.14
7.61	6.66	7.57	8.12	8.12	7.23	3.13	4.122	8.145	3.15
7.69	6.78	7.58	8.11	8.13	7.24	3.14	4.123	8.17	3.16
7.77	6.9	7.59	8.1	8.14	7.25	3.15	4.124	8.195	3.17
7.85	7.02	7.6	8.09	8.15	7.26	3.16	4.125	8.22	3.18
7.93	7.14	7.61	8.08	8.16	7.27	3.17	4.126	8.245	3.19
8.01	7.26	7.62	8.07	8.17	7.28	3.18	4.127	8.27	3.2
8.09	7.38	7.63	8.06	8.18	7.29	3.19	4.128	8.295	3.21
8.17	7.5	7.64	8.05	8.19	7.3	3.2	4.129	8.32	3.22
8.25	7.62	7.65	8.04	8.2	7.31	3.21	4.13	8.345	3.23
8.33	7.74	7.66	8.03	8.21	7.32	3.22	4.131	8.37	3.24
8.41	7.86	7.67	8.02	8.22	7.33	3.23	4.132	8.395	3.25
8.49	7.98	7.68	8.01	8.23	7.34	3.24	4.133	8.42	3.26
8.57	8.1	7.69	8	8.24	7.35	3.25	4.134	8.445	3.27
8.65	8.22	7.7	7.99	8.25	7.36	3.26	4.135	8.47	3.28
8.73	8.34	7.71	7.98	8.26	7.37	3.27	4.136	8.495	3.29
8.81	8.46	7.72	7.97	8.27	7.38	3.28	4.137	8.52	3.3
8.89	8.58	7.73	7.96	8.28	7.39	3.29	4.138	8.545	3.31
8.97	8.7	7.74	7.95	8.29	7.4	3.3	4.139	8.57	3.32
9.05	8.82	7.75	7.94	8.3	7.41	3.31	4.14	8.595	3.33
9.13	8.94	7.76	7.93	8.31	7.42	3.32	4.141	8.62	3.34
9.21	9.06	7.22	7.92	8.32	7.43	3.33	4.142	8.645	3.35
8.12	9.18	7.23	7.91	8.33	7.44	3.34	4.143	8.67	3.36
8.11	7.26	7.24	7.9	8.34	7.45	3.35	4.12	8.695	3.37
8.1	7.18	7.25	7.89	8.35	7.46	3.36	4.13	8.72	3.38
8.09	7.1	7.26	7.88	8.36	7.47	3.37	4.14	8.745	3.39
8.08	7.02	7.27	7.87	8.37	7.48	3.38	4.15	8.77	3.4
8.07	6.94	7.28	7.86	5.26	7.49	3.39	4.16	8.795	3.41
8.06	6.86	7.29	7.85	5.18	7.5	3.4	4.17	8.82	3.42
8.05	6.78	7.3	7.84	5.1	7.51	3.41	4.18	8.845	3.43
8.04	6.7	7.31	7.83	5.02	7.52	3.42	4.19	8.87	3.44
8.03	6.62	7.32	7.82	4.94	7.53	3.43	4.2	8.895	3.45
8.02	6.54	7.33	7.81	4.86	7.54	3.44	4.21	8.92	3.46
8.01	6.46	7.34	7.8	4.78	7.55	3.45	4.22	8.945	3.47
8	6.38	7.35	7.79	4.7	7.56	3.46	4.23	8.97	3.48
7.99	6.3	7.36	7.78	4.62	7.57	3.47	4.24	8.995	3.49
7.98	6.22	7.37	7.77	4.54	7.58	3.48	4.25	9.02	3.5
7.97	6.14	7.38	7.76	4.46	7.59	3.49	4.26	9.045	3.51
7.96	6.06	7.39	7.75	4.38	7.6	3.5	4.27	9.07	3.52
7.95	5.98	7.4	7.74	4.3	7.61	3.51	4.28	9.095	3.53
7.94	5.9	7.41	7.73	4.22	7.62	3.52	4.29	9.12	3.54
7.93	5.82	7.42	7.72	4.14	7.63	3.53	4.3	9.145	3.55

(continued on next page)

Table 9.22 Solar radiation data of different city—cont'd

London	Perris	Delhi	Sagar	Bhopal	Ahmedabad	Indore	Pune	Perth	Bangalore
7.92	5.74	7.43	7.71	4.06	7.64	3.54	4.31	9.17	3.56
7.91	5.66	7.44	7.7	3.98	7.65	3.55	4.32	9.195	3.57
7.9	5.58	7.45	7.69	3.9	7.66	3.56	4.33	9.22	3.58
7.89	5.5	7.46	7.68	3.82	7.67	3.57	4.34	9.245	3.59
7.88	5.42	7.47	7.67	3.74	7.68	3.58	4.35	9.27	3.6
7.87	5.34	7.48	7.66	3.66	7.69	3.59	4.36	9.295	3.61
7.86	5.26	7.49	7.65	3.58	7.7	3.6	4.37	9.32	3.62
7.85	5.18	7.5	7.64	3.5	7.71	3.61	4.38	9.345	3.63
7.84	5.1	7.51	7.63	3.42	7.72	3.62	4.39	9.37	3.64
7.83	5.02	7.52	7.62	3.34	7.73	3.63	4.4	9.395	3.65
7.82	4.94	7.53	7.61	3.26	7.74	3.64	4.41	9.42	3.66
7.81	4.86	7.54	7.6	3.18	7.75	3.65	4.42	9.445	3.67
7.8	4.78	7.55	7.59	3.1	7.76	3.66	4.43	9.47	3.68
7.79	4.7	7.56	7.58	3.02	7.77	3.67	4.44	9.495	3.69
7.78	4.62	7.57	7.57	2.94	7.78	3.68	4.45	9.52	3.7
7.77	4.54	7.58	7.56	2.86	7.79	3.69	4.46	9.545	3.71
7.76	4.46	7.59	7.55	2.78	7.8	3.7	4.47	9.57	3.72
7.75	4.38	7.6	7.54	2.7	7.81	3.71	4.48	9.595	3.73
7.74	4.3	7.61	7.53	2.62	7.82	3.72	4.49	9.62	3.74
7.73	4.22	7.62	7.52	2.54	7.83	3.73	4.5	9.645	3.75
7.72	4.14	7.63	7.51	2.46	7.84	3.74	4.51	9.67	3.76
7.71	4.06	7.64	7.5	2.38	7.89	3.75	4.52	9.695	3.77
7.7	3.98	7.65	7.49	2.3	7.88	3.76	4.53	9.72	3.78
7.69	3.9	7.66	7.48	2.22	7.87	3.77	4.54	9.745	3.79
7.68	3.82	7.67	7.47	2.14	7.86	3.78	4.55	9.77	3.8
7.67	3.74	7.68	7.46	2.06	7.85	3.79	4.56	9.795	3.81
7.66	3.66	7.69	7.45	1.98	7.84	3.8	4.57	9.82	3.82
7.65	3.58	7.7	7.44	1.9	7.83	3.81	4.58	9.845	3.83
7.64	3.5	7.71	7.43	1.82	7.82	3.82	4.59	9.87	3.84
7.63	3.42	7.72	7.42	1.74	7.81	3.83	4.6	9.895	3.85
7.62	3.34	7.73	7.41	1.66	7.8	3.84	4.61	9.92	3.86
7.61	3.26	7.74	7.4	1.58	7.79	3.85	4.62	9.945	3.87
7.6	3.18	7.75	7.39	1.5	7.78	3.86	4.63	9.97	3.88
7.59	3.1	7.76	7.38	1.42	7.77	3.87	4.64	9.995	3.89
7.58	3.02	7.77	7.37	1.34	7.76	3.88	4.65	10.02	3.9
7.57	2.94	7.78	7.36	1.26	7.75	3.89	4.66	10.045	3.91
7.56	2.86	7.79	3.45	1.18	7.74	3.9	4.67	10.07	3.92
7.55	2.78	7.8	3.46	1.1	7.73	3.91	4.68	10.095	3.93
7.54	2.7	7.81	3.47	1.02	7.72	3.92	4.69	10.12	3.94
7.53	2.62	7.82	3.48	0.94	7.71	3.93	4.7	10.145	3.95
7.52	2.54	7.83	3.49	0.86	7.7	3.94	4.71	10.17	3.96
7.51	2.46	7.84	3.5	0.78	7.69	3.95	4.72	10.195	3.97
7.5	2.38	7.89	3.51	0.7	7.68	3.96	4.73	10.22	3.98

(continued on next page)

Table 9.22 Solar radiation data of different city—cont'd

London	Perris	Delhi	Sagar	Bhopal	Ahmedabad	Indore	Pune	Perth	Bangalore
7.49	2.3	7.88	3.52	0.62	7.67	3.97	4.74	10.245	3.99
7.48	2.22	7.87	3.53	0.54	7.66	3.98	4.75	10.27	4
7.47	2.14	7.86	3.54	0.52	7.65	3.99	4.76	10.295	4.01
7.46	2.06	7.85	3.55	0.53	7.64	4	4.77	10.32	4.02
7.45	1.98	7.84	3.56	0.54	7.63	4.01	4.78	10.345	4.03
7.44	1.9	7.83	3.57	0.55	7.62	4.02	4.79	10.37	4.04
7.43	1.82	7.82	3.58	0.56	7.61	4.03	4.8	10.395	4.05
7.42	1.74	7.81	3.59	0.57	7.6	4.04	4.81	10.42	4.06
7.41	1.66	7.8	3.6	0.58	7.59	4.05	4.82	10.445	4.07
7.4	1.58	7.79	3.61	0.59	7.58	4.06	4.83	10.47	4.08
7.39	1.5	7.78	3.62	0.6	7.57	4.07	4.84	10.495	4.09
7.38	1.42	7.77	3.63	0.61	7.56	4.08	4.85	10.52	4.1
7.37	1.34	7.76	3.64	0.62	7.55	4.09	4.86	10.545	4.11
7.36	1.26	7.75	3.65	0.63	7.54	4.1	4.87	10.57	4.12
3.45	1.18	7.74	3.66	0.64	7.53	4.11	4.88	10.595	4.13
3.46	1.1	7.73	3.67	0.65	7.52	4.12	4.89	10.62	4.14
3.47	1.02	7.72	3.68	0.66	7.51	4.13	4.9	10.645	4.15
3.48	0.94	7.71	3.69	0.67	7.5	4.14	4.91	10.67	4.16
3.49	0.86	7.7	3.7	0.68	7.49	4.15	4.92	10.695	4.17
3.5	0.78	7.69	3.71	0.69	7.48	4.16	4.93	10.72	4.18
3.51	0.7	7.68	3.72	0.7	7.47	4.17	4.94	10.745	4.19
3.52	0.62	7.67	3.73	0.71	7.46	4.18	4.95	10.77	4.2
3.53	0.54	7.66	3.74	0.72	7.45	4.19	4.96	10.795	4.21
3.54	0.52	7.65	3.75	0.73	7.44	4.2	4.97	10.82	4.22
3.55	0.53	7.64	3.76	0.74	7.43	4.21	4.98	10.845	4.23
3.56	0.54	7.63	3.77	0.75	7.42	4.22	4.99	10.87	4.24
3.57	0.55	7.62	3.78	0.76	7.41	4.23	5	10.895	4.25
3.58	0.56	7.61	3.79	0.77	7.4	4.24	5.01	10.92	4.26
3.59	0.57	7.6	3.8	0.78	7.39	4.25	5.02	10.945	4.27
3.6	0.58	7.59	3.81	0.79	7.38	4.26	5.03	10.97	4.28
3.61	0.59	7.58	3.82	0.8	7.37	4.27	5.04	10.995	4.29
3.62	0.6	7.57	3.83	0.81	7.36	4.28	5.05	11.02	4.3
3.63	0.61	7.56	3.84	0.82	7.35	4.29	5.06	11.045	4.31
3.64	0.62	7.55	3.85	0.83	7.34	4.3	5.07	11.07	4.32
3.65	0.63	7.54	3.86	0.84	7.33	4.31	5.08	11.095	4.33
3.66	0.64	7.53	3.87	0.85	7.32	4.32	5.09	11.12	4.34
3.67	0.65	7.52	3.88	0.86	7.31	4.33	5.1	11.145	4.35
3.68	0.66	7.51	3.89	0.87	7.3	4.34	5.11	11.17	4.36
3.69	0.67	7.5	3.9	0.88	7.29	4.35	5.12	11.195	4.37
3.7	0.68	7.49	3.91	0.89	7.28	4.36	5.13	11.22	4.38
3.71	0.69	7.48	3.92	0.9	7.27	4.37	5.14	11.245	4.39
3.72	0.7	7.47	3.93	0.91	7.26	4.38	5.15	11.27	4.4

(continued on next page)

Table 9.22 Solar radiation data of different city—cont'd

London	Perris	Delhi	Sagar	Bhopal	Ahmedabad	Indore	Pune	Perth	Bangalore
3.73	0.71	7.46	3.94	0.92	7.25	4.39	5.16	11.295	4.41
3.74	0.72	7.45	3.95	0.93	7.24	4.4	5.17	11.32	4.42
3.75	0.73	7.44	3.96	0.94	7.23	4.41	5.18	11.345	4.43
3.76	0.74	7.43	3.97	0.95	7.22	4.42	5.19	11.37	4.44
3.77	0.75	7.42	3.98	0.96	7.21	4.43	5.2	11.395	4.45
3.78	0.76	7.41	3.99	0.97	7.2	4.44	5.21	11.42	4.46
3.79	0.77	7.4	4	0.98	7.19	4.45	5.22	11.445	4.47
3.8	0.78	7.39	4.01	0.99	7.18	4.46	5.23	11.47	4.48
3.81	0.79	7.38	4.02	1	7.17	4.47	5.24	11.495	4.49
3.82	0.8	7.37	4.03	1.01	7.16	4.48	5.25	11.52	4.5
3.83	0.81	7.36	4.04	1.02	7.15	4.49	5.26	11.545	4.51
3.84	0.82	7.35	4.05	1.03	7.14	4.5	5.27	11.57	4.52
3.85	0.83	7.34	4.06	1.04	7.13	4.51	5.28	11.595	4.53
3.86	0.84	7.33	4.07	1.05	7.12	4.52	5.29	11.62	4.54
3.87	0.85	7.32	4.08	1.06	7.11	4.53	5.3	11.645	4.55
3.88	0.86	7.31	4.09	1.07	7.1	4.54	5.31	11.67	4.56
3.89	0.87	7.3	4.1	1.08	7.09	4.55	5.32	11.695	4.57
3.9	0.88	7.29	4.11	1.09	7.08	4.56	5.33	11.72	4.58
3.91	0.89	7.28	4.12	1.1	7.07	4.57	5.34	11.745	4.59
3.92	0.9	7.27	4.13	1.11	7.06	4.58	5.35	11.77	4.6
3.93	0.91	7.26	4.14	1.12	7.05	4.59	5.36	11.795	4.61
3.94	0.92	7.25	4.15	1.13	7.04	4.6	5.37	11.82	4.62
3.95	0.93	7.24	4.16	1.14	7.03	4.61	5.38	11.845	4.63
3.96	0.94	7.23	4.17	1.15	7.02	4.62	5.39	11.87	4.64
3.97	0.95	7.22	4.18	1.16	7.01	4.63	5.4	11.895	4.65
3.98	0.96	7.21	4.19	1.17	7	4.64	5.41	11.92	4.66
3.99	0.97	7.2	4.2	1.18	6.99	4.65	5.42	11.945	4.67
4	0.98	7.19	4.21	1.19	6.98	4.66	5.43	11.97	4.68
4.01	0.99	7.18	4.22	1.2	6.97	4.67	5.44	11.995	4.69
4.02	1	7.17	4.23	1.21	6.96	4.68	5.45	12.02	4.7
4.03	1.01	7.16	4.24	1.22	6.95	4.69	5.46	12.045	4.71
4.04	1.02	7.15	4.25	1.23	6.94	4.7	5.47	12.07	4.72
4.05	1.03	7.14	4.26	1.24	6.93	4.71	5.48	12.095	4.73
4.06	1.04	7.13	4.27	1.25	6.92	4.72	5.49	12.12	4.74
4.07	1.05	7.12	4.28	1.26	6.91	4.73	5.5	12.145	4.75
4.08	1.06	7.11	4.29	1.27	6.9	4.74	5.51	12.17	4.76
4.09	1.07	7.1	4.3	1.28	6.89	4.75	5.52	12.195	4.77
4.1	1.08	7.09	4.31	1.29	6.88	4.76	5.53	12.22	4.78
4.11	1.09	7.08	4.32	1.3	6.87	4.77	5.54	12.245	4.79
4.12	1.1	7.07	4.33	1.31	6.86	4.78	5.55	12.27	4.8
4.13	1.11	7.06	4.34	1.32	6.85	4.79	5.56	12.295	4.81
4.14	1.12	7.05	4.35	1.33	6.84	4.8	5.57	12.32	4.82

(continued on next page)

Table 9.22 Solar radiation data of different city—cont'd

London	Perris	Delhi	Sagar	Bhopal	Ahmedabad	Indore	Pune	Perth	Bangalore
4.15	1.13	7.04	4.36	1.34	6.83	4.81	5.58	12.345	4.83
4.16	1.14	7.03	4.37	1.35	6.82	4.82	5.59	12.37	4.84
4.17	1.15	7.02	4.38	1.36	6.81	4.83	5.6	12.395	4.85
4.18	1.16	7.01	4.39	1.37	6.8	4.84	5.61	12.42	4.86
4.19	1.17	7	4.4	1.38	6.79	4.85	5.62	12.445	4.87
4.2	1.18	6.99	4.41	1.39	6.78	4.86	5.63	11.45	4.88
6.7	2.45	6.98	4.42	1.4	6.77	4.87	5.64	11.46	4.89
6.71	2.56	6.97	4.43	1.41	6.76	4.88	5.65	11.47	4.9
6.72	2.67	6.96	4.44	1.42	6.75	4.89	5.66	11.48	4.91
6.73	2.78	6.95	4.45	1.43	6.74	4.9	5.67	11.49	4.92
6.74	2.89	6.94	4.46	1.44	6.73	4.91	5.68	11.5	4.93
6.75	3	6.93	4.47	1.45	6.72	4.92	5.69	11.51	4.94
6.76	3.11	6.92	4.48	1.46	6.71	4.93	5.7	11.52	4.95
6.77	3.22	6.91	4.49	1.47	6.7	4.94	5.71	11.53	4.96
6.78	3.33	6.9	4.5	1.48	6.69	4.95	5.72	11.54	4.97
6.79	3.44	6.89	4.51	1.49	6.68	4.96	5.73	11.55	4.98
6.8	3.55	6.88	4.52	1.5	6.67	4.97	5.74	11.56	4.99
6.81	3.66	6.87	4.53	1.51	6.66	4.98	5.75	11.57	5
6.82	3.77	6.86	4.54	1.52	6.65	4.99	5.76	11.58	5.01
6.83	3.88	6.85	4.55	1.53	6.64	5	5.77	11.59	5.02
6.84	3.99	6.84	4.56	1.54	6.63	5.01	5.78	11.6	5.03
6.85	4.1	6.83	4.57	1.55	6.62	5.02	5.79	11.61	5.04
6.86	4.21	6.82	4.58	1.56	6.61	5.03	5.8	11.62	5.05
6.87	4.32	6.81	4.59	1.57	6.6	5.04	5.81	11.63	5.06
6.88	4.43	6.8	4.6	1.58	6.59	5.05	5.82	11.64	5.07
6.89	4.54	6.79	4.61	1.59	6.58	5.06	5.83	11.65	5.08
6.9	4.65	6.78	4.62	1.6	6.57	5.07	5.84	11.66	5.09
6.91	4.76	6.77	4.63	1.61	6.56	5.08	5.85	11.67	5.1
6.92	4.87	6.76	4.64	1.62	6.55	5.09	5.86	11.68	5.11
6.93	4.98	6.75	4.65	1.63	6.54	5.1	5.87	11.69	5.12

- **i.** Histogram plot of solar radiation of London city.
- **ii.** Histogram plot of solar radiation of Delhi city.
- **iii.** Bar plot of solar radiation of Bhopal city.
- **iv.** Bar plot of solar radiation of Indore city.
- **v.** Line plot of solar radiation of Ahmedabad city.
- **vi.** Line plot of solar radiation of Pune city.
- **vii.** Scatter plot of solar radiation of Ahmedabad city and Bhopal city.

2. Apply the Python programming for linear regression analysis as a dependent variable (solar radiation of Perris) and independent variable (Solar radiation of Sagar) (**Data refer to** Table 9.22).

Table 9.23 Dataset for logistic regression.

Temperature	Clearness index	Solar radiation	Power generation
16	0.49	5.5	0
20	0.62	6.3	0
28	0.71	7.2	1
31	0.72	8	1
35	0.81	10.1	1
37	0.81	12.1	1
41	0.83	12	1
18	0.6	7	0
17	0.63	6.5	0
21	0.73	6.2	0
22	0.7	8	1
42	0.83	10.1	1
43	0.85	11.5	1
41	0.83	12	1
38	0.79	10	1
20	0.55	6	0
36	0.71	9	1
21	0.42	7	0
36	0.81	11	1

3. Apply the Python programming for linear regression analysis as a dependent variable (solar radiation of Delhi) and independent variable (Solar radiation of Pune) (**Data refer to** Table 9.22).
4. Apply the Python programming for logistic regression analysis for the **Data refer to** Table 9.23.
5. Apply the Python programming for Naïve Bayes analysis for the **Data refer to** Table 9.23.
6. Apply the Python programming for Random Forest analysis for the **Data refer to** Table 9.23.

References

[1] F. Milano, An open source power system analysis toolbox, IEEE Trans. Power Syst. 20 (3) (Aug 2005) 1199–1206.

[2] L. Thurner, A. Scheidler, F. Schäfer, J.-.H. Menke, J. Dollichon, F. Meier, S. Meinecke and M. Braun 2017 Pandapower—an Open Source Python Tool for convenient modeling, analysis and optimization of electric power systems. https://arxiv.org/abs/1709.06743. Accessed date: 1 February 2022.

[3] S. Pfenninger, Dealing with multiple decades of hourly wind and PV time series in energy models: a comparison of methods to reduce time resolution and the planning implications of inter-annual variability, Appl. Energy 197 (2017) 1–13.

[4] A. Greenhall, R. Christie, Minpower: a power systems optimization toolkit, in: 2012 IEEE Power and Energy Society General Meeting, July 2012, pp. 1–6.
[5] C.E. Murillo-Sanchez, R.D. Zimmerman, C.L. Anderson, R.J. Thomas, Secure planning and operations of systems with stochastic sources, energy storage and active demand, IEEE Trans. Smart Grid 4 (2013) 2220–2229.

Index

Page numbers followed by "*f*" and "*t*" indicate, figures and tables respectively.

Printed in the United States
by Baker & Taylor Publisher Services